Basics of Bioinformatics

Rui Jiang • Xuegong Zhang • Michael Q. Zhang
Editors

Basics of Bioinformatics

Lecture Notes of the Graduate Summer
School on Bioinformatics of China

Editors
Rui Jiang
Xuegong Zhang
Department of Automation
Tsinghua University
Beijing
China, People's Republic

Michael Q. Zhang
Department of Molecular and Cell Biology
The University of Texas at Dallas
Richardson, TX, USA

Tsinghua National Laboratory
    for Information Science and Technology
Tsinghua University
Beijing, China, People's Republic

ISBN 978-3-662-52325-4   ISBN 978-3-642-38951-1 (eBook)
DOI 10.1007/978-3-642-38951-1
Springer Heidelberg New York Dordrecht London

Jointly published with Tsinghua University Press, Beijing
ISBN: 978-7-302-32359-4 Tsinghua University Press, Beijing

Foreword

This ambitious volume is the result of the successful 2007 Graduate Summer School on Bioinformatics of China held at Tsinghua University. It is remarkable for its range of topics as well as the depth of coverage. Bioinformatics draws on many subjects for analysis of the data generated by the biological sciences and biotechnology. This foreword will describe briefly each of the 12 chapters and close with additional general comments about the field. Many of the chapters overlap and include useful introductions to concepts such as gene or Bayesian methods. This is a valuable aspect of the volume allowing a student various angles of approach to a new topic.

Chapter 1, "Basics for Bioinformatics," defines bioinformatics as "the storage, manipulation and interpretation of biological data especially data of nucleic acids and amino acids, and studies molecular rules and systems that govern or affect the structure, function and evolution of various forms of life from computational approaches." Thus, the first subject they turn to is molecular biology, a subject that has had an enormous development in the last decades and shows no signs of slowing down. Without a basic knowledge of biology, the bioinformatics student is greatly handicapped. From basic biology the authors turn to biotechnology, in particular, methods for DNA sequencing, microarrays, and proteomics. DNA sequencing is undergoing a revolution. The mass of data collected in a decade of the Human Genome Project from 1990 to 2001 can be generated in 1 day in 2010. This is changing the science of biology at the same time. A 1,000 genome project became a 10,000 genome project 2 years later, and one expects another zero any time now. Chromatin Immunoprecipitation or ChIP allows access to DNA bound by proteins and thus to a large number of important biological processes. Another topic under the umbrella of biological sciences is genetics, the study of heredity and inherited characteristics (phenotypes). Heredity is encoded in DNA and thus is closely related to the goals of bioinformatics. This whole area of genetics beginning with Mendel's laws deserves careful attention, and genetics is a key aspect of the so-called genetic mapping and other techniques where the chromosomal locations of disease genes are sought.

Chapter 2, "Basic Statistics for Bioinformatics," presents important material for the understanding and analysis of data. Probability and statistics are basic to bioinformatics, and this chapter begins with the fundamentals including many classical distributions (including the binomial, Poisson, and normal). Usually the observation of complete populations such as "all people in China over 35 years old" is not practical to obtain. Instead random samples of the population of interest are obtained and then inferences about parameters of the population are made. Statistics guides us in making those inferences and gaining information about the quality of the estimates. The chapter describes techniques such as method of moments, maximum likelihood, and Bayesian methods. Bayesian methods have become indispensable in the era of powerful computing machines. The chapter treats hypothesis testing which is less used than parameter estimation, but hypothesis testing provides understanding of p-values which are ubiquitous in bioinformatics and data analysis. Classical testing situations reveal useful statistics such as the t-statistic. Analysis of variance and regression analysis are crucial for testing and fitting large data sets. All of these methods and many more are included in the free open-source package called R.

Chapter 3, "Topics in Computational Genomics," takes us on a tour of important topics that arise when complete genome information is available. The subject did not begin until nearly 2000 when complete genome sequences became a possibility. The authors present us with a list of questions, some of which are listed next. What are the genes of an organism? How are they turned off and on? How do they interact with each other? How are introns and exons organized and expressed in RNA transcripts? What are the gene products, both structure and function? How has a genome evolved? This last question has to be asked with other genomes and with members of the population comprising the species. Then the authors treat some of the questions in detail. They describe "finding protein coding genes," "identifying promoters," "genomic arrays and a CGH/CNP analysis," "modeling regulatory elements," "predicting transcription factor binding sites," and motif enrichment and analysis. Within this last topic, for example, various word counting methods are employed including the Bayesian methods of expectation maximization and Gibbs sampling.

An alert reader will have noticed the prominence of Bayesian methods in the preceding paragraphs. Chapter 4, "Statistical Methods in Bioinformatics," in this collection focuses on this subject. There is a nice discussion of statistical modeling and then Bayesian inference. Dynamic programming, a recursive method of optimization, is introduced and then employed in the development of Hidden Markov Models (HMMs). Of course the basics of Markov chains must also be covered. The Metropolis-Hastings algorithm, Monte Carlo Markov chains (MCMC), and Gibbs sampling are carefully presented. Then these ideas find application in the analysis of microarray data. Here the challenging aspects of multiple hypothesis testing appear, and false discovery rate analysis is described. Hierarchical clustering and bi-clustering appear naturally in the context of microarray analysis. Then the issues of sequence analysis (especially multiple sequence analysis) are approached using these HHM and Bayesian methods along with pattern discovery in the sequences.

Discovering regulatory sequence patterns is an especially important topic in this section. The topics of this chapter appear in computer science as "machine learning" or under "data mining"; here the subject is called statistical or Bayesian methods. Whatever it is named, this is an essential area for bioinformatics.

The next chapter (Chap. 5), "Algorithms in Computational Biology," takes up the formal computational approach to our biological problems. It should be pointed out that the previous chapters contained algorithmic content, but there it was less acknowledged. It is my belief that the statistical and algorithmic approaches go hand in hand. Even with the Euclid's algorithm example of the present chapter, there are statistical issues nearby. For example, the three descriptions of Euclid's algorithm are analyzed for time complexity. It is easy to ask how efficient the algorithms are on randomly chosen pairs of integers. What is the expected running time of the algorithms? What is the variance? Amazingly these questions have answers which are rather deep. The authors soon turn to dynamic programming (DP), and once again they present clear illustrative examples, in this case Fibonacci numbers. Designing DP algorithms for sequence alignment is covered. Then a more recently developed area of genome rearrangements is described along with some of the impressive (and deep) results from the area. This topic is relevant to whole genome analysis as chromosomes evolve on a larger scale than just alterations of individual letters as covered by sequence alignment.

In Chap. 6, "Multivariate Statistical Methods in Bioinformatics Research," we have a thorough excursion into multivariate statistics. This can be viewed as the third statistical chapter in this volume. Here the multivariate normal distribution is studied in its many rich incarnations. This is justified by the ubiquitous nature of the normal distribution. Just as with the bell-shaped curve which appears in one dimension due to the central limit theorem (add up enough independent random variables and suitably normalized, one gets the normal under quite general conditions), there is also a multivariate central limit theorem. Here detailed properties are described as well as related distributions such as the Wishart distribution (the analog of the chi-square). Estimation is relevant as is a multivariate t-test. Principal component analysis, factor analysis, and linear discriminant analysis are all covered with some nice examples to illustrate the power of approaches. Then classification problems and variable selection both give platforms to further illustrate and develop the methods on important bioinformatics application areas.

Chapter 7, "Association Analysis for Human Diseases: Methods and Examples," gives us the opportunity to look more deeply into aspects of genetics. While this chapter emphasizes statistics, be aware that computational issues also drive much of the research and cannot be ignored. Population genetics is introduced and then the important subjects of genetic linkage analysis and association studies. Genomic information such as single-nucleotide polymorphisms (SNPs) provide voluminous data for many of these studies, where multiple hypothesis testing is a critical issue.

Chapter 8, "Data Mining and Knowledge Discovery Methods with Case Examples," deals with the area of knowledge discovery and data mining. To quote the authors, this area "has emerged as an important research direction for extracting useful information from vast repositories of data of various types. The basic concepts, problems and challenges deals with the area of knowledge discovery and data mining that has emerged as an important research direction for extracting useful information from vast repositories of data of various types. The basic concepts, problems and challenges are first briefly discussed. Some of the major data mining tasks like classification, clustering and association rule mining are then described in some detail. This is followed by a description of some tools that are frequently used for data mining. Two case examples of supervised and unsupervised classification for satellite image analysis are presented. Finally an extensive bibliography is provided."

The valuable chapter on Applied Bioinformatics Tools (Chap. 9) provides a step-by-step description of the application tools used in the course and data sources as well as a list of the problems. It should be strongly emphasized that no one learns this material without actually having hands-on experience with the derivations and the applications. This is not a subject for contemplation only!

Protein structure and function is a vast and critically important topic. In this collection it is covered by Chap. 10, "Foundations for the Study of Structure and Function of Proteins." There the detailed structure of amino acids is presented with their role in the various levels of protein structure (including amino acid sequence, secondary structure, tertiary structure, and spatial arrangements of the subunits). The geometry of the polypeptide chain is key to these studies as are the forces causing the three-dimensional structures (including electrostatic and van der Waals forces). Secondary structural units are classified into α-helix, β-sheets, and β-turns. Structural motifs and folds are described. Protein structure prediction is an active field, and various approaches are described including homology modeling and machine learning.

Systems biology is a recently described approach to combining system-wide data of biology in order to gain a global understanding of a biological system, such as a bacterial cell. The science is far from succeeding in this endeavor in general, let alone having powerful techniques to understand the biology of multicellular organisms. It is a grand challenge goal at this time. The fascinating chapter on Computational Systems Biology Approaches for Deciphering Traditional Chinese Medicine (Chap. 11) seeks to apply the computational systems biology (CSB) approach to traditional Chinese medicine (TCM). The chapter sets up parallel concepts between CSB and CTM. In Sect. 11.3.2 the main focus is "on a CSB-based case study for TCM ZHENG—a systems biology approach with the combination of computational analysis and animal experiment to investigate Cold ZHENG and Hot ZHENG in the context of the neuro-endocrine-immune (NEI) system." With increasing emphasis on the so-called nontraditional medicine, these studies have great potential to unlock new understandings for both CSB and TCM.

Finally I close with a few remarks about this general area. Biology is a major science for our new century; perhaps it will be the major science of the twenty-first century. However, if someone is not excited by biology, then they should find a subject that does excite them. I have almost continuously found the new discoveries such as introns or microRNA absolutely amazing. It is such a young science when such profound wonders keep showing up. Clearly no one analysis subject can solve all the problems arising in modern computational molecular biology. Statistics alone, computer science alone, experimental molecular biology alone, none of these are sufficient in isolation. Protein structure studies require an entire additional set of tools such as classical mechanics. And as systems biology comes into play, systems of differential equations and scientific computing will surely be important. None of us can learn everything, but everyone working in this area needs a set of well-understood tools. We all learn new techniques as we proceed, learning things required to solve the problems. This requires people who evolve with the subject. This is exciting, but I admit it is hard work too. Bioinformatics will evolve as it confronts new data created by the latest biotechnology and biological sciences.

University of Southern California   Michael S. Waterman
Los Angeles, USA
March 2, 2013

Contents

Contents   xix

Chapter 1
Basics for Bioinformatics

Xuegong Zhang, Xueya Zhou, and Xiaowo Wang

1.1 What Is Bioinformatics

Bioinformatics has become a hot research topic in recent years, a hot topic in several disciplines that were not so closely linked with biology previously. A side evidence of this is the fact that the 2007 Graduate Summer School on Bioinformatics of China had received more than 800 applications from graduate students from all over the nation and from a wide collection of disciplines in biological sciences, mathematics and statistics, automation and electrical engineering, computer science and engineering, medical sciences, environmental sciences, and even social sciences. So what is bioinformatics?

It is always challenging to define a new term, especially a term like bioinformatics that has many meanings. As an emerging discipline, it covers a lot of topics from the storage of DNA data and the mathematical modeling of biological sequences, to the analysis of possible mechanisms behind complex human diseases, to the understanding and modeling of the evolutionary history of life, etc.

Another term that often goes together or close with bioinformatics is computational molecular biology, and also computational systems biology in recent years, or computational biology as a more general term. People sometimes use these terms to mean different things, but sometimes use them in exchangeable manners. In our personal understanding, computational biology is a broad term, which covers all efforts of scientific investigations on or related with biology that involve mathematics and computation. Computational molecular biology, on the other hand, concentrates on the molecular aspects of biology in computational biology, which therefore has more or less the same meaning with bioinformatics.

X. Zhang (✉) • X. Zhou • X. Wang
MOE Key Laboratory of Bioinformatics and Bioinformatics Division, TNLIST/Department
of Automation, Tsinghua University, Beijing 100084, China
e-mail: zhangxg@tsinghua.edu

R. Jiang et al. (eds.), *Basics of Bioinformatics: Lecture Notes of the Graduate Summer School on Bioinformatics of China*, DOI 10.1007/978-3-642-38951-1_1,
© Tsinghua University Press, Beijing and Springer-Verlag Berlin Heidelberg 2013

Bioinformatics studies the storage, manipulation, and interpretation of biological data, especially data of nucleic acids and amino acids, and studies molecular rules and systems that govern or affect the structure, function, and evolution of various forms of life from computational approaches. The word "computational" does not only mean "with computers," but it refers to data analysis with mathematical, statistical, and algorithmic methods, most of which need to be implemented with computer programs. As computational biology or bioinformatics studies biology with quantitative data, people also call it as quantitative biology.

Most molecules do not work independently in living cells, and most biological functions are accomplished by the harmonic interaction of multiple molecules. In recent years, the new term systems biology came into being. Systems biology studies cells and organisms as systems of multiple molecules and their interactions with the environment. Bioinformatics plays key roles in analyzing such systems. People have invented the term computational systems biology, which, from a general viewpoint, can be seen as a branch of bioinformatics that focuses more on systems rather than individual elements.

For a certain period, people regarded bioinformatics as the development of software tools that help to store, manipulate, and analyze biological data. While this is still an important role of bioinformatics, more and more scientists realize that bioinformatics can and should do more. As the advancement of modern biochemistry, biophysics, and biotechnologies is enabling people to accumulate massive data of multiple aspects of biology in an exponential manner, scientists begin to believe that bioinformatics and computational biology must play a key role for understanding biology.

People are studying bioinformatics in different ways. Some people are devoted to developing new computational tools, both from software and hardware viewpoints, for the better handling and processing of biological data. They develop new models and new algorithms for existing questions and propose and tackle new questions when new experimental techniques bring in new data. Other people take the study of bioinformatics as the study of biology with the viewpoint of informatics and systems. These people also develop tools when needed, but they are more interested in understanding biological procedures and mechanisms. They do not restrict their research to computational study, but try to integrate computational and experimental investigations.

1.2   Some Basic Biology

No matter what type of bioinformatics one is interested in, basic understanding of existing knowledge of biology especially molecular biology is a must. This chapter was designed as the first course in the summer school to provide students with non-biology backgrounds very basic and abstractive understanding of molecular biology. It can also give biology students a clue how biology is understood by researchers from other disciplines, which may help them to better communicate with bioinformaticians.

1.2.1   Scale and Time

Biology is the science about things that live in nature. There are many forms of life on the earth. Some forms are visible to human naked eyes, like animals and plants. Some can only be observed under light microscope or electron microscope, like many types of cells in the scale of 1.100 μm and some virus in the scale of 100 nm. The basic components of those life forms are molecules of various types, which scale around 1.10 nm. Because of the difficulty of direct observation at those tiny scales, scientists have to invent various types of techniques that can measure some aspects of the molecules and cells. These techniques produce a large amount of data, from which biologists and bioinformaticians infer the complex mechanisms underlying various life procedures.

Life has a long history. The earliest form of life appeared on the earth about 4 billion years ago, not long after the forming of the earth. Since then, life has experienced a long way of evolution to reach today's variety and complexity. If the entire history of the earth is scaled to a 30-day month, the origin of life happened during days 3–4, but there has been abundant life only since day 27. A lot of higher organisms appeared in the last few days: first land plants and first land animals all appeared on day 28, mammals began to exist on day 29, and birds and flowering plants came into being on the last day. Modern humans, which are named homo sapiens in biology, appeared in the last 10 min of the last day. If we consider the recorded human history, it takes up only the last 30 s of the last day. The process that life gradually changes into different and often more complex or higher forms is called evolution. When studying the biology of a particular organism, it is important to realize that it is one leaf or branch on the huge tree of evolution. Comparison between related species is one major approach when investigating the unknown.

1.2.2   Cells

The basic component of all organisms is the cell. Many organisms are unicellular, which means one cell itself is an organism. However, for higher species like animals and plants, an organism can contain thousands of billions of cells.

Cells are of two major types: prokaryotic cells and eukaryotic cells. Eukaryotic cells are cells with real nucleus, while prokaryotic cells do not have nucleus. Living organisms are also categorized as two major groups: prokaryotes and eukaryotes according to whether their cells have nucleus. Prokaryotes are the earlier forms of life on the earth, which includes bacteria and archaea. All higher organisms are eukaryote, including unicellular organisms like yeasts and higher organisms like plants and animals. The bacteria *E. coli* is a widely studied prokaryote. Figure 1.1 shows the structure of an *E. coli* cell, as a representative of prokaryotic cells.

Eukaryotic cells have more complex structures, as shown in the example of a human plasma cell in Fig. 1.2. In eukaryotic cells, the key genetic materials, DNA, live in nucleus, in the form of chromatin or chromosomes. When a cell is

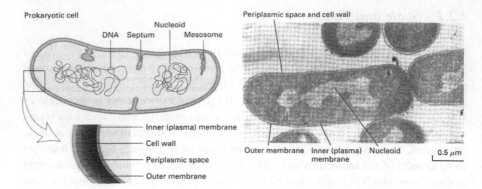

Fig. 1.1 A prokaryotic cell

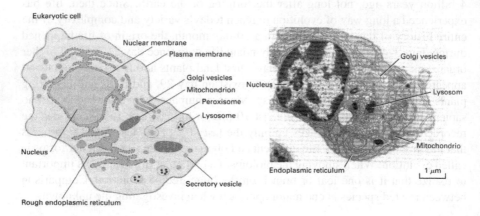

Fig. 1.2 An eukaryotic cell

not dividing, the nuclear DNA and proteins are aggregated as chromatin, which is dispersed throughout the nucleus. The chromatin in a dividing cell is packed into dense bodies called chromosomes. Chromosomes are of two parts, called the P-arm and Q-arm, or the shorter arm and longer arm, separated by the centromere.

1.2.3 DNA and Chromosome

DNA is the short name for deoxyribonucleic acid, which is the molecule that stores the major genetic information in cells. A nucleotide consists of three parts: a phosphate group, a pentose sugar (ribose sugar), and a base. The bases are of four types: adenine (A), guanine (G), cytosine (C), and thymine (T). A and G are purines with two fused rings. C and T are pyrimidines with one single ring. Besides DNA,

5'-A T T A C G G T A C C G T-3'
3'-T A A T G C C A T G G C A-5'

Fig. 1.3 An example segment of a double-strand DNA sequence

there is another type of nucleotide called RNA or ribonucleic acid. For RNA, the bases are also of these four types except that the T is replaced by the uracil (U) in RNA.

DNA usually consists of two strands running in opposite directions. The backbone of each strand is a series of pentose and phosphate groups. Hydrogen bonds between purines and pyrimidines hold the two strands of DNA together, forming the famous double helix. In the hydrogen bonds, a base A always pairs with a base T on the other stand and a G always with a C. This mechanism is called base pairing. RNA is usually a single strand. When an RNA strand pairs with a DNA strand, the base-pairing rule becomes A-U, T-A, G-C, and C-G.

The ribose sugar is called pentose sugar because it contains five carbons, numbered as $1'$–$5'$, respectively. The definition of the direction of a DNA or RNA strand is also based on this numbering, so that the two ends of a DNA or RNA strand are called the $5'$ end and the $3'$ end. The series of bases along the strand is called the DNA or RNA sequence and can be viewed as character strings composed with the alphabet of "A," "C," "G," and "T" ("U" for RNA). We always read a sequence from the $5'$ end to the $3'$ end. On a DNA double helix, the two strands run oppositely. Figure 1.3 is an example of a segment of double-strand DNA sequence. Because of the DNA base-pairing rule, we only need to save one strand of the sequence.

DNA molecules have very complicated structures. A DNA molecule binds with histones to form a vast number of nucleosomes, which look like "beads" on DNA "string." Nucleosomes pack into a coil that twists into another larger coil and so forth, producing condensed supercoiled chromatin fibers. The coils fold to form loops, which coil further to form a chromosome. The length of all the DNA in a single human cell is about 2 m, but with the complicated packing, they fit into the nucleus with diameter around 5 μm.

1.2.4 The Central Dogma

The central dogma in genetics describes the typical mechanism by which the information saved in DNA sequences fulfills its job: information coded in DNA sequence is passed on to a type of RNA called messenger RNA (mRNA). Information in mRNA is then passed on to proteins. The former step is called transcription, and latter step is called translation.

Transcription is governed by the rule of complementary base pairing between the DNA base and the transcribed RNA base. That is, an A in the DNA is transcribed to a U in the RNA, a T to an A, a G to a C, and vice versa.

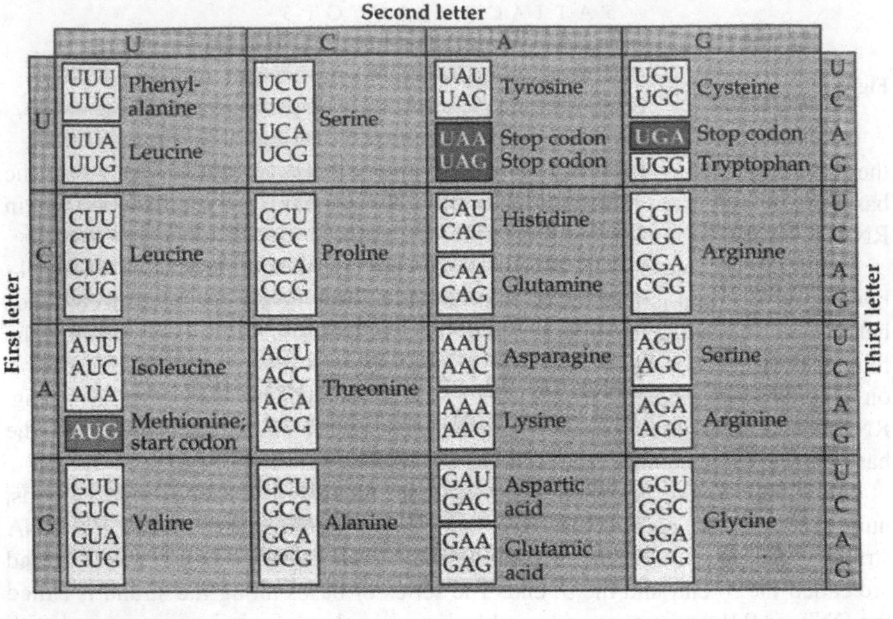

Second letter

		U		C		A		G		
U		UUU UUC	Phenyl-alanine	UCU UCC	Serine	UAU UAC	Tyrosine	UGU UGC	Cysteine	U C	
		UUA UUG	Leucine	UCA UCG		UAA UAG	Stop codon Stop codon	UGA UGG	Stop codon Tryptophan	A G	
C		CUU CUC CUA CUG	Leucine	CCU CCC CCA CCG	Proline	CAU CAC	Histidine	CGU CGC CGA CGG	Arginine	U C A G	
						CAA CAG	Glutamine				
A		AUU AUC AUA	Isoleucine	ACU ACC ACA ACG	Threonine	AAU AAC	Asparagine	AGU AGC	Serine	U C	
		AUG	Methionine; start codon			AAA AAG	Lysine	AGA AGG	Arginine	A G	
G		GUU GUC GUA GUG	Valine	GCU GCC GCA GCG	Alanine	GAU GAC	Aspartic acid	GGU GGC GGA GGG	Glycine	U C A G	
						GAA GAG	Glutamic acid				

Fig. 1.4 The genetic codes

Proteins are chains of amino acids. There are 20 types of standard amino acids used in lives. The procedure of translation converts the information from the language of nucleotides to the language of amino acids. The translation is done by a special dictionary: the genetic codes or codon. Figure 1.4 shows the codon table. Every three nucleotides code for one particular amino acid. The three nucleotides are called a triplet. Because three nucleotides can encode 64 unique items, there are redundancies in this coding scheme, as shown in Fig. 1.4. Many amino acids are coded by more than one codon. For the redundant codons, usually their first and second nucleotides are consistent, but some variation in the third nucleotide is tolerated. AUG is the start codon that starts a protein sequence, and there are three stop codons CAA, CAG, and UGA that stop the sequence.

Figure 1.5a illustrates the central dogma in prokaryotes. First, DNA double helix is opened and one strand of the double helix is used as a template to transcribe the mRNA. The mRNA is then translated to protein in ribosome with the help of tRNAs.

Figure 1.5b illustrates the central dogma in eukaryotes. There are several differences with the prokaryote case. In eukaryotic cells, DNAs live in the nucleus, where they are transcribed to mRNA similar to the prokaryote case. However, this mRNA is only the preliminary form of message RNA or pre-mRNA. Pre-mRNA is processed in several steps: parts are removed (called spicing), and ends of 150–200 As (called poly-A tail) are added. The processed mRNA is exported outside the nucleus to the cytoplasm, where it is translated to protein.

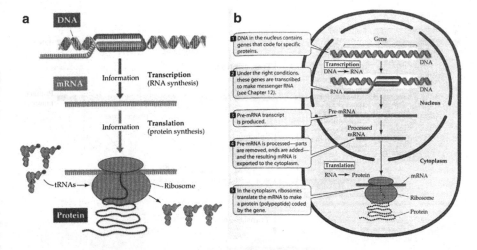

Fig. 1.5 The central dogma

The procedure that genes are transcribed to mRNAs which are then translated to proteins is called the expression of genes. And the abundance of the mRNA molecules of a gene is usually called the expression value (level) of that gene, or simply the expression of the gene.

1.2.5 Genes and the Genome

We believe that the Chinese translation "基因"of the term "gene" is one of the best scientific term ever translated. Besides that the pronunciation is very close to the English version, the literal meaning of the two characters is also very close to the definition of the term: basic elements. Genes are the basic genetic elements that, together with interaction with environment, are decisive for the phenotypes.

Armed with knowledge of central dogma and genetic code, people had long taken the concept of a gene as the fragments of the DNA sequence that finally produce some protein products. This is still true in many contexts today. More strictly, these DNA segments should be called protein-coding genes, as scientists have found that there are some or many other parts on the genome that do not involve in protein products but also play important genetic roles. Some people call them as nonprotein-coding genes or noncoding genes for short. One important type of noncoding genes is the so-called microRNAs or miRNAs. There are several other types of known noncoding genes and may be more unknown. In most current literature, people still use gene to refer to protein-coding genes and add attributes like "noncoding" and "miRNA" when referring to other types of genes. We also follow this convention in this chapter.

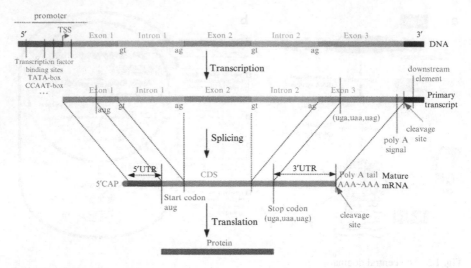

Fig. 1.6 The structure of a gene

The length of a DNA segment is often counted by the number of nucleotides (nt) in the segment. Because DNAs usually stay as double helix, we can also use the number of base pairs (bp) as the measurement of the length. For convenience, people usually use "k" to represent "1,000." For example, 10 kb means that the sequence is of 10,000 bp. A protein-coding gene stretches from several hundreds of bp to several k bp in the DNA sequence. Figure 1.6 shows an example structure of a gene in high eukaryotes.

The site on the DNA sequence where a gene is started to be transcribed is called the transcription start site or TSS. The sequences around (especially the upstream) the TSS contain several elements that play important roles in the regulation of the transcription. These elements are called cis-elements. Transcription factors bind to such factors to start, enhance, or repress the transcription procedure. Therefore, sequences upstream the TSS are called promoters. Promoter is a loosely defined concept, and it can be divided into three parts: (1) a core promoter which is about 100 bp long around the TSS containing binding sites for RNA polymerase II (Pol II) and general transcription factors, (2) a proximal promoter of several hundred base pairs long containing primary specific regulatory elements located at the immediately upstream of the core promoter, and (3) a distal promoter up to thousands of base pairs long providing additional regulatory information. In eukaryotes, the preliminary transcript of a gene undergoes a processing step called splicing, during which some parts are cut off and remaining parts are joined. The remaining part is called exon, and the cut part is called intron. There can be multiple exons and introns in a gene. After introns are removed, the exons are connected to form the processed mRNA. Only the processed mRNAs are exported to the cytoplasm, and only parts of the mRNAs are translated to proteins. There may be untranslated regions (UTRs) at both ends of the mRNA: one at the TSS end is called 5′-UTR, and the other at the tail end is called 3′-UTR. The parts of exons that are translated are called CDS or coding DNA sequences. Usually exons constitute only a small part in the sequence of a gene.

In higher eukaryotes, a single gene can have more than one exon-intron settings. Such genes will have multiple forms of protein products (called isoforms). One isoform may contain only parts of the exons, and the stretch of some exons may also differ among isoforms. This phenomenon is called alternative splicing. It is an important mechanism to increase the diversity of protein products without increasing the number of genes.

The term "genome" literally means the set of all genes of an organism. For prokaryotes and some low eukaryotes, majority of their genome is composed of protein-coding genes. However, as more and more knowledge about genes and DNA sequences in human and other high eukaryotes became available, people learned that protein-coding genes only take a small proportion of all the DNA sequences in the eukaryotic genome. Now people tend to use "genome" to refer all the DNA sequences of an organism or a cell. (The genomes of most cell types in an organism are the same.)

The human genome is arranged in 24 chromosomes, with the total length of about 3 billion base pairs (3×10^9 bp). There are 22 autosomes (Chr.1.22) and 2 sex chromosomes (X and Y). The 22 autosomes are ordered by their lengths (with the exception that Chr.21 is slightly shorter than Chr.22): Chr.1 is the longest chromosome and Chr.21 is the shortest autosome. A normal human somatic cell contains 23 pairs of chromosomes: two copies of chromosomes 1.22 and two copies of X chromosome in females or one copy of X and one copy of Y in males. The largest human chromosome (Chr.1) has about 250 million bp, and the smallest human chromosome (Chr.Y) has about 50 million bp.

There are about 20,000–25,000 protein-coding genes in the human genome, spanning about 1/3 of the genome. The average human gene consists of some 3,000 base pairs, but sizes vary greatly, from several hundred to several million base pairs. The protein-coding part only takes about 1.5–2 % of the whole genome. Besides these regions, there are regulatory sequences like promoters, intronic sequences, and intergenic (between-gene) regions. Recent high-throughput transcriptomic (the study of all RNA transcripts) study revealed that more than half of the human genomes are transcribed, although only a very small part of them are processed to mature mRNAs. Among the transcripts are the well-known microRNAs and some other types of noncoding RNAs. The functional roles played by majority of the transcripts are still largely unknown. There are many repetitive sequences in the genome, and they have not been observed to have direct functions.

Human is regarded as the most advanced form of life on the earth, but the human genome is not the largest. Bacteria like *E. coli* has genomes of several million bp, yeast has about 15 million bp, Drosophila (fruit fly) has about 3 million bp, and some plants can have genomes as large as 100 billion bp. The number of genes in a genome is also not directly correlated with the complexity of the organism's complexity. The unicellular organism yeast has about 6,000 genes, fruit fly has about 15,000 genes, and the rice that we eat everyday has about 40,000 genes. In lower species, protein-coding genes are more densely distributed on the genome. The human genome also has a much greater portion (50 %) of repeat sequences than the lower organisms like the worm (7 %) and the fly (3 %).

1.2.6 Measurements Along the Central Dogma

For many years, molecular biology can only study one or a small number of objects (genes, mRNAs, or proteins) at a time. This picture was changed since the development of a series of high-throughput technologies. They are called high throughput because they can obtain measurement of thousands of objects in one experiment in a short time. The emergence of massive genomic and proteomic data generated with these high-throughput technologies was actually a major motivation that promotes the birth and development of bioinformatics as a scientific discipline. In some sense, what bioinformatics does is manipulating and analyzing massive biological data and aiding scientific reasoning based on such data. It is therefore crucial to have the basic understanding of how the data are generated and what the data are for.

1.2.7 DNA Sequencing

The sequencing reaction is a key technique that enables the completion of sequencing the human genome. Figure 1.7 illustrates the principle of the widely used Sanger sequencing technique.

The technique is based on the complementary base-pairing property of DNA. When a single-strand DNA fragment is isolated and places with primers, DNA polymerase, and the four types of deoxyribonucleoside triphosphate (dNTP), a new DNA strand complementary to the existing one will be synthesized. In the DNA sequencing reaction, dideoxyribonucleoside triphosphate (ddNTP) is added besides the above components, and the four types of ddNTPs are bound to four different fluorescent dyes. The synthesis of a new strand will stop when a ddNTP instead of a dNTP is added. Therefore, with abundant template single-strand DNA fragments, we'll be able to get a set of complementary DNA segments of all different lengths, each one stopped by a colored ddNTP. Under electrophoresis, these segments of different lengths will run at different speeds, with the shortest segments running the fastest and the longest segments running the slowest. By scanning the color of all segments ordered by their length, we'll be able to read the nucleotide at each position of the complementary sequence and therefore read the original template sequence. This technique is implemented in the first generation of sequencing machines.

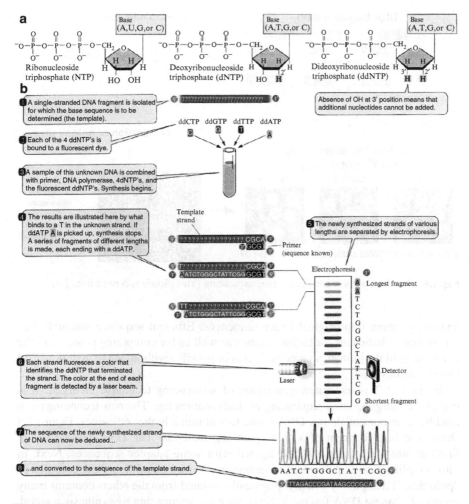

Fig. 1.7 Sequencing reaction

The sequencing reaction can only measure sequence fragments of up to ~800 nt (it's very difficult to separate larger DNA fragments that have only one nucleotide difference in length by current capillary electrophoresis). For the whole human genome, scientists have been able to mark the long genome with DNA sequence tags whose genomic position can be uniquely identified and cut the DNA into large fragments (~million bp). These fragments are still too long for the sequencing machine. Scientists invented the so-called shotgun strategy to sequence those long DNA fragments. The DNA is randomly broken into shorter fragments of 500–800 bp, which can be sequenced by the sequencing machine to obtain reads. Multiple overlapping reads are gathered by several rounds of fragmentation followed by sequencing. Computer programs piece together those overlapping reads, align,

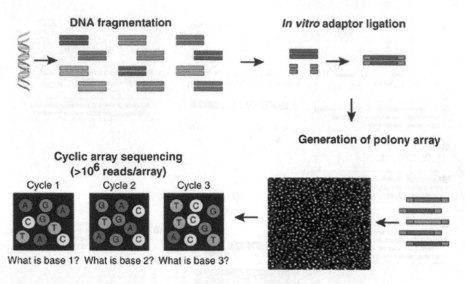

Fig. 1.8 Flowchart of the second-generation sequencing (This picture is derived from [1])

and merge them into original larger sequences. Efficient sequence assembly has raised many challenges for bioinformatics as well as for computing power, and the completion of the human genome project is impossible without the help of powerful bioinformatics tools.

In the last 2 years, a new generation of sequencing technique emerged. It is called second-generation sequencing or deep sequencing. The new technology can read huge amount of shorter DNA sequences at much higher efficiency. Figure 1.8 shows the brief concept of such sequencing methods: The DNA fragments are first cut into short fragments and ligated with some adaptor sequences. Next, in vitro amplification is performed to generate an array of million PCR colonies or "polonies." Each polony which is physically isolated from the others contains many copies of a single DNA fragment. Next, with the sequencing by synthesis method, serial extension of primed templates is performed, and fluorescent labels incorporated with each extension are captured by high-resolution image-based detection system. The nucleotide synthesis (complement to the template DNA fragment) of each polony at each cycle is recalled by processing the serial image data. Compared with the Sanger sequencing technology, the deep sequencing method is highly parallel and can produce gigas of sequence data in a single day. Up to the end of 2007, there are mainly three deep sequencing platforms: 454, Solexa, and SOLiD. The 454 system which is designed based on the pyrosequencing technology could produce about 100 Mb sequences in a single instrument run with reads length up to 200–400 nt. Solexa and SOLiD could produce about one to two Gb sequences in a single run but with reads length only up to about 36 nt. The key advantage of the 454 system is its longer reads length, and its application focuses on de novo genome sequencing (sequence the unknown genome). In contrast the Solexa and SOLiD platforms are mainly used for genome resequencing (like SNP detection), transcriptomic analysis, ChIP-seq analysis, etc. With the rapid development of new technologies, sequencing personal genomes is becoming a realistic goal. In October 2006, the X Prize Foundation established the Archon X Prize, which intends to award $10 million to "the first Team that can build a device and use it to sequence 100 human genomes within 10 days or less, with an accuracy of no more than one error in every 100,000 bases sequenced, with sequences accurately covering at least 98 % of the genome, and at a recurring cost of no more than $10,000 (US) per genome."

1.2.8 Transcriptomics and DNA Microarrays

The genome can be viewed as the original blueprint of the cell. When a gene is to take effect, it is transcribed into mRNAs as the acting copy, according to which proteins are made. This procedure is called the expression of a gene. When more proteins of some type are needed, more mRNAs will be made. Therefore, the abundance of the mRNA in the cell indicates the expression level of the corresponding gene. Sometimes people call it the expression of the gene for simplicity.

The genomic information in all or most cells of an organism is the same, but genes express differently at different developmental stage and in different tissues. There is an estimation that only about one third of all genes are expressed at the same time in a certain tissue. There are genes that perform basic functions in cells and therefore are expressed in all tissues. They are called housekeeping genes. On the other hand, many genes show distinctive tissue-specific expression patterns. That means they may be expressed highly in one type of cells but not in other cells. Different cell types in a multicellular organism express different sets of genes at different time and with different quantities. Basic cellular processes are realized by tightly regulated gene expression programs.

Therefore, it is important to study the expression profiles of the whole repertoire of genes. The study of all transcripts is called transcriptomes. The DNA microarray is a key high-throughput technique in transcriptomic investigations. It can simultaneously measure the abundance of mRNAs of thousands or more genes. As mRNAs often degrade rapidly, usually complementary DNAs (cDNAs) reverse transcribed from the mRNAs are used in the measurement.

The basic principle of microarrays is also the complementary base-pairing hybridization of DNAs. Pieces of different DNA fragments (called probes) are placed on a small chip. The probes were designed in ways that they can represent individual genes. When the samples' cDNAs are applied on the chip, they'll hybridize with the probes whose sequences are complementary to the cDNA sequences, and those DNAs that do not hybridize to any probe will be washed off. With proper fluorescence labeling on the cDNAs, their abundances can be "read" from the fluorescence intensities at each probe locations. These readings measure the expression levels of the genes represented by the probes.

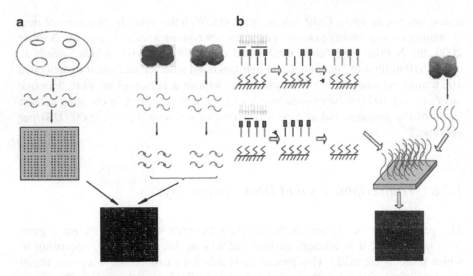

Fig. 1.9 DNA microarrays. (**a**) Printed cDNA microarray. (**b**) Oligonucleotide microarray

There are two different types of DNA microarrays: the printed cDNA microarray (cDNA microarray for short) and the oligonucleotide microarray. The major difference is their ways of preparing the probes. Figure 1.9 illustrates the basic principle of the two types of methods.

In cDNA microarrays, probes are relatively long segments of genes made from cloned cDNA libraries. They are spotted on the chip with techniques similar to jet printers. Different labs can prepare their own probes according to the genes they want to study. However, this advantage of flexibility also brings the disadvantage that the quantity of each probe can hardly be controlled precisely. Therefore, data reproducibility and comparison between the data from two labs can be a problem. To tackle this problem, usually two samples of identical amount labeled with different fluorescences are applied to the chip. If a gene is expressed at different abundances in the two samples, the two fluorescences will have different intensities as the result of competitive hybridization, and the ratio of the two intensities will reflect the ratio of the gene's expression in the two samples. This strategy can partially eliminate effect of possible variance in the quantity of probes. The two samples can be one patient and one control. When studying multiple samples, each of them can be compared with a common control or a matched control. Different experiment designs have different characteristics and can have different implications on the bioinformatics processing of the data.

Oligonucleotide microarrays use much shorter probes (about ~25nt), and multiple probes may be used for a single gene to increase the specificity. Taking the Affymetrix GeneChip as an example, the oligonucleotide probes are grown on the chip with light-directed oligonucleotide synthesis. The quantity of the probes can be precisely controlled. Usually only one sample is applied on the chip, and the reading is the expression level of each gene in the sample instead of the ratio of two expressions. These types of chips have higher density than cDNA arrays as the probes are much shorter. The latest Affymetrix expression microarray contains probes for all known human genes. The data obtained at two labs with the same system can be better compared. The disadvantage of oligonucleotide microarrays is that the chips are factory made and are therefore less flexible: individual labs cannot design their own chips, and ordering customized chips can be much more expensive. The data quality of oligonucleotide microarrays is in general regarded as of better quality than cDNA microarrays.

No matter what type of microarray it is, the original data form is scanned images. From reading the intensities on the images to getting expression of genes, there are many problems that bioinformatics has played major roles. After getting the expression of multiple genes in different samples, bioinformatics then becomes the major player in analyzing the data.

A typical microarray-based study is to compare the expression of many genes between two groups of samples, or across certain time course. For example, in studying the molecular features of two subtypes of the same cancer, patient samples of the two subtypes are collected and microarrays are used to measure the gene expression. A vector of the expression of all genes is obtained for each sample, and a gene expression matrix is obtained on all the samples, with columns representing samples and rows representing genes. A typical bioinformatics task is to identify the genes underlying the distinction between the two subtypes of the cancer.

Only a small proportion of the genome contains protein-coding genes; however, many other parts of the genome are also transcribed. The microarrays introduced above are now often called gene expression microarrays. Many other types of microarrays have emerged in recent years, following the same general principle. For example, microRNAs (miRNAs) are a type of small noncoding RNAs that play important regulatory roles in cells. By using probes for those microRNAs, microarrays can be designed for measuring the expression of microRNAs in the sample. As the density of the microarray chips increases, scientists have developed the so-called tiling arrays that have probes tiling the whole genome at high resolution. With such tiling arrays, we can measure the abundance of all transcribed parts of the genome, no matter they are known protein-coding genes, microRNAs, or transcripts that are previously unknown. It is with such technique that scientists have found that most parts of the human genome are transcribed. However, the noise level of high-density microarrays is still very high, which raises more questions to bioinformatics for processing the data.

In early days, the term "transcriptome" often meant the study of the expression of all genes at mRNA level, which is similar to "expression profiling." However, with more and more noncoding transcripts discovered, the term is closer to its literal meaning: the study of all or many transcripts.

It should be noted that with the development of second-generation sequencing, microarrays are no longer the only choice for transcriptomic study. Deep sequencing can be applied to count the fragments of cDNAs so that the expression of RNAs can be measured in digital.

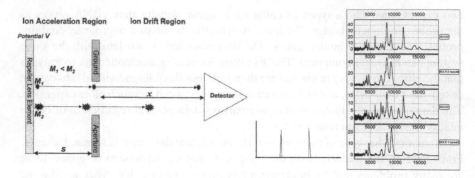

Fig. 1.10 TOF-MS

1.2.9 Proteomics and Mass Spectrometry

For the protein-coding genes, mRNAs are only the intermediate product. Proteins are even more complicated than DNA/RNA. For many genes, the abundance of their protein products is not linearly correlated with the expression at mRNA level. Therefore, studying the expression of proteins is important for understanding the molecular system of cells. Proteomics is the study of all or many proteins. Due to the mechanisms like alternative splicing and posttranslational protein modification, the diversity of proteins is much larger than that of genes. People even don't have the common understanding about the approximate number of all types of proteins in human. It may be several magnitudes larger than the number of genes.

Two key features for identifying proteins are their electrical charge and their molecular mass. Scientists developed techniques to separate protein mixtures according to these factors. A representative technique is the 2D gel electrophoresis (2D gel, for short), which separates protein mixtures first according to the isoelectric focusing (IEF) and then by the mass.

Mass spectrometry is a technique that is widely used in proteomics study. The key principle is the time-of-flight mass spectrometry (TOF-MS): ionized proteins are located on certain surface or matrix, and an electricity field is applied. The charged protein or peptides (protein segments) will fly in the electricity field and reach a detector. The time of the flight before reaching the detector depends on the mass-to-charge ratio of the protein, and the strength of the signal at the detector is proportional to the accumulation of the molecules which reflects the quantity of the proteins. Figure 1.10 illustrates the basic principle of TOF-MS.

Mass spectrometry has three typical types of applications in proteomics study. The first one is to identify proteins/peptides in a mixture. This is rather straightforward: on the mass spectrum of a mixture, a peak corresponds to the abundance of certain protein or peptide. By searching for proteins with the same or very close molecular weight as the peak location in protein databases, the protein can be identified if it has been reported before.

The second type of application is de novo amino acid sequencing. A protein segment is broken into all possible pieces before applying to the MS machine. On the mass spectrum, multiple peaks can be detected corresponding to peptide segments of different lengths. Different amino acid sequence segments will result in different molecular masses corresponding to peaks at different locations. Therefore, from all the peaks, it is theoretically possible to resolve the sequence. However, this is a combinatorial problem and is a challenging task for bioinformatics algorithms.

Usually tandem mass spectrometry (or MS/MS) is adopted in such applications. Tandem mass spectrometry is two (or more) rounds of mass spectrometry. For example, the first round can isolate one peptide from the protein mixture, and the second round is used to resolve the sequence.

Another typical application of mass spectrometry is to study the expression of multiple proteins in the samples, like the microarrays are used for measuring mRNA abundances. The mass spectrum of each sample provides the expression profile of all the proteins in the sample. By aligning the peaks between different samples, we can detect the proteins that are differentially expressed between groups of samples, and we can also study the different patterns of multiple peaks between the compared samples.

1.2.10   ChIP-Chip and ChIP-Seq

Chromatin Immunoprecipitation or ChIP is an experimental technique that can capture the DNA segments bound by certain proteins. The ChIP-chip technique combines ChIP with DNA microarrays (especially tiling arrays) to detect the DNAs that are bound by specific proteins like transcription factors. This technique is widely used in the study of transcription factor binding sites and histone modification states. ChIP-sequencing or ChIP-seq is a more recent developed technology which shares the same idea with ChIP but replaces the microarrays by deep sequencing. Figure 1.11 illustrates the basic principle of ChIP-seq method. In the first step, the DNA-binding proteins (transcription factors or histones) are cross-linked with the DNA site they bind to in an in vivo environment. Then, the total DNA is extracted and cut into fragments of hundreds of base pairs in length by sonication or some nuclease. Next, the DNA-protein complex are selected by using an antibody specific to the protein of interest, and the DNA fragments that do not bind by this protein are washed away. In the next step, the DNA-protein complexes are reverse cross-linked and the remaining DNA is purified. After size selection, all the resulting ChIP-DNA fragments are sequenced simultaneously using deep sequencing machines like Solexa Genome Analyzer which can generate millions of short reads with length of about 36 bp in a single sequencing run. Finally, the sequenced reads are mapped to the genome by some high efficient reads mapping programs like ELAND, RMAP, or ZOOM, and the genomic loci with high reads intensity are identified as putative binding sites of the protein of interest.

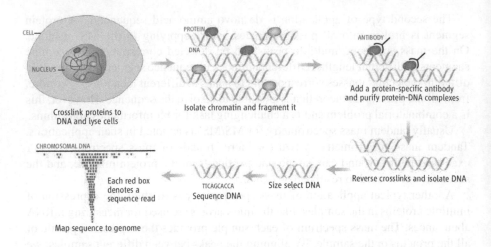

CELL—
NUCLEUS—
Crosslink proteins to
DNA and lyse cells

PROTEIN
DNA
Isolate chromatin and fragment it

ANTIBODY
Add a protein-specific antibody
and purify protein-DNA complexes

CHROMOSOMAL DNA
Each red box
denotes a
sequence read
Map sequence to genome

TTCAGCACCA
Sequence DNA

Size select DNA

Reverse crosslinks and isolate DNA

Fig. 1.11 Summary of ChIP-seq method (This figure is adapted from [2])

1.3 Example Topics of Bioinformatics

With the quick review of some warm-up biological knowledge, far from being adequate for building any solid background, we quickly move to introduce some typical examples of bioinformatics research. Some of the topics will be deeply covered by following chapters. The purpose of this section is to give readers an overall feeling about what types of questions can be and should be answered by bioinformatics and computational biology.

1.3.1 Examples of Algorithmatic Topics

The birth and growth of bioinformatics as a discipline was accompanied by the generation and accumulation of data in molecular biology. When data size increases, even a simple task in manipulating the data may become non-straightforward. We need special algorithms to do that. "Algorithm" means a step-by-step computational procedure for solving a problem.

For example, when we get many DNA sequence data, storing them in computer databases seems trivial. However, when we find a new sequence segment (called a query) and ask whether this sequence has already been deposited in the database, it becomes less trivial, or even challenging, when the database is huge. When we are looking not only at the sequence segments that are exactly the same as the query but at those sequences that look similar with the query, it becomes even more difficult. This is what sequence analysis is about.

Fig. 1.12 An example of pairwise sequence alignment

```
ATGGTGCACCTGACTGATGCTGAGAAGGCTGCTGT
ATGGTGCACCTGACTCCT---GAGGAGAAGTCTGC
```

The sequence database query problem can be boiled down to finding the best local alignment between two sequences, or two strings in computer science jargon. Figure 1.12 presents a very simple illustration of the problem. It was a breakthrough when the dynamic programming approach for such problems was proposed by Temple Smith and Michael Waterman in 1981, although at that time only few people realized how important the work was. Sequence alignment is a very basic problem in bioinformatics. The question has many variations, and it is the foundation for many other topics.

From the brief introduction of the shotgun sequencing method, we could realize that assembling the long sequence from many short reads is a challenging task for algorithms. It is like to solve a huge jigsaw puzzle problem. With the availability of massive deep sequencing data, a related problem is how to efficiently map the short sequence reads back to the genome.

Multiple sequence alignment brings another dimension of complexity to the problem. Comparative genomics is based on multiple sequence alignment. The genomes of multiple organisms can be compared to infer the evolutionary history of the species. Building the phylogenetic tree is an important challenge for algorithms.

Besides sequence-related problems, there are also many other types of algorithmatic problems in bioinformatics, such as finding hidden patterns in a noisy microarray matrix, inferring the amino acid sequences from possible combinations, and analyzing biological network graphs.

1.3.2   Examples of Statistical Topics

When we search for a short query sequence in a long genome segment, we need to design powerful algorithms to find matched targets efficiently. But when we want to infer biological conclusions from the searching result, we need to ask questions like "what is the probability of finding the matched targets in the candidate sequence under a certain biological context?" This is one type of questions that statistics help to answer.

From data point of view, there are two types of bioinformatics tasks: one is the processing of the data themselves, and the other is inferring answers to biological questions from the data. Most, if not all, biological data can be viewed as noisy sample generated by some underlying probabilistic rules. Statistical inference is a discipline to infer the underlying rules from data. The key concept is the so called p-value, which gives an estimation of the probability to have the observed data when a hypothesized rule does not apply. For example, when a particular short sequence pattern (called motif) is found in the promoters of a set of genes that tend to express in a coordinated manner, one will ask the probability of observing the multiple appearance of such a sequence pattern by chance. The question can be answered with some statistical models about the DNA sequences. If the probability is small enough, then one will tend to believe that the sequence motif has some role in the coordination of the genes. This example is a simplified description of the motif discovery problem which plays a key role in many bioinformatics and functional genomics study.

In microarray study of cancers, a basic question is which genes are differentially expressed between cancer and normal samples. This is a typical statistical question, and many standard statistical methods can be applied. However, due to the special characteristics of microarray data, new methods are also needed.

The complexity nature of many types of biological data raises many new challenges to established statistical models. How to build proper statistical models based on biological knowledge and make inferences from data is a key question in many bioinformatics research. For example, in gene recognition tasks, scientists have built very sophisticated hidden Markov models that incorporate existing knowledge about gene structure.

1.3.3 Machine Learning and Pattern Recognition Examples

Building statistical models is one way to describe the data and make predictions. Another approach is to build a prediction machine directly from the data. This approach is called machine learning, which is an important topic in the field of intelligent information processing. When the target to be predicted is discrete classes, the task is called pattern recognition or pattern classification.

Machine learning has been widely used in bioinformatics. For example, recognizing genes and other functional elements on the genome is an important topic in bioinformatics and genomics. Scientists have developed machine learning methods such as artificial neural networks and support vector machines for these types of tasks. A learning machine is actually also a model, but not necessarily a statistical one, and data reported with biological experiments are used to train the model. HMM can also be regarded as a machine learning method. It uses a sequential statistical model to describe the data, and parameters in the model also need to be trained with known data.

Another typical example is using microarray data or proteomics expression data to classify cancers. For each patient, the gene expressions measured by microarrays compose a vector. They can be viewed as the original features for classifying the samples. One can select a smaller number of genes to classify a certain type of cancer with normal cells or to classify subtypes of the cancer. It seems like a standard pattern recognition task. However, microarray data has several unique properties: the data dimension can be very high (tens of thousands of dimension), but the sample size is usually small (in hundreds or less). Some traditional machine programs cannot work in such extreme scenario. Many people developed new or improved machine methods for this type of questions.

Besides supervised machine learning problems, unsupervised machine learning also has broad application in bioinformatics. Among the many other examples, hierarchical clustering can be used to cluster genes into groups with possible function correlation according to their expression profiles and can be used to cluster samples into groups based on their gene expressions.

1.3.4 Basic Principles of Genetics

Up to this point, we have compiled an incomplete compendium of research areas of modern biology from bioinformaticians' perspective. One of the important areas worth a separate discussion here is genetics. As we elaborated at the beginning of this chapter, it is hard to quantify the precise scope of bioinformatics, as a result of its multidisciplinary nature. Genetics, however, has seldom been taken as a part of bioinformatics. It sounds surprising, since both fields are entrenched on the shared methodological background: statistics and algorithm. But it is understandable that while the large body of bioinformatics is focused on a single representative sequence of the genome, the principal concept of genetics is interindividual variation which makes it quite detached from the result of biology. On the other hand, we emphasize that the development of modern genetics cannot be possible without the advancement in biotechnology and aids from bioinformatics; bioinformaticians should be acquainted with the basic principles of genetics in order to better communicate with geneticists. In this section, we take a historical approach to distill the essential concepts of genetics within the context of disease gene mapping.

1.3.4.1 Mendel and Morgan's Legacy

The dawn of the modern genetics is unanimously attributed to Mendel's seminal work on pea plant. More than 140 years ago, Mendel observed that crossing purebred peas with one binary trait (e.g., yellow and green seed color) resulted in one trait (yellow seeds) rather than a mixture of two; after selfing of F1 generation, seed color (yellow/green) exhibited 3:1 ratio. Similarly when crossing two binary traits (e.g., purple or white flower color plus spherical or wrinkled seed shape), 9:3:3:1 ratio was observed among the F2 generation for all combination of traits. Mendel postulated that each individual's binary trait was controlled by a distinct factor (later called *genes*), which had two different forms (*alleles*), recessive and dominant. Genes normally occur in pairs in a normal body cell: one is maternal derived and the other paternal derived. Within an individual, if two alleles are identical, then the individual is called *homozygous* for that gene; otherwise, the individual is called *heterozygous*. Individual's appearance is determined by the set of alleles it happens to possess (*genotype*) and environment. In case of heterozygote, dominant allele will hide the effect of recessive allele. During the formation of sex cells (*gametes*), two alleles of a gene will segregate and pass on to eggs or sperms, each of which receives one randomly chosen allele copy (law of segregation). And alleles of different genes will pass on independently to each other to the offspring, so there is no relation between, for example, seed shape and color of flower (law of independent assortment).

The significance Mendel's work was the proposition of the concept of gene as the discrete hereditary unit whose different alleles control different traits. It took another 40 years until the importance of Mendel's idea was recognized. Soon after geneticists rediscovered Mendel's law, they found that the independent assortment for different traits was not always the case. Instead, they observed that there are groups of traits tended to be inherited together (*linked*) by the offspring rather than assorted independently (*unlinked*). The dependence of inheritance (*linkage*) led Morgan et al. to the chromosome theory of inheritance in which chromosomes were thought to harbor genetic material. In diploid organism, chromosomes come in pairs; each *homolog* comes from one parent. During *meiosis*, the process to produce gametes, one parent provides one chromosome from each homologous pair. During first round division of meiosis, several *crossover* events will take place between homologous positions of two parental chromosomes, such that the transmitted chromosome consists alternating segments from two parental alleles. Chromosome theory elucidated the biological basis for Mendel's law of segregation and also reconciled the contradiction between linked traits and the violation to law of independent assortment. It turned out that genes controlling Mendel's pea traits were either on different chromosomes or located far apart on the same chromosome where an obligatory crossover in between must occur. Chromosome theory postulated that genes are arranged linearly along the chromosomes; the combination of nearby alleles along the same chromosome (*haplotype*) tended to be transmitted jointly unless they are shuffled by crossover.

The distance separating two genes on the same chromosome determines the frequency of their recombinant (*genetic distance*) and the probability that corresponding traits will be inherited together by offspring. By analyzing co-inheritance pattern of many linked traits from experimental crosses or family pedigrees, it is possible to place corresponding genes in order and estimate genetic distances between neighboring genes. Rigorous statistical methods were developed to construct such genetic maps. It is truly remarkable in retrospect that early-day geneticists were able to know where genes were and their relative positions even they had no idea about molecular structure of genes.

1.3.4.2 Disease Gene Mapping in the Genomic Era

The early day practice taking gene as a polymorphic landmark naturally spawned the concept of *genetic markers* (or locus) in the genomic era. Alleles giving rise to different Mendel's pea traits are just coding variants that produce different protein *isoforms* among individuals (called *non-synonymous* variants; also recall that alternative splicing creates protein isoforms within the same individuals). There are many more types of variations whose different forms (also termed *alleles*), coding or noncoding, can be directly assayed from DNA level. While some alleles may cause changes in phenotypes, for example, increasing the risk to diseases, most are *neutral* (little phenotypic consequences) and commonly occurring within human population. Among them, two types of variations have shown greatest practical utility: single base-pair change (*single nucleotide polymorphism, SNP*) and short sequence of 1.6 bp repeated in tandem (*microsatellite*).

A microsatellite locus typically has tens of alleles (copy numbers of repeating unit), which can be determined via PCR amplification from unique flanking sequences. Highly variable alleles among human individuals make microsatellite the ideal markers to construct human genetic map from extended pedigrees. A map of ordered DNA markers had huge practical values. It allowed geneticists to localize *loci* (e.g., protein-coding genes and regulatory elements) whose mutations therein are responsible for the trait of our interest (e.g., diseases status and crop yield) on to the grid of prearranged genomic landmarks, a process known as *gene mapping*. The idea of *gene mapping* via *linkage analysis* is not new, inheriting the legacy from Mendel and Morgan: both DNA tags and traits loci are taken as genetic markers; and their relative orders are determined by tracing co-inheritance pattern of traits with markers in families or experimental crosses. Linkage studies using human pedigrees during the past 30 years have led to the mapping of thousands of genes within which some single mutations cause severe disorders (*Mendelian disease*), like Tay-Sachs diseases and cystic fibrosis, among others (see Online Mendelian Inheritance in Man for a complete compendium).

Encouraged by the huge success of mapping genes for rare Mendelian disease, geneticists were eager to apply the linkage analysis to common and complex diseases (like hypertension, diabetes), which also exhibit familial aggregation. But this time, they fell short of luck. At least two distinct features of common diseases are known to compromise the power of linkage analysis: first, the risk of getting the diseases for the carriers of causal variants is much lower than in Mendelian cases. Second, there may be multiple genes that, possibly through their interaction with environment, influence the disease susceptibility.

An alternative way emerged during mid-1990s. Rather than tracing the segregation patterns within families, we can pinpoint disease mutations by systematically testing each common genetic variation for their allele frequency differences between unrelated cases and controls sampled from population (*association mapping*). Aside from the practical tractability, the focus on common variants is based on the "common disease-common variants" (CDCV) hypothesis, which proposes that variants conferring susceptibility to common diseases occur commonly in population (with allele frequency >5 % as an operational criteria). While idea of association study is absolutely simple, transforming this blueprint into practices awaits for more than a decade.

As a first step toward this goal, great efforts were made in parallel with human genome project to compile a comprehensive catalog of sequence variations and map them to the reference genome backbone. SNPs are the most abundant form of variants. In contrast to high variability of microsatellites, they typically have two alleles at each locus which can be measured by hybridization (*genotyping*). Two homologous chromosomes within an individual differ on average 1 in every 1,000 bases in their aligned regions (*heterozygosity*). And more than 95 % of those heterozygous loci will have >5 % *minor allele frequencies* within population. Up to now, it has been estimated that more than 70 % of total 10 million common SNPs have been discovered and deposited in the public databases. Other forms of variations including those altering copy numbers of large DNA chunks have also been mapped in an accelerated pace recently. Nevertheless, high abundance and easy to genotype make SNPs the primal choice for association study. Meanwhile, off-the-shelf SNP genotyping microarrays nowadays can simultaneously genotype more than half million SNPs in one individual with more than 99 % accuracy. With both genomic resources and cutting-edge technologies at hand, genome-wide association study seemed tantalizing.

But question remained: do we really need to type all the variants in the genome-wide association study (which is still infeasible)? Even provided that we could type all common SNPs, but if the disease-causing variant is not SNPs, are we still able to find them? To answer these questions, we need to take on an evolutionary perspective.

Variations do not come out of nowhere. All the variations that we observe in the current day population result from historical mutations that happen on the chromosomes that are passed on to the next generation. Each SNP is typically biallelic due to a unique point mutation event earlier in the human history (because point mutation rate is very low, 10^{-8} per site per generation, recurrent mutation is negligible). As we mentioned above, most of the variation is *neutral*, so the frequencies of newly arisen alleles will subject to random fluctuation because population size is finite (*genetic drift*). As time goes by, most of the newly arisen alleles will be removed from the population, while some of them will happen to spread across the entire population (*fixation*). So the polymorphisms we observe are those old mutations that have neither become extinct nor reached fixation until today. Some of the mutations can influence individual's fitness to the environment, for example, causing severe disorder in the early age. In such cases, the probability for this allele being transmitted to the next generation will be reduced, since the carrier may unlikely to survive until reproductive age. The frequencies of such deleterious alleles, including those causing Mendelian diseases, will be kept low as a consequence of *purifying selection*. Most common diseases, however, have only mild impact on individual's reproduction. So the variants that predispose individuals to common diseases can rise to moderate frequencies, consistent with but not proving the CDCV hypothesis.

Variations do not come alone. Whenever a new allele was born, it must be embedded on the particular background of a specific combination of existing alleles (*haplotype*) at that time. In subsequent generations, the haplotype background of that specific allele will be reshuffled by the meiotic crossovers. Because nearby markers undergo fewer crossovers, alleles of closely linked loci (be it SNPs, indels, copy number variations, etc.) exhibit allelic associations with each other (termed *linkage disequilibrium*, abbreviated as *LD*). It suggests that even if the disease-causing mutations are not directly typed and tested for association, they can still be "tagged" by the alleles of nearby SNPs. And by properly selecting markers based on the LD patterns of human population, genome-wide association studies can be made in a cost-effective way. Both the marker selection and result interpretation therefore require the knowledge about the interrelationship between variants.

International HapMap Project has been completed to achieve this goal, with the priority given to the common SNPs. We now know that there are regions of tens or even hundreds of kilobases long, where diversity of SNP haplotypes is limited. These "haplotype blocks" are separated by sharp breakdown of LD as a result of punctuated distribution of crossover events (with ~80 % of crossovers happen within *recombination hotspots*). Within blocks, a reduced number of common SNPs can serve as a proxy to predict allelic status of remaining common SNPs or even other common genetic variations (like copy number gain or loss). Half million SNPs can provide adequate power in association study to test most of the common SNPs in East Asia and European populations. These findings, together with the maturity of technology and statistical methodology, have paved the way for the first wave of association study during the past 2 years. More than a hundred loci have now been identified to be bona fide reproducibly associated with common forms of human diseases.

Never satisfied by the initial success, geneticists want to extend the power of the association mapping to rare variants. To this end, they call for a map that catalogs and describes the relationships among almost all the variants, be it common and rare. Armed with cutting-edge sequencers, the 1000 Genomes Project has been launched with this ambition. Geneticists and expertise from other disciplines are now working in an ever closer manner.

References

1. Shendure J, Ji H (2008) Next-generation DNA sequencing. Nat Biotechnol 26:1135–1145
2. Fields S (2007) Site-seeing by sequencing. Science 316(5830):1441–1442

Chapter 2
Basic Statistics for Bioinformatics

Yuanlie Lin and Rui Jiang

2.1 Introduction

Statistics is a branch of mathematics that targets on the collection, organization, and interpretation of numerical data, especially on the analysis of population characteristics by inferences from random sampling. Many research topics in computational biology and bioinformatics heavily rely on the application of probabilistic models and statistical methods. It is therefore necessary for students in bioinformatics programs to take introductory statistics as their first course. In this chapter, the basics of statistics are introduced from the following clue: foundations of statistics, point estimation, hypothesis testing, interval estimation, analysis of variance (ANOVA), and regression models. Besides, the free and open-source statistical computing environment R is also briefly introduced. Students can refer to the book by George Casella and Roger L. Berger [1] and other related textbooks for further studies.

2.2 Foundations of Statistics

2.2.1 Probabilities

One of the main objectives of researches in bioinformatics is to draw conclusions about a population of objects by conducting experiments. If the outcome of an experiment cannot be predicted but all possible outcomes of the experiment can be

Y. Lin (✉)
Department of Mathematical Sciences, Tsinghua University, Beijing 100084, China
e-mail: ylin@math.tsinghua.edu.cn

R. Jiang
MOE Key Laboratory of Bioinformatics and Bioinformatics Division, TNLIST/Department of Automation, Tsinghua University, Beijing 100084, China

R. Jiang et al. (eds.), *Basics of Bioinformatics: Lecture Notes of the Graduate Summer School on Bioinformatics of China*, DOI 10.1007/978-3-642-38951-1_2,
© Tsinghua University Press, Beijing and Springer-Verlag Berlin Heidelberg 2013

pointed out in advance, the experiment is called a random experiment, or experiment for simplicity. The set, S, of all possible outcomes of a particular experiment is called the sample space for the experiment. A collection of some possible outcomes of an experiment is called an event, which can be equivalently defined as a subset of S (including S itself). Since events are sets of outcomes, the relationships between sets (such as union, intersection, and complementation) and the operations about sets (such as commutativity, associativity, distributive laws, and De Morgan's laws) are also applicable to events. Two events are disjoint if the intersection of them is the empty set. A group of events are pairwise disjoint (or mutually exclusive) if any two of these events are disjoint. If the union of a group of mutually exclusive events is the sample space S, the group of events forms a partition of S.

A collection of subsets of S is called a *sigma algebra* (or *Borel field*), denoted by $\mathcal{B}$, if it satisfies:

1. $\emptyset \in \mathcal{B}$ (the empty set is an element of $\mathcal{B}$).
2. If $A \in \mathcal{B}$, then $A^c \in \mathcal{B}$ ($\mathcal{B}$ is closed under complementation).
3. If $A_1, A_2, \ldots \in \mathcal{B}$, then $\bigcup_{i=1}^{\infty} A_i \in \mathcal{B}$ ($\mathcal{B}$ is closed under countable unions).

Given a sample space S and an associated sigma algebra $\mathcal{B}$, a *probability function* is a function P with domain $\mathcal{B}$ that satisfies:

1. $P(A) \geq 0$ for all $A \in \mathcal{B}$ (nonnegativity).
2. $P(S) = 1$ (normality).
3. If $A_1, A_2, \ldots \in \mathcal{B}$ are mutually exclusive, then $P\left(\bigcup_{i=1}^{\infty} A_i\right) = \sum_{i=1}^{\infty} P(A_i)$ (countable additivity).

These three properties are usually referred to as the *Axioms of Probability*, or *Kolmogorov Axioms*. Any function that satisfies the Axioms of Probability is a probability function.

Let $S = \{s_1, \ldots, s_n\}$ be a finite set and $\mathcal{B}$ be any sigma algebra of subset S. Let $p_1, \ldots, p_n$ be nonnegative numbers that sum to 1. For any $A \in \mathcal{B}$, define $P(A)$ by

$$P(A) = \sum_{\{i:s_i \in A\}} p_i.$$

Then P is a probability function on $\mathcal{B}$. If $p_1 = p_2 = \cdots = p_n = 1/n$, we have an equivalent form as the *classical definition of probabilities*.

Many properties of the probability function can be derived from the Axioms of Probability. For example, for a single event A, we have:

1. $P(\emptyset) = 0$.
2. $P(A) \leq 1$.
3. $P(A^c) = 1 - P(A)$.

For any two events A and B, we have:

1. $P(B \cap A^c) = P(B) - P(A \cap B)$.
2. $P(A \cup B) = P(A) + P(B) - P(A \cap B)$.
3. $P(A) \leq P(B)$ if $A \subset B$.

For an event A and any partition $C_1, C_2, \ldots$ of the sample space, we have

$$P(A) = \sum_{i=1}^{\infty} P(A \cap C_i).$$

For any sets $A_1, A_2, \ldots$, we have

$$P\left(\bigcup_{i=1}^{\infty} A_i\right) \leq \sum_{i=1}^{\infty} P(A_i).$$

For any two events A and B in the sample space S, the conditional probability of A given B, written $P(A|B)$, is defined by

$$P(A|B) = \frac{P(A \cap B)}{P(B)},$$

if $P(B) > 0$. In the above conditional probability calculation, the sample space has been updated from S to B. If A and B are disjoint events, $P(A|B) = P(B|A) = 0$, because $P(A \cap B) = 0$.

From the definition of the conditional probability, we have

$$P(A \cap B) = P(B|A)P(A) = P(A|B)P(B).$$

This is known as the multiplication rule of probabilities. If $C_1, C_2, \ldots$ is a partition of the sample space S, we have

$$P(A) = \sum_{i=1}^{\infty} P(A \cap C_i) = \sum_{i=1}^{\infty} P(A|C_i)P(C_i).$$

This is referred to as the law of total probability. Furthermore, for $i = 1, 2, \ldots$

$$P(C_i|A) = \frac{P(C_i \cap A)}{P(A)} = \frac{P(A|C_i)P(C_i)}{\sum_{k=1}^{\infty} P(A|C_k)P(C_k)}.$$

This is well known as the Bayes rule.

In general, $P(A|B)$ is not necessarily equal to $P(A)$. If $P(A|B) = P(A)$, or equivalently $P(A \cap B) = P(A)P(B)$, we say that A and B are *statistically independent*. If $P(A \cap B) - P(A)P(B) > 0$, we say that A and B are *positively correlated*. If $P(A \cap B) - P(A)P(B) < 0$, we say that A and B are *negatively correlated*. A collection of events $A_1, \ldots, A_n$ are mutually independent if for any subcollection $A_{i_1}, \ldots, A_{i_k}$, $P\left(\bigcap_{j=1}^{k} A_{i_j}\right) = \prod_{j=1}^{k} P(A_{i_j})$.

2.2.2 Random Variables

In many experiments it is easier to deal with a summary variable than with the original outcomes. The summary variable can also greatly reduce the size of the sample space. In these cases, we can define a mapping (a function) from the original sample space to a new (typically much smaller or well structured) sample space. These mapping functions are typically real valued and generally referred to as random variables.

A *random variable* is a function from a sample space S into the real number space $\Re$. In defining a random variable, a new sample space (the range of the random variable) and a new probability function on this sample space are defined.

We are interested in all possible values of a random variable, but we are more interested in the probabilities that the random variable takes these values. For any random variable X, we associate a function $F_X(x)$, defined by

$$F_X(x) = P(X \leq x), \quad \text{for all } x.$$

This function is referred to the *cumulative distribution function* or *cdf* of X. For a cdf $F(x)$:

1. $\lim_{x \to -\infty} F(x) = 0$ and $\lim_{x \to \infty} F(x) = 1$.
2. $F(x)$ is a nondecreasing function of x.
3. $F(x)$ is right-continuous, say, for every number x_0, $\lim_{x \to x_0^+} F(x) = F(x_0)$.

These are sufficient and necessary conditions for a cdf. We say that X is continuous if $F_X(x)$ is a *continuous* function, and X is *discrete* if $F_X(x)$ is a step function. If two cdfs take equal values for all possible points in their common domains, we say that the corresponding random variables are *identically distributed*.

For a discrete random variable X, the *probability mass function* (pmf) is defined by

$$f_X(x) = P(X = x), \quad \text{for all } x.$$

By doing this, we have

$$P(a \leq X \leq b) = \sum_{a \leq x \leq b} f_X(x).$$

Particularly,

$$F_X(x) = P(X \leq b) = \sum_{x \leq b} f_X(x).$$

For a continuous random variable X, the probability mass is equal to 0 for every x, that is, $P(X = a) = 0$ for any $a \in \Re$. Therefore,

$$P(a < X < b) = P(a < X \leq b) = P(a \leq X < b) = P(a \leq X \leq b).$$

To circumvent this situation, we define a *probability density function* (pdf) $f_X(x)$ that satisfies

$$F_X(x) = \int_{-\infty}^{x} f_X(t)dt, \quad \text{for all } x.$$

By doing this, we have

$$P(a < X < b) = \int_{a}^{b} f_X(x)dx.$$

Of course, both pmf and pdf should be nonnegative and sum (integrate) to 1 for all possible values in their domains.

In some cases we know the distribution of a random variable but are interested in some other quantities that can be mapped from this random variable. To facilitate the manipulation, we introduce functions of random variables. For a random variable X, the function $Y = g(X)$ is again a random variable, with the sample space and probability function being updated. In order to derive the distribution of the new random variable, we need to do a *transformation* from the original random variable.

According to the definition of cdf, if the transformation function $Y = g(X)$ is an increasing function, then $F_Y(y) = F_X\left(g^{-1}(y)\right)$. On the other hand, if $Y = g(X)$ is an decreasing function, then $F_Y(y) = 1 - F_X\left(g^{-1}(y)\right)$. Differentiating the cdf, we have the pdf for the transformation as

$$f_Y(y) = f_X\left(g^{-1}(y)\right)\left|\frac{d}{dy}g^{-1}(y)\right|.$$

If the transformation function $g(X)$ is not monotone, the situation would be complicated. An expedient is to separate the domain of $g(X)$ to monotone pieces, find pdf for each piece, and then sum up the resulting pdf to obtain that for the transformation.

Sometimes we are interested in seeking for a typical value as the representative of a random variable. A reasonable choice is the weighted average of all possible values of the random variable, with the weight being their probabilities of occurrence. This summarization value is called the expected value or expectation of the random variable. Formally, given the pdf or pmf $f(x)$ of a random variable X, the *expected value* or *mean* of a random variable $g(X)$, denoted by $Eg(X)$, is

$$Eg(X) = \begin{cases} \int_{-\infty}^{\infty} g(x)f(x)dx & \text{if } X \text{ is continuous} \\ \sum_{x \in \mathcal{X}} g(x)f(x) = \sum_{x \in \mathcal{X}} g(x)P(X = x) & \text{if } X \text{ is discrete} \end{cases}$$

providing that the integral or sum exists. If $E|g(X)| = \infty$, we say that $Eg(X)$ does not exist. For each integer n, $\mu'_n = EX^n$ is called the nth moment of X and $\mu_n = E(X - \mu)^n$ the nth central moment of X, where $\mu = \mu'_1 = EX$ is the first moment and called the *mean*. The second central moment $\mathrm{Var}X = \sigma^2 = E(X - EX)^2$ is called the *variance*, and the positive square root of $\mathrm{Var}X$ is called the *standard deviation*. Obviously, we have $E(aX + b) = aEX + b$, $\mathrm{Var}(aX + b) = a^2\mathrm{Var}X$, and $\mathrm{Var}X = EX^2 - (EX)^2$. The fact that $\min_{c \in \Re} E(X - c)^2 = E(X - EX)^2$ suggests that EX is the best guess of X if measured in terms of the mean squared error.

2.2.3   Multiple Random Variables

In some cases, we are interested in multiple numeric values of the objects in study. More generally, we infer the property of interest of a population of objects via multiple observations of the objects. In these cases, we need to deal with multiple random variables.

An *n-dimensional random vector* is a function from a sample space S into $\Re^n$, the n-dimensional Euclidean space. The simplest case is $n = 2$, resulting in *bivariate random vector*. Studies regarding multivariate random vectors are done by investigating the bivariate case and then generalizing to multivariate case.

Let (X, Y) be a discrete bivariate random vector. Then the function $f(x, y)$ from $\Re^2$ into $\Re$ defined by $f_{X,Y}(x, y) = P(X = x, Y = y)$ is called the *joint probability mass function* (joint pmf) of (X, Y). For a certain region A in $\Re^2$, $P((x, y) \in A) = \sum_{(x,y) \in A} f_{X,Y}(x, y)$. Summing up the joint pmf over all possible values in one dimension gives us the *marginal pmf* of another dimension, as

$$f_X(x) = \sum_{y \in \Re} f_{X,Y}(x, y) \quad \text{and} \quad f_Y(y) = \sum_{x \in \Re} f_{X,Y}(x, y).$$

Parallel to the definition of conditional probabilities, dividing the joint pmf by the marginal pmf of one dimension gives us the *conditional pmf* of another dimension, as

$$f(x|y) = \frac{f_{X,Y}(x, y)}{f_Y(y)} \quad \text{and} \quad f(y|x) = \frac{f_{X,Y}(x, y)}{f_X(x)},$$

provided that $f_X(x) > 0$ and $f_Y(y) > 0$.

For a continuous bivariate random vector (X, Y), the probability mass is equal to 0 for every (x, y). We then use the following integral to define the *joint probability density function* (joint pdf)

$$P((x, y) \in A) = \int\int_A f_{X,Y}(x, y)\mathrm{d}x\mathrm{d}y.$$

The marginal pdf and conditional pdf are defined using similar formula as the discrete case. For example,

$$f_X(x) = \int_{-\infty}^{\infty} f_{X,Y}(x,y)dy \quad \text{and} \quad f_Y(y) = \int_{-\infty}^{\infty} f_{X,Y}(x,y)dx.$$

In general, the marginal pmf (or pdf) $f_X(x)$ is not necessarily equal to the conditional pmf (or pdf) $f(x|y)$ for every possible x. If they are equal for every x, or equivalently

$$f_{X,Y}(x,y) = f_X(x)f_Y(y),$$

we say that the two variables X and Y are independent. A sufficient and necessary condition for two variables X and Y being independent is that the joint pmf (or pdf) can be factorized to two functions, one is only of X and the other is only of Y, as

$$f_{X,Y}(x,y) = g(x)h(y).$$

Parallel to the univariate case, a function of a bivariate random vector is also a bivariate random vector. In order to derive the distribution functions of the new random vector, we need to do transformations. The general situation for this kind of transformations would be complicated, but if the function is a one-to-one transformation, a way similar to the univariate case exists. Formally, for a continuous bivariate random vector (X, Y), if $U = g(X, Y)$ and $V = h(X, Y)$, or equivalently, $X = \varphi(U, V)$ and $Y = \psi(U, V)$ when the function is one-to-one, then the joint pdf of (U, V) is given by

$$f_{U,V}(u,v) = f_{X,Y}(\varphi(x,y), \psi(x,y)) |J|,$$

where J is the *Jacobian of the transformation* and is the determinant of the a matrix of partial derivatives, given by

$$J = \begin{vmatrix} \dfrac{\partial \varphi}{\partial u} & \dfrac{\partial \varphi}{\partial v} \\[2mm] \dfrac{\partial \psi}{\partial u} & \dfrac{\partial \psi}{\partial v} \end{vmatrix} = \dfrac{\partial \varphi}{\partial u}\dfrac{\partial \psi}{\partial v} - \dfrac{\partial \varphi}{\partial v}\dfrac{\partial \psi}{\partial u}.$$

The expected value of a bivariate random vector $g(X, Y)$, with respect to the joint pmf (or pdf) of (X, Y), $f(x, y)$ is defined as

$$Eg(X,Y) = \sum_{(x,y)\in\Re} g(x,y)f(x,y) \quad \text{and} \quad Eg(X,Y) = \int_{-\infty}^{\infty}\int_{-\infty}^{\infty} g(x,y)f(x,y)dxdy$$

for the discrete and continuous case, respectively. Since the conditional pmf (or pdf) $f(x|y)$ depends on the value of y, for $h(X)$, a function of only X, the expectation $E[h(X)|y] = \sum_{x \in \Re} h(x)f(x|y)$ in the discrete case, or $E[h(X)|y] = \int_{-\infty}^{\infty} h(x)f(x|y)\mathrm{d}x$ in the continuous case, is referred to as the *conditional expectation* of $h(X)$, given that $Y = y$.

For two random variables X and Y, their *covariance* is the number defined by

$$\mathrm{Cov}(X, Y) = E[(X - EX)(Y - EY)] = E[(X - \mu_X)(Y - \mu_Y)],$$

and their *correlation coefficient* is the number defined by

$$\rho_{XY} = \frac{\mathrm{Cov}(X, Y)}{\sqrt{\mathrm{Var}X\,\mathrm{Var}Y}} = \frac{\mathrm{Cov}(X, Y)}{\sigma_X \sigma_Y}.$$

Obviously, $\mathrm{Cov}(X, Y) = EXY - \mu_X \mu_Y$. The correlation coefficient ρ_{XY} satisfies $-1 \le \rho_{XY} \le 1$, and $|\rho_{XY}| = 1$ if and only if X and Y have a linear relationship with probability 1; that is, $P(Y = aX + b) = 1$ for some $a \ne 0$ and b.

2.2.4 Distributions

A numeric property of interest that is associated with a population of objects in study is often modeled as a random variable and described using its distribution functions. Typically, these distribution functions depend on some constants, called *parameters*. With different values of the parameters, the distribution functions represent different distributions, though the mathematical formations of the functions remain unchanged. In this sense, a distribution function with parameters represents a family of distributions.

The simplest random experiment has only two possible outcomes, say, "success" and "failure," and usually referred to as a Bernoulli trial; if we define "success" as the value 1 and "failure" as 0, we have a *Bernoulli* random variable, whose pmf is

$$P(X = 1) = p \quad \text{and} \quad P(X = 0) = 1 - p,$$

given that the probability of X being 1 is p.

The number of successes in a serial of n independent Bernoulli (p) trials is a *binomial* (n, p) random variable, whose pmf is

$$P(X = x|n, p) = \binom{n}{x} p^x (1 - p)^{n-x}, \quad x = 0, 1, \ldots, n.$$

The limitation of a binomial variable is a Poisson variable, whose pmf is

$$P(X = x|\lambda) = \frac{e^{-\lambda}\lambda^x}{x!}, \quad x = 0, 1, \ldots.$$

A Poisson distribution is often used to model the number of occurrence of a particular event in a period of time.

From another point of view, considering the number of failures when the rth success occurs in multiple independent Bernoulli (p) trials leads to the following *negative binomial* (r, p) distribution:

$$P(X = x|r, p) = \binom{r + x - 1}{x} p^r (1 - p)^x, \quad x = 0, 1, \ldots.$$

Obviously, $r + x$ is the total number of Bernoulli trials performed. In the case that $r = 1$, the total number of trials $y = x + 1$ has a *geometric* (p) distribution, whose pmf is

$$P(Y = y|p) = p(1 - p)^{y-1}, \quad y = 1, 2, \ldots.$$

Bernoulli trials can be generalized to have more than two outcomes. If an experiment has multiple, say m, possible outcomes, it is often referred to as a *multinomial trial*. Let these outcomes be labeled as $1, 2, \ldots, m$ and the corresponding probability of observing the kth outcome be p_k. We have the pmf for a multinomial trial variable, as

$$P(X = k) = p_k, \quad k = 1, 2, \ldots, m.$$

In the case that all p_k are equal to $1/m$, we have a *discrete uniform* distribution, whose pmf is

$$P(X = k) = \frac{1}{m}, \quad k = 1, 2, \ldots, m.$$

The numbers of occurrences of the m possible outcomes in a serial of n independent multinomial trials has a multinomial distribution, whose pmf is

$$P(X_1 = x_1, \ldots, X_m = x_m | n, m, p_1, \ldots, p_m) = \frac{n!}{x_1! \cdots x_m!} \prod_{i=1}^{m} p_i^{x_i},$$

where $\sum_{i=1}^{m} x_i = n$ and $\sum_{i=1}^{m} p_i = 1$. The multinomial family is a generalization of the binomial family and has been widely used to model biopolymer sequences such as the DNA and the protein.

36   Y. Lin and R. Jiang

Suppose there is an urn containing m red and n green balls and we select k balls at random from the total of $m + n$ balls. The probability of obtaining exactly x red balls would be

$$P(X = x|m,n,k) = \frac{\binom{m}{x}\binom{n}{k-x}}{\binom{m+n}{k}}, \quad x = 0,1,\ldots,k.$$

This is the pmf of a hypergeometric (m,n,k) distribution.

The simplest family of continuous distributions is the *uniform* family, whose probability density keeps constant in the interval $[a,b]$, formally as

$$f(x|a,b) = \frac{1}{b-a}, \quad x \in [a,b].$$

The most widely used family of continuous distributions is the *normal* family, $N(\mu,\sigma^2)$, whose pdf is

$$f(x|\mu,\sigma^2) = \frac{1}{\sqrt{2\pi}\sigma}e^{-\frac{(x-\mu)^2}{2\sigma^2}}.$$

$N(0,1)$ is often referred to as the *standard normal distribution*, and normal distributions are also referred to as Gaussian distributions.

The ratio of two independent normal random variables has a Cauchy distribution, whose pdf is

$$f(x|\theta) = \frac{1}{\pi}\frac{1}{1+(x-\theta)^2}.$$

The Cauchy family is famous for its pathological behavior, because its mean does not exist.

The most versatile family of continuous distributions is the *Gamma* family, whose pdf is

$$f(x|\alpha,\beta) = \frac{1}{\Gamma(\alpha)\beta^\alpha}x^{\alpha-1}e^{-\frac{x}{\beta}}, 0 < x < \infty, \alpha > 0, \beta > 0,$$

where $\Gamma(\alpha) = \int_0^\infty t^{\alpha-1}e^{-t}dt$ is the Gamma function, who has many interesting properties, such as $\Gamma(\alpha+1) = \alpha\Gamma(\alpha)$, $\Gamma(n) = (n-1)!$, $\Gamma(1) = 1$, and $\Gamma(1/2) = \sqrt{2\pi}$. The Gamma family is the headstream of many other important families. For example, a Gamma$(p/2,2)$ random variable is a χ^2 random variable with p degrees of freedom. A Gamma$(1,\lambda)$ random variable is an exponential random variable with the parameter λ. A student's t random variable with n degrees of freedom is the quotient of a standard normal variable divided by the square root of a χ^2 random variable with the same degrees of freedom. A Snedecor's F random variable with n and m degrees of freedom is the quotient of two χ^2 random variables, where the numerator has n degrees of freedom and the denominator has m degrees of freedom.

One of the few common named continuous families that give probability density to the interval $[0, 1]$ is the *Beta* family, whose pdf is

$$f(x|\alpha, \beta) = \frac{\Gamma(\alpha + \beta)}{\Gamma(\alpha)\Gamma(\beta)} x^{\alpha-1}(1 - x)^{\beta-1}, 0 \leq x \leq 1, \alpha > 0, \beta > 0.$$

Because the beta family is trapped in the same range as probabilities, it is often used to model the distributions of probabilities.

The multivariate version of the Beta distribution is the Dirichlet distribution, whose pdf is

$$f(x_1, \ldots, x_m | \alpha_1, \ldots, \alpha_m) = \frac{\Gamma\left(\sum_{i=1}^{m} \alpha_i\right)}{\prod_{i=1}^{m} \Gamma(\alpha_i)} \prod_{i=1}^{m} x_i^{\alpha_i - 1}, \quad \alpha_i \geq 0, m \geq 2$$

where every $x_i \geq 0$ and all x_i sum up to 1. The Dirichlet family is a generalization of the Beta family and has been widely used to model the prior distributions of probabilities in Bayesian inference and Bayesian networks.

2.2.5   Random Sampling

Inferences about a numeric property of interest are typically conducted by selecting at random some objects and observing their properties. Each observation is equivalent to a random experiment of randomly selecting an object from the sample space of all objects, and in this experiment we can define a random variable whose domain includes all the objects and whose range contains all possible values of the property of interest. After a number of observations being performed, the same number of random variables is obtained, and all of these random variables have the same domain, the same range, and the same distribution. Conceptually, the collection of all of the objects forms a population in study. Because we are interested in the property of the objects, in statistics, the word *population* is used to refer to the common distribution of the random variables that are obtained in the process of the multiple observations. Since by experiment design the random variables are mutually independent and have identical distribution, they are often referred to as *independent and identically distributed* (iid) random variables. A group of n such kind of iid random variables is called a *random sample* of size n. A random sample is often simplified as a sample.

For a random sample $X_1, \ldots, X_n$ of size n, a real-valued or vector-valued multivariate function $Y = T(X_1, \ldots, X_n)$ is called a *statistic*. The distribution of a statistic Y is called the sampling distribution of Y.

Two of the most widely used statistics are the sample mean

$$\overline{X} = \frac{X_1 + \cdots + X_n}{n} = \frac{1}{n}\sum_{i=1}^{n} X_i,$$

and the sample variance

$$S^2 = \frac{1}{n-1}\sum_{i=1}^{n}\left(X_i - \overline{X}\right)^2.$$

Their properties include $\mathrm{E}\overline{X} = \mu$, $\mathrm{Var}\overline{X} = \sigma^2/n$, and $\mathrm{E}S^2 = \sigma^2$, where μ and σ^2 are population mean and population variance, respectively.

Let $Z = \sqrt{n}(\overline{X} - \mu)/\sigma$. Z is often referred to as the *Z-score*, *Z-value*, *standard score*, or *normal score*. Let $X_{(i)}$ as the ith smallest random variable in a sample $X_1, \ldots, X_n$. $X_{(i)}$ is called the ith *order statistic* in the sample. Particularly, $R = X_{(n)} - X_{(1)}$ is called the *sample range*. $M = X_{((n+1)/2)}$ when n is odd, together with $M = \left(X_{(n/2)} + X_{(n/2+1)}\right)/2$ when n is even, is called the *sample median*.

When a sample $X_1, \ldots, X_n$ comes from a normal population $\mathrm{N}(\mu, \sigma^2)$, the sample mean $\overline{X}$ and the sample variance S^2 are independent random variables. The quotient $\sqrt{n}(\overline{X} - \mu)/\sigma$ has a standard normal distribution. The squared sum $\sum_{i=1}^{n}(X_i - \mu)^2/\sigma^2$ has a χ^2 distribution with n degrees of freedom. The squared sum $\sum_{i=1}^{n}(X_i - \overline{X})^2/\sigma^2 = (n-1)S^2/\sigma^2$ has a χ^2 distribution with $n-1$ degrees of freedom. When the population variance σ^2 is unknown, the quotient $\sqrt{n}(\overline{X} - \mu)/S$ has a student's t distribution with $n - 1$ degrees of freedom.

When a paired sample $(X_1, Y_1), \ldots, (X_n, Y_n)$ comes from a bivariate normal population $\mathrm{N}(\mu_X, \mu_Y, \sigma_X^2, \sigma_Y^2, \rho)$, $\sqrt{n}\left[(\overline{X} - \overline{Y}) - (\mu_X - \mu_Y)\right]/\sigma_{X-Y}$ has a standard normal distribution, and $\sqrt{n}\left[(\overline{X} - \overline{Y}) - (\mu_X - \mu_Y)\right]/S_{X-Y}$ has a t distribution with $n - 1$ degrees of freedom, where $\sigma_{X-Y}^2 = \sigma_X^2 - 2\rho\sigma_X\sigma_Y + \sigma_Y^2$ and $S_{X-Y}^2 = \sum_{i=1}^{n}\left[(X_i - Y_i) - (\overline{X} - \overline{Y})\right]^2/(n - 1)$.

When two samples $X_1, \ldots, X_m$ and $Y_1, \ldots, Y_n$ come from two independent normal populations $\mathrm{N}(\mu_X, \sigma_X^2)$ and $\mathrm{N}(\mu_Y, \sigma_Y^2)$ ($\sigma_X = \sigma_Y = \sigma$), respectively, $\sqrt{n'}\left[(\overline{X} - \overline{Y}) - (\mu_X - \mu_Y)\right]/\sigma$ has a standard normal distribution, and $\sqrt{n'}\left[(\overline{X} - \overline{Y}) - (\mu_X - \mu_Y)\right]/S_p$ has a student's t distribution with $m + n - 2$ degrees of freedom, where $n' = mn/(m + n)$ and $S_p = \left[(n - 1)S_X^2 + (m - 1)S_Y^2\right]/(m + n - 2)$.

A sequence of random variables, $X_1, X_2, \ldots$, converges *in probability* to a random variable X, if for every $\varepsilon > 0$

$$\lim_{n \to \infty} P\left(|X_n - X| \geq \varepsilon\right) = 0 \quad \text{or,} \quad \text{equivalently,} \quad \lim_{n \to \infty} P\left(|X_n - X| < \varepsilon\right) = 1.$$

The Chebychev's inequality states that for a random variable X,

$$P\left(g(X) \geq r\right) \leq \frac{\mathrm{E}g(X)}{r},$$

where $g(x)$ is a nonnegative function and $r > 0$. A direct result of this inequality is the so-called Weak Law of Large Numbers. Let $X_1, X_2, \ldots$ be a sequence of iid random variables with $EX_i = \mu$ and $VarX_i = \sigma^2 < \infty$. Define $\overline{X}_n = \sum_{i=1}^{n} X_i / n$. Then for every $\varepsilon > 0$,

$$\lim_{n \to \infty} P \left(\left| \overline{X}_n - \mu \right| < \varepsilon \right) = 1;$$

that is, $\overline{X}_n$ converges in probability to μ. The weak law of large numbers states that, under general condition, the sample mean approaches the population mean as $n \to \infty$, regardless of the distribution of each X_i.

A sequence of random variables, $X_1, X_2, \ldots$, converges *in distribution* to a random variable X, if

$$\lim_{n \to \infty} F_{X_n}(x) = F_X(x)$$

at all points x where $F_X(x)$ is continuous. With this concept, we have the following *Central Limit Theorem*. Let $X_1, X_2, \ldots$ be a sequence of iid random variables with $EX_i = \mu$ and $VarX_i = \sigma^2 < \infty$. Define $\overline{X}_n = \sum_{i=1}^{n} X_i / n$. Let $G_n(x)$ denote the cdf of $\sqrt{n} \left(\overline{X}_n - \mu \right) / \sigma$. Then, for any x, $-\infty < x < \infty$,

$$\lim_{n \to \infty} G_n(x) = \int_{-\infty}^{x} \frac{1}{\sqrt{2\pi}} e^{-\frac{t^2}{2}} dt;$$

that is, $\sqrt{n} \left(\overline{X}_n - \mu \right) / \sigma$ has a limiting standard normal distribution. The Central Limit Theorem states that, under general condition, the standard score has a standard normal distribution as $n \to \infty$, regardless of the distribution of each X_i.

2.2.6 Sufficient Statistics

Statistical inference about a population is done via a sample of the population, more precisely, via some statistics of the sample. In principle, a statistic summarizes the information in a sample by determining a key feature of the sample values and thus defines a form of data reduction or data summary. Particularly, we are interested in methods of data reduction that do not discard important information about the population and methods that can successfully discard information that is irrelevant to the characteristics of interest. We typically use three principles of data reduction: the sufficiency principle, the likelihood principle, and the equivariance principle. Here we will briefly introduce the sufficient principle.

A *sufficient statistic* for a parameter θ of a population is a statistic that captures all the information about the parameter contained in the sample. This consideration yields to the following sufficiency principle: if $T(\mathbf{X})$ is a sufficient statistic for θ, then any inference about θ should depend on the sample $\mathbf{X}$ only through the value $T(\mathbf{X})$. In other words, if $\mathbf{x}$ and $\mathbf{y}$ are two sample points such that $T(\mathbf{x}) = T(\mathbf{y})$, then the inference about θ should be the same whether $\mathbf{X} = \mathbf{x}$ or $\mathbf{X} = \mathbf{y}$ is observed. With this understanding, a statistic $T(\mathbf{X})$ is a sufficient statistic for a parameter θ if the conditional distribution of the sample $\mathbf{X}$ given the value of $T(\mathbf{X})$ does not depend on θ. Mathematically, if $p(\mathbf{x}|\theta)$ is the joint pdf or pmf of a sample $\mathbf{X}$ and $q(t|\theta)$ is the pdf or pmf of $T(\mathbf{X})$ is a sufficient statistic for θ if for every $\mathbf{x}$ in the sample space, the ratio $p(\mathbf{x}|\theta)/q(T(\mathbf{x})|\theta)$ is constant as a function of θ. More practically, let $f(\mathbf{x}|\theta)$ denote the joint pdf or pmf of a sample $\mathbf{X}$. A statistic $T(\mathbf{X})$ is a sufficient statistic for θ if and only if there exist function $g(t|\theta)$ and $h(\mathbf{x})$ such that for all sample points $\mathbf{x}$ and all parameter points θ,

$$f(\mathbf{x}|\theta) = g(T(\mathbf{x})|\theta)h(\mathbf{x}).$$

This sufficient and necessary condition of a sufficient statistic is usually referred to as the *factorization theorem*.

It should be noted that for any parameter of any population, sufficient statistic is not unique. In fact, the complete sample $\mathbf{X}$ is always a sufficient statistic, though no data reduction is done by this trivial statistic. In general, any one-to-one function of a sufficient statistic is also a sufficient statistic. Since the purpose of a sufficient statistic is to achieve data reduction without loss of information about the parameters of the given population, a sufficient statistic that can achieve the most data reduction might be considered as the most preferred one. This consideration yields the following definition of the minimal sufficient statistic. A sufficient statistic $T(\mathbf{X})$ is called a *minimal sufficient statistic* if for any other sufficient statistic $T'(\mathbf{X})$, $T(\mathbf{X})$ is a function of $T'(\mathbf{X})$. A minimal sufficient statistic can be found by the following simple way. Let $f(\mathbf{x}|\theta)$ be the pmf or pdf of a sample $\mathbf{X}$. Suppose there exists a function $T(\mathbf{X})$ such that for every two sample points $\mathbf{x}$ and $\mathbf{y}$, the ratio $f(\mathbf{x}|\theta)/f(\mathbf{y}|\theta)$ is constant as a function of θ if and only if $T(\mathbf{x}) = T(\mathbf{y})$, then $T(\mathbf{X})$ is a minimal sufficient statistic.

It should be noted that a minimal sufficient statistic is not unique. Any one-to-one function of a minimal sufficient statistic is also a minimal sufficient statistic.

2.3   Point Estimation

A random sample is typically from a population that is described by a pdf or pmf $f(x|\theta)$, where θ denotes any parameter associated with the population. Since knowledge of θ can provide knowledge of the entire population, it is reasonable that we derive a representative statistic, a *point estimator*, of the parameter from the sample. There are in general three methods for deriving point estimators: method of moments, maximum likelihood, and Bayes inference. With a point estimator being derived, we can calculate its mean squared error, which is composed of the variance and the bias of the estimator. We prefer unbiased estimators, and we like to seek for estimators with the minimum variance.

2.3.1 Method of Moments

Let $\mathbf{X} = (X_1,\ldots,X_n)$ be a sample of size n from a population with pdf or pmf $f(\mathbf{x}|\boldsymbol{\theta})$, where $\boldsymbol{\theta} = (\theta_1,\ldots,\theta_k)$ is the set of parameters. One can find a method of moments estimators by equating the first k population moments to the corresponding sample moments, and then solving the resulting equations. Mathematically, define

$$\mu'_1 = \mathrm{E}X^1,\ m_1 = \frac{1}{n}\sum_{i=1}^{n}X_i^1,$$
$$\mu'_2 = \mathrm{E}X^2,\ m_2 = \frac{1}{n}\sum_{i=1}^{n}X_i^2,$$
$$\vdots \qquad\qquad \vdots$$
$$\mu'_k = \mathrm{E}X^k,\ m_k = \frac{1}{n}\sum_{i=1}^{n}X_i^k.$$

The method of moments estimator $\tilde{\boldsymbol{\theta}} = \left(\tilde{\theta}_1,\ldots,\tilde{\theta}_k\right)$ of $\boldsymbol{\theta} = (\theta_1,\ldots,\theta_k)$ can then be obtained by solving

$$\begin{cases}\mu'_1(\theta_1,\ldots,\theta_k) = m_1\\ \mu'_2(\theta_1,\ldots,\theta_k) = m_2\\ \qquad\vdots\\ \mu'_k(\theta_1,\ldots,\theta_k) = m_k\end{cases}$$

Applying the method of moments to $X_1,\ldots,X_n$, a sample of size n from a normal population $\mathrm{N}(\mu,\sigma^2)$, we can obtain

$$\tilde{\mu} = \bar{X} \quad\text{and}\quad \tilde{\sigma}^2 = \frac{1}{n}\sum_{i=1}^{n}(X_i - \bar{X})^2.$$

$\tilde{\mu}^2$ and $\tilde{\sigma}^2$ are therefore the method of moments estimator for the parameters of the normal population.

2.3.2 Maximum Likelihood Estimators

Maximum likelihood is the most popular technique for finding estimators. Let $\mathbf{X} = (X_1,\ldots,X_n)$ be a sample of size n from a population with pdf or pmf $f(x|\boldsymbol{\theta})$, where $\boldsymbol{\theta} = (\theta_1,\ldots,\theta_k)$ is the set of parameters. One can find $\hat{\boldsymbol{\theta}}(\mathbf{x})$, the maximum likelihood estimate of $\boldsymbol{\theta}$, by maximizing its likelihood function

$$L(\boldsymbol{\theta}|\mathbf{x}) = \prod_{i=1}^{n}f(x_i|\theta_1,\ldots,\theta_k).$$

Replace sample point **x** with the random sample **X**, we obtain the maximum likelihood estimator of θ as $\hat{\theta}$ (**x**).

For example, let $X_1, \ldots, X_n$ be a sample of size n from a normal population $N\left(\mu, \sigma^2\right)$, where both μ and σ^2 are unknown. The likelihood function of the parameters is then

$$L(\mu, \sigma^2|\mathbf{x}) = \frac{1}{(2\pi\sigma^2)^{n/2}} \exp\left[-\frac{1}{2}\sum_{i=1}^{n}\frac{(x_i - \theta)^2}{\sigma^2}\right].$$

Taking logarithm of this likelihood function, we have

$$l(\mu, \sigma^2|\mathbf{x}) = \log L(\mu, \sigma^2|\mathbf{x}) = -\frac{n}{2}\log 2\pi - \frac{n}{2}\log\sigma^2 - \frac{1}{2}\sum_{i=1}^{n}\frac{(x_i - \theta)^2}{\sigma^2}.$$

Maximizing this function and replace **x** with **X**, we obtain the maximum likelihood estimator of $\left(\mu, \sigma^2\right)$ as

$$\tilde{\mu} = \bar{X} \quad \text{and} \quad \tilde{\sigma}^2 = \frac{1}{n}\sum_{i=1}^{n}\left(X_i - \bar{X}\right)^2.$$

It is worth noting that we need to get the global maximum of the likelihood function, and sometimes boundary conditions need to be verified.

2.3.3 Bayes Estimators

The derivation of Bayes estimators is based on the Bayesian approach, where parameters are also treated as random variables. The goal of the Bayesian approach is therefore to derive the distribution of the parameters, given the observed values of the random sample. In the Bayesian approach, the distribution of the parameters without the observation is called the *prior* distribution, typically denoted by $\pi(\theta)$, while the distribution updated with the observation is called the *posterior* distribution, denoted by $\pi(\theta|\mathbf{x})$. The relationship between the prior and the posterior distributions is given by the Bayes rule, as

$$\pi(\theta|\mathbf{x}) = \frac{\pi(\theta)\, p(\mathbf{x}|\theta)}{p(\mathbf{x})} = \frac{\pi(\theta)\, p(\mathbf{x}|\theta)}{\int \pi(\theta)\, p(\mathbf{x}|\theta)\, d\theta},$$

where $p(\mathbf{x}|\theta)$ is the pdf or pmf of the random sample and is equivalent to the likelihood function of θ and $p(\mathbf{x})$ is the marginal distribution of **X**, typically referred to as the marginal likelihood (evidence). The above relationship can therefore be intuitively written as

$$\text{Posterior} = \frac{\text{Prior} \times \text{likelihood}}{\text{Marginal likelihood}}.$$

In order to derive a Bayes estimator, we need to (1) assume a prior distribution, (2) derive the likelihood, and (3) calculate the posterior distribution. Typically, the prior distribution is chosen in such a family that the posterior distribution is also in that family. This class of prior is referred to as *conjugate family* for the population. For example, the beta family is the conjugate family of the binomial family and the Dirichlet family is the conjugate family of the multinomial family.

As an example, let $X_1, \ldots, X_n$ be a sample of size n from a Bernoulli (θ) population and $Y = \sum_{i=1}^{n} X_i$. Y has a binomial (n, θ) distribution. Assume that θ has a beta (α, β) distribution, that is

$$\pi(\theta) = \frac{\Gamma(\alpha + \beta)}{\Gamma(\alpha)\Gamma(\beta)} \theta^{\alpha-1}(1-\theta)^{\beta-1}.$$

The likelihood is then

$$p(y|\theta) = \binom{n}{y} \theta^y (1-\theta)^{n-y},$$

and the evidence is

$$p(x) = \int_0^1 \frac{\Gamma(\alpha + \beta)}{\Gamma(\alpha)\Gamma(\beta)} \theta^{\alpha-1}(1-\theta)^{\beta-1} \binom{n}{y} \theta^y (1-\theta)^{n-y} d\theta$$

$$= \binom{n}{y} \frac{\Gamma(\alpha + \beta)}{\Gamma(\alpha)\Gamma(\beta)} \frac{\Gamma(y + \alpha)\Gamma(n - y + \beta)}{\Gamma(n + \alpha + \beta)}.$$

Putting together, we have that the posterior distribution

$$\pi(\theta|y) = \frac{\Gamma(n + \alpha + \beta)}{\Gamma(y + \alpha)\Gamma(n - y + \beta)} \theta^{y+\alpha-1}(1-\theta)^{n-y+\beta-1},$$

which is also a beta distribution with parameters updated from (α, β) to $(y + \alpha, n - y + \beta)$.

With the posterior distribution available, we can reasonably choose the mean of this distribution as a point estimate of the parameters. In the above example, the point estimate of θ is

$$\hat{\theta} = \frac{y + \alpha}{n + \alpha + \beta}.$$

Note that we can write $\hat{\theta}$ as

$$\hat{\theta} = \left(\frac{\alpha + \beta}{n + \alpha + \beta}\right)\left(\frac{\alpha}{\alpha + \beta}\right) + \left(\frac{n}{n + \alpha + \beta}\right)\left(\frac{y}{n}\right),$$

which shows that the posterior mean is a linear combination of the prior mean $\alpha/(\alpha + \beta)$ and the sample mean y/n.

2.3.4 Mean Squared Error

The mean squared error (MSE) of an estimator W of a parameter θ, defined as $\text{MSE} = E_\theta(W - \theta)^2$, is often used to measure the goodness of an estimator. MSE can be written as

$$\text{MSE} = E_\theta(W - \theta)^2 = \text{Var}_\theta W + (E_\theta W - \theta)^2,$$

where $\text{Var}_\theta W$ is the variance of the estimator and $E_\theta W - \theta$ is defined as the bias of the estimator W. If $\text{Bias}_\theta W = E_\theta W - \theta$ is identically equal to 0 as a function of θ, the estimator W is called unbiased. Certainly, for an unbiased estimator W, $E_\theta W = \theta$ for all θ, and $\text{MSE} = \text{Var}_\theta W$. In the family of all unbiased estimator, the one has the smallest variance for all θ is called the *best unbiased estimator*, also referred to as the *uniform minimum variance unbiased estimator* (UMVUE). Finding a best unbiased estimator is typically not trivial.

2.4 Hypothesis Testing

In point estimation, we infer population parameters by giving the most likely (representative) values of the parameters. In some other situation, the inference is done by assuming certain characteristics of the parameters and then assessing whether the assumption is appropriate or not. In this case, we use another statistic inference method called *hypothesis testing*.

In statistics, a *hypothesis* is a statement about a population or about the parameters associated with the population. There are typically two complementary hypotheses in a hypothesis testing problem: the *null hypothesis* (H_0) and the *alternative hypothesis* (H_1). The two hypotheses partition the parameter space Θ into two complementary sets, Θ_0 for H_0 and Θ_0^c for H_1, where $\Theta_0 \cup \Theta_0^c = \Theta$ and $\Theta_0 \cap \Theta_0^c = \emptyset$. The hypothesis testing problem can thus be mathematically represented as

$$H_0 : \theta \in \Theta_0 \text{ versus } H_1 : \theta \in \Theta_0^c.$$

We should make decisions to either accept H_0 or reject H_0; this procedure is called a *hypothesis testing procedure* or *hypothesis test*. More specifically, in a hypothesis testing procedure, we need to determine for which subset of the sample space (*acceptance region*) the null hypothesis H_0 should be accepted and for which subset of the sample space (*rejection region*) H_0 should be rejected.

In general, we are more interested in the hypothesis that the experiment data can support. This hypothesis is sometimes referred to as the *research hypothesis* and is typically used as the alternative hypothesis. Because the rejection of the null hypothesis means the acceptance of the alternative hypothesis, yielding positive conclusion of the experiment data, we also treat the rejection of H_0 as the claim of positive, and the acceptance of H_0 as the claim of negative.

A hypothesis test is typically done by first finding a test statistic $W(\mathbf{X}) = W(X_1, \ldots, X_n)$ and then determining the rejection region, an interval/region, of this test statistic. We will introduce likelihood ratio tests for this purpose. The goodness of a hypothesis test is evaluated using the power function, which measures the probability of making wrong decisions. As a general test statistic, p-values are widely used to provide more flexibility on the determination of the rejection region.

2.4.1 Likelihood Ratio Tests

For a hypothesis testing problem

$$H_0 : \theta \in \Theta_0 \text{ versus } H_1 : \theta \in \Theta_0^c,$$

the likelihood ratio test statistic is defined as

$$\lambda(\mathbf{x}) = \frac{\sup_{\Theta_0} L(\theta | \mathbf{x})}{\sup_{\Theta} L(\theta | \mathbf{x})},$$

and a likelihood ratio test (LRT) is a test that has a rejection region $\{\mathbf{x} : \lambda(\mathbf{x}) \leq c\}$, where $0 \leq c \leq 1$ is a constant. In other words, a likelihood ratio test will reject the null hypothesis if the upper bound of the likelihood for all possible parameters in the null hypothesis is significantly less than that for the entire parameter space.

Recall the maximum likelihood estimates of the population parameter, a likelihood ratio test can be determined via the following steps:

1. Determine the entire parameter space and that of the null hypothesis.
2. Perform a restricted region MLE to determine the upper bound of the likelihood for the null hypothesis.
3. Perform an unrestricted MLE to determine the upper bound of the likelihood for the entire parameter space.
4. Calculate the likelihood ratio test statistic using results of the above two steps.
5. Determine the rejection region of the likelihood ratio test.

For example, let $\mathbf{X} = (X_1, \ldots, X_n)$ be a sample of size n from a normal population $N(\mu, \sigma^2)$, where both μ and σ^2 are unknown. For the hypothesis testing problem

$$H_0 : \mu = \mu_0 \text{ versus } H_1 : \mu \neq \mu_0,$$

the unrestricted MLE gives us

$$\hat{\mu} = \bar{X} \quad \text{and} \quad \hat{\sigma}^2 = \frac{1}{n} \sum_{i=1}^{n} (X_i - \bar{X})^2,$$

while the restricted region MLE gives us

$$\hat{\mu} = \mu_0 \quad \text{and} \quad \hat{\sigma}^2 = \frac{1}{n} \sum_{i=1}^{n} (X_i - \mu_0)^2.$$

The likelihood ratio test statistic is then

$$\lambda(\mathbf{X}) = \left[\frac{\sum_{i=1}^{n} (X_i - \bar{X})^2}{\sum_{i=1}^{n} (X_i - \bar{X})^2 + n(\bar{X} - \mu_0)^2} \right]^{\frac{n}{2}}$$

$$= \left[\frac{1}{1 + n(\bar{X} - \mu_0)^2 / \sum_{i=1}^{n} (X_i - \bar{X})^2} \right]^{\frac{n}{2}},$$

and the likelihood ratio test is to reject the null hypothesis if

$$\frac{1}{1 + n(\bar{x} - \mu_0)^2 / \sum_{i=1}^{n} (x_i - \bar{x})^2} \leq c^{2/n},$$

which is equivalent to

$$\frac{|\bar{x} - \mu_0|}{s / \sqrt{n}} > c',$$

where c' is a constant to be determined using the criteria discussed in the next section. This test is called the *one-sample t test*, which is one of the most widely used tests in bioinformatics.

2.4.2 Error Probabilities and the Power Function

A hypothesis test has two possible decisions: either reject H_0 or accept H_0. The true population parameter θ has also two possible states, either $\theta \in \Theta_0$ or $\theta \in \Theta_0^c$. If the true parameter is in the subset that is given by the alternative hypothesis ($\theta \in \Theta_0^c$) and the decision is to reject H_0, we make a correct decision of claiming positive. Similarly, a correct decision of claiming negative is made if the true parameter is in the subset that is given by the null hypothesis ($\theta \in \Theta_0$), and the decision is to accept H_0. On the other hand, if the true parameter $\theta \in \Theta_0$ but the decision is to reject H_0,

Table 2.1 Two types of errors in hypothesis testing

		Truth	
		$\theta \in \Theta_0^c$	$\theta \in \Theta_0$
Decision	Reject H_0 ($\mathbf{X} \in R$)	Correct decision	Type I error
	Accept H_0 ($\mathbf{X} \in R^c$)	Type II error	Correct decision

we make a wrong decision of claiming positive, yielding a *type I error*; if the true parameter $\theta \in \Theta_0^c$ but the decision is to accept H_0, we make a wrong decision of claiming negative, yielding a *type II error*. Therefore, type I error is always resulted by rejecting H_0, and type II error is always resulted by accepting H_0.

From the machine learning point of view, a type I error is equivalent to a false positive, and a type II error is equivalent to a false negative. Table 2.1 illustrates the two types of errors in hypothesis testing.

Let R be the rejection region and R^c be the acceptance region. For $\theta \in \Theta_0$, the probability of a type I error is then $P_\theta (\mathbf{X} \in R)$. For $\theta \in \Theta_0^c$, the probability of a type II error is $P_\theta (\mathbf{X} \in R^c) = 1 - P_\theta (\mathbf{X} \in R)$. Putting together, we have

$$P_\theta (\mathbf{X} \in R) = \begin{cases} P \text{ (a type I error)} & \text{if } \theta \in \Theta_0 \\ 1 - P \text{ (a type II error)} & \text{if } \theta \in \Theta_0^c \end{cases}$$

Define $\beta(\theta) = P_\theta(\mathbf{X} \in R)$ as the power function of a hypothesis test with rejection region R. $\beta(\theta)$ then characterizes the probability of making wrong decisions. With the power function, we can find good tests by controlling the probability of its type I error at a certain level and then seek for the tests that can minimize the type II error.

The control of type I error probability is via the *significance level* of a test. Let $0 \le \alpha \le 1$ be a real number, a test with power function $\beta(\theta)$ is a level α test if $\sup_{\theta \in \Theta_0} \beta(\theta) \le \alpha$. In other words, a test with significance level α can guarantee that the type I error probability is less than or equal to α for all possible population parameters. Typical values of α include 0.05, 0.01, and 0.001. A test is called unbiased of for every $\theta' \in \Theta_0^c$ and $\theta'' \in \Theta_0$, $\beta(\theta') \ge \beta(\theta'')$. In other words, an unbiased test can make sure that correct decisions are more likely to happen than errors.

With the type I error being controlled by the significance level, the control of type II error probability is via seeking *uniformly most powerful* tests and then determining a suitable sample size. A uniformly most powerful test always has larger power values than any other tests for every $\theta \in \Theta_0^c$, that is, it can achieve the minimum type II error among all possible tests. The following example demonstrates the selection of the constant in the likelihood ratio test and the selection of the sample size.

Let $\mathbf{X} = (X_1, \ldots, X_n)$ be a sample of size n from a normal population $N(\mu, \sigma^2)$, where σ^2 is already known. For the hypothesis testing problem

$$H_0 : \mu \le \mu_0 \text{ versus } H_1 : \mu > \mu_0,$$

the likelihood ratio test can be derived as reject the null hypothesis if

$$\frac{\bar{X} - \mu_0}{\sigma/\sqrt{n}} > c,$$

where c is a constant to be determined. The power function of this test is

$$\beta(\mu) = P_\mu\left(\frac{\bar{X} - \mu_0}{\sigma/\sqrt{n}} > c\right) = P_\mu\left(\frac{\bar{X} - \mu}{\sigma/\sqrt{n}} > c - \frac{\mu - \mu_0}{\sigma/\sqrt{n}}\right) = 1 - \Phi\left(c - \frac{\mu - \mu_0}{\sigma/\sqrt{n}}\right),$$

where $\Phi(\cdot)$ is the normal cdf. In order to ensure the significance level of this test being $\alpha = 0.05$, we need to have $\beta(\mu) \leq \alpha = 0.05$ for all $\mu \leq \mu_0$. This yield the choose of c as $\Phi^{-1}(1 - \alpha) = \Phi^{-1}(0.95) \approx 1.64$. If we further require that the type II error probability must be less than or equal to $\delta = 0.05$ for all $\mu \geq \mu_0 + \sigma$, we need to select such a sample size n that $\max_{\mu \geq \mu_0 + \sigma}\{1 - \beta(\mu)\} = \delta = 0.05$. This yields the selection of n as $[c - \Phi^{-1}(\delta)]^2 = [\Phi^{-1}(0.95) - \Phi^{-1}(0.05)]^2 \approx 10.82$. Because n must be an integer, we can choose $n = 11$.

2.4.3 p-Values

A hypothesis test is done by report the rejection region in some statistically significance level, which controls the probability of type I error and makes the test meaningful. The use of significance level α in the control of type I error is a kind of *hard* decision behavior, in the sense that the null hypothesis is either rejected or accepted at this level. Ideally, people prefer another kind of *soft* decision behavior, where the degree of meaningful, or the significance level, can be flexibly selected by the user instead of the investigator. This is typically done by reporting the results of a hypothesis test via the means of a certain kind of test statistic called a *p-value*.

A *p*-value $p(\mathbf{X})$ is a test statistic that satisfies $0 \leq p(\mathbf{x}) \leq 1$ for every possible sample point $\mathbf{x}$. Smaller *p*-values give stronger evidence that the alternative hypothesis H_1 is true, while larger *p*-values give stronger evidence that the null hypothesis H_0 is true. A *p*-value is valid if, for every possible population parameter $\theta \in \Theta_0$ and every real number $0 \leq \alpha \leq 1$,

$$P_\theta\left(p(\mathbf{X}) \leq \alpha\right) \leq \alpha.$$

In other words, a test that rejects H_0 if and only if $p(\mathbf{X}) \leq \alpha$ is a test having significance level α. Obviously, the reported *p*-values provide a means of soft decision, in that people can easily construct a level α test by simply set a *p*-value cutoff of α.

A more common and usable way to define a valid *p*-value is as follows. Let $W(\mathbf{X})$ be such a test statistic that larger values of W give stronger evidence that the alternative hypothesis H_1 is true. For every sample point $\mathbf{x}$,

$$p(\mathbf{x}) = \sup_{\theta \in \Theta_0} P_\theta \left(W(\mathbf{X}) \geq W(\mathbf{x}) \right)$$

is a valid p-value. Similarly, in the case that smaller values of W give stronger evidence that the alternative hypothesis is true, a valid p-value can be found by defining

$$p(\mathbf{x}) = \sup_{\theta \in \Theta_0} P_\theta \left(W(\mathbf{X}) \leq W(\mathbf{x}) \right).$$

An interesting characteristic of p-values is that a valid p-value is uniformly distributed (more precisely, stochastically greater than or equal to a uniform $(0, 1)$ distribution) under the null hypothesis.

For example, Let $\mathbf{X} = (X_1, \ldots, X_n)$ be a sample of size n from a normal population $N(\mu, \sigma^2)$, where σ^2 is yet unknown. For the hypothesis testing problem

$$H_0 : \mu \leq \mu_0 \text{ versus } H_1 : \mu > \mu_0,$$

the likelihood ratio test can be derived as rejecting the null hypothesis if

$$W(\mathbf{X}) = \frac{\bar{X} - \mu_0}{S/\sqrt{n}} > c,$$

where c is a constant to be determined. Obviously, larger values of W give stronger evidence for rejecting the null hypothesis H_0, or H_1 is true. Because

$$P_{\mu, \sigma^2} \left(W(\mathbf{X}) \geq W(\mathbf{x}) \right) = P_{\mu, \sigma^2} \left(\frac{\bar{X} - \mu_0}{S/\sqrt{n}} \geq W(\mathbf{x}) \right)$$

$$= P_{\mu, \sigma^2} \left(\frac{\bar{X} - \mu}{S/\sqrt{n}} \geq W(\mathbf{x}) - \frac{\mu - \mu_0}{S/\sqrt{n}} \right),$$

$$\leq P \left(T_{n-1} \geq W(\mathbf{x}) \right)$$

where T_{n-1} is a random variable having the T distribution with degree of freedom $n - 1$, a valid p-value can be defined as

$$p(\mathbf{x}) = P \left(T_{n-1} \geq W(\mathbf{x}) \right) = P \left(T_{n-1} \geq \frac{\bar{X} - \mu_0}{S/\sqrt{n}} \right).$$

This test is called the one-sample t test (one-sided).

p-Value is one of the most important concepts in hypothesis testing. With the use of p-values, rejection region of different formats and scale can be transferred into a single uniformed style and scale, thereby providing great flexibility for investigators to select their preferred cutoffs.

2.4.4   Some Widely Used Tests

In this section, we will summarize some widely used tests, including tests for mean, median, symmetric point, variance, equality of two or more proportions, and goodness of fit of any two distributions. Almost all of these tests can be derived from the likelihood ratio test, together with the central limit theorem (CLT) and the definition of the p-value.

One-sample t tests. Let $\mathbf{X} = (X_1, \ldots, X_n)$ be a sample of size n from a normal population $N(\mu, \sigma^2)$, where σ^2 is unknown. We want to test the following hypotheses:

Two-sided: $H_0: \mu = \mu_0$ versus $H_1: \mu \neq \mu_0$;
One-sided: $H_0: \mu \leq \mu_0$ versus $H_1: \mu > \mu_0$ or
              $H_0: \mu = \mu_0$ versus $H_1: \mu > \mu_0$;
One-sided: $H_0: \mu \geq \mu_0$ versus $H_1: \mu < \mu_0$ or
              $H_0: \mu = \mu_0$ versus $H_1: \mu < \mu_0$.

Using one-sample t test, the test statistic for all of the above tests is

$$T(\mathbf{X}) = \frac{\bar{X} - \mu_0}{S/\sqrt{n}}.$$

Under the null hypothesis, this statistic has a T distribution with $n - 1$ degrees of freedom. The rejection regions and p-values can then be determined accordingly.

Two-sample t tests. Let $\mathbf{X} = (X_1, \ldots, X_n)$ and $\mathbf{Y} = (Y_1, \ldots, Y_m)$ be two sample of size n and m, respectively, from two independent normal populations $N(\mu_X, \sigma_X^2)$ and $N(\mu_Y, \sigma_Y^2)$, where both σ_X^2 and σ_Y^2 are unknown but are equal. We want to test the following hypotheses:

Two-sided: $H_0: \mu_X = \mu_Y$ versus $H_1: \mu_X \neq \mu_Y$;
One-sided: $H_0: \mu_X \leq \mu_Y$ versus $H_1: \mu_X > \mu_Y$ or
              $H_0: \mu_X = \mu_Y$ versus $H_1: \mu_X < \mu_Y$;
One-sided: $H_0: \mu_X \geq \mu_Y$ versus $H_1: \mu_X < \mu_Y$ or
              $H_0: \mu_X = \mu_Y$ versus $H_1: \mu_X < \mu_Y$.

Using one-sample t test, the test statistic for all of the above tests is

$$T(\mathbf{X}, \mathbf{Y}) = \frac{\bar{X} - \bar{Y}}{S_p \sqrt{1/n + 1/m}},$$

where

$$S_p^2 = \frac{(n-1)S_X^2 + (m-1)S_X^2}{n+m-2}$$

is the pool variance estimate. Under the null hypothesis, this statistic has a T distribution with $n + m - 2$ degrees of freedom. The rejection regions and p-values can then be determined.

Paired-sample t tests. Let $(X_1, Y_1), \ldots, (X_n, Y_n)$ be a sample of size n from a bivariate normal populations $N\left(\mu_X, \mu_Y, \sigma_X^2, \sigma_Y^2, \rho\right)$, where σ_X^2, σ_Y^2, and ρ are unknown. We want to test the following hypotheses:

Two-sided: $H_0: \mu_X = \mu_Y$ versus $H_1: \mu_X \neq \mu_Y$;
One-sided: $H_0: \mu_X \leq \mu_Y$ versus $H_1: \mu_X > \mu_Y$ or
             $H_0: \mu_X = \mu_Y$ versus $H_1: \mu_X > \mu_Y$;
One-sided: $H_0: \mu_X \geq \mu_Y$ versus $H_1: \mu_X < \mu_Y$ or
             $H_0: \mu_X = \mu_Y$ versus $H_1: \mu_X < \mu_Y$.

Using one-sample t test, the test statistic for all of the above tests is

$$T(\mathbf{X}, \mathbf{Y}) = \frac{\bar{X} - \bar{Y}}{S_{X-Y}/\sqrt{n}},$$

where

$$S_{X-Y}^2 = \frac{1}{n-1} \sum_{i=1}^{n} \left[(X_i - Y_i) - (\bar{X} - \bar{Y})\right]^2$$

is sample variance corresponding to the bivariate normal population. Under the null hypothesis, this statistic has a T distribution with $n - 1$ degrees of freedom. The rejection regions and p-values can then be determined.

The paired-sample t tests are different from the two-sample t tests in that the paired sample is from a bivariate normal population, while the two samples are from two independent normal populations.

One-sample binomial exact tests. Let $\mathbf{X} = (X_1, \ldots, X_n)$ be a sample of size n from a Bernoulli population $B(\theta)$. We want to test the following hypotheses:

Two-sided: $H_0: \theta = \theta_0$ versus $H_1: \theta \neq \theta_0$
One-sided: $H_0: \theta = \theta_0$ versus $H_1: \theta > \theta_0$
One-sided: $H_0: \theta = \theta_0$ versus $H_1: \theta < \theta_0$

Using one-sample Binomial exact tests, the test statistic for all of the above tests is

$$B(\mathbf{X}) = \sum_{i=1}^{n} X_i.$$

Under the null hypothesis, this statistic has a binomial distribution with parameters n and θ_0. The rejection regions and p-values can then be determined.

Define

$$\hat{\theta} = \frac{1}{n} \sum_{i=1}^{n} X_i \quad \text{and} \quad \hat{\sigma}^2 = \hat{\theta}(1 - \hat{\theta}),$$

With the use of the central limit theorem (CLT), we have

$$Z_n = \frac{\hat{\theta} - \theta_0}{\sqrt{\hat{\theta}(1 - \hat{\theta})/n}} \to N(0, 1).$$

This suggests a test with statistic Z_n, which asymptotically has a standard normal distribution. The rejection regions and p-values can then be determined. Because the square of a standard normal random variable is a χ^2 random variable with one degree of freedom, the above normal approximation can also be equivalent to χ^2 approximations.

Two-sample Fisher exact tests. Let $\mathbf{X} = (X_1, \ldots, X_{n_X})$ and $\mathbf{Y} = (Y_1, \ldots, Y_{n_Y})$ be two sample of size n_X and n_Y, respectively, from two Bernoulli populations $B(\theta_1)$ and $B(\theta_2)$, respectively. We are interested in testing the following hypothesis:

One-sided: $H_0 : \theta_1 = \theta_2$ versus $H_1 : \theta_1 > \theta_2$.

Using two-sample Fisher exact test, the p-value is calculated by

$$p = P(S \geq s_X | s_X + s_Y) = \sum_{k=s_X}^{\min\{n_X, s_X + s_Y\}} \binom{n_X}{k} \binom{n_Y}{s_X + s_Y - k} \bigg/ \binom{n_X + n_Y}{s_X + s_Y},$$

where s_X and s_Y are the number of one's observed in the two samples, respectively. For other one-sided and the two-sided case, similar results held. Similar to the one proportion case, with the use of the central limit theorem, the normal approximation and the χ^2 approximation exist for testing the equality of two proportions.

2.5   Interval Estimation

In point estimation, the statistic inference about the population is made by estimating a single representative value of the population parameter. In hypothesis testing, the inference is made by assuming certain characteristics (hypotheses) of the population parameter and then making decisions by either rejecting or accepting the null hypothesis. In this section, we will introduce another kind of statistic inference method that estimates an interval of the population parameter. Obviously, point estimation gives us the most precise possible knowledge about the population parameter, while the interval estimation gives us lower precision of the parameter. As a gain, the probability that an interval estimator covers the true parameter increases over the point estimator. The sacrifice in precision results in the gain in coverage probability.

An interval estimate of a population parameter θ is a pair of functions $L(\mathbf{x})$ and $U(\mathbf{x})$ that satisfies $L(\mathbf{x}) \leq U(\mathbf{x})$ for all sample points. When a sample point $\mathbf{x}$ is observed, the inference $\theta \in [L(\mathbf{x}), U(\mathbf{x})]$ is made, resulting in an interval estimator $[L(\mathbf{X}), U(\mathbf{X})]$. Sometimes we use one-sided interval estimators such as $[L(\mathbf{x}), \infty)$ and $(-\infty, U(\mathbf{x})]$. Sometimes the interval can also be open intervals such as $(L(\mathbf{x}), U(\mathbf{x}))$, $(L(\mathbf{x}), \infty)$, and $(-\infty, U(\mathbf{x}))$. It is worth noting that in classical statistics, a parameter is a fixed number, while an interval estimator is typically a pair of random variables. In terminology, we can say that an interval estimator covers the true parameter, but we cannot say that the true parameter locates in an interval. For this reason, the *coverage probability* of an interval estimator $[L(\mathbf{X}), U(\mathbf{X})]$ is defined as the probability that the random interval $[L(\mathbf{X}), U(\mathbf{X})]$ covers the true parameter θ, denoted by $P_\theta (\theta \in [L(\mathbf{X}), U(\mathbf{X})])$. The *confidence coefficient* of the interval estimator is the infimum of the coverage probability, that is, $\inf_\theta P_\theta (\theta \in [L(\mathbf{X}), U(\mathbf{X})])$. An interval estimator, together with its confidence coefficient, is called a *confidence interval*. An interval estimator with confidence coefficient $1 - \alpha$ is called a $1 - \alpha$ *confidence interval*.

The confidence coefficient of an interval estimator is closely related to a serious of hypothesis testing problems. Consequently, we can find an interval estimator by inverting a test statistic, as the following example.

Let $\mathbf{X} = (X_1, \ldots, X_n)$ be a sample of size n from a normal population $\mathrm{N}(\mu, \sigma^2)$, where σ^2 is unknown. Applying the one-sample t test for the two-sided hypothesis testing problem

$$H_0 : \mu = \mu_0 \text{ versus } H_1 : \mu \neq \mu_0$$

yields the test statistic

$$T(\mathbf{X}) = \frac{\bar{X} - \mu_0}{S/\sqrt{n}}.$$

The acceptance region for a test with significance level α is

$$R = \left\{ \mathbf{x} : \left| \frac{\bar{x} - \mu_0}{S/\sqrt{n}} \right| \leq t_{n-1,\alpha/2} \right\}.$$

Or equivalently

$$\bar{x} - t_{n-1,\alpha/2} \frac{S}{\sqrt{n}} \leq \mu_0 \leq \bar{x} + t_{n-1,\alpha/2} \frac{S}{\sqrt{n}}.$$

Since the test has significance level α, we have

$$P\left(\bar{x} - t_{n-1,\alpha/2}\frac{S}{\sqrt{n}} \leq \mu_0 \leq \bar{x} + t_{n-1,\alpha/2}\frac{S}{\sqrt{n}} \,\Big|\, \mu = \mu_0\right) = 1 - \alpha.$$

Considering this is true for every μ_0, we have that

$$P\left(\bar{x} - t_{n-1,\alpha/2}\frac{S}{\sqrt{n}} \leq \mu \leq \bar{x} + t_{n-1,\alpha/2}\frac{S}{\sqrt{n}}\right) = 1 - \alpha.$$

This is to say that the interval estimator $\left[\bar{x} - t_{n-1,\alpha/2}S/\sqrt{n}, \bar{x} + t_{n-1,\alpha/2}S/\sqrt{n}\right]$ is a $1 - \alpha$ confidence interval.

From this example, we can see that by inverting a test statistic, interval estimators with certain confidence coefficient can be found.

Another means of deriving interval estimators is via the use of *pivotal quantities*. In general, a distribution depends on certain parameters θ. If a formula $Q(\mathbf{X}, \theta)$ of a random sample and the population parameters is free of any parameter, Q is called a pivotal quantity. For example, for a normal sample with size n,

$$T = \frac{\bar{X} - \mu}{S/\sqrt{n}}$$

has a t distribution with $n - 1$ degrees of freedom and does not depend on the parameters μ and σ^2. T is therefore a pivotal quantity. Because student t distributions are symmetric, we can easily derive that in order to have

$$P\left(a \leq \frac{\bar{X} - \mu}{S/\sqrt{n}} \leq b\right) = 1 - \alpha$$

a and b should satisfy $a = b = t_{n-1,\alpha/2}$. Hence

$$P\left(\bar{x} - t_{n-1,\alpha/2}\frac{S}{\sqrt{n}} \leq \mu \leq \bar{x} + t_{n-1,\alpha/2}\frac{S}{\sqrt{n}}\right) = 1 - \alpha.$$

This gives us a $1-\alpha$ confidence interval $\left[\bar{x} - t_{n-1,\alpha/2}S/\sqrt{n}, \bar{x} + t_{n-1,\alpha/2}S/\sqrt{n}\right]$.

2.6   Analysis of Variance

Using two-sample t tests, one can test whether two normal samples have equal means or not. A generalization is to test whether three or more normal samples have equal means or not. An intuitive solution is to do a series of pair-wise two-sample

Table 2.2 One-way
ANOVA data

	1	2	...	k
1	Y_{11}	Y_{21}	...	Y_{k1}
2	Y_{12}	Y_{22}	...	Y_{k2}
3	Y_{13}	Y_{23}	...	Y_{k3}
4	...	...	...	Y_{k4}
	Y_{1n_1}	...	...	...
		Y_{2n_2}	...	Y_{kn_k}

tests. Although this strategy is feasible, there are two problems involved: (1) the increase in the computational burden and (2) the multiple testing correction problems. In detail, for a number of N normal samples, we need to do $N(N-1)/2$ two-sample tests, and obviously the increase in the computational burden is not trivial. Furthermore, because we need to do $N(N-1)/2$ tests, we need to adjust the resulting p-values in order to ensure that the type I error can meet our requirement. The discussion about the multiple testing correction problem is beyond the scope of this introductory course. What we want to have is a simple strategy to test whether more than two normal means are equal or not—the so-called *analysis of variance* (ANOVA) methodology. We will skip the detailed derivation of ANOVA and focus on the calculation involved in ANOVA.

2.6.1 One-Way Analysis of Variance

The data for one-way ANOVA is shown in Table 2.2. Briefly, there are k random samples, indexed by 1, 2, ..., k, and the size of the kth sample is n_k. We assume that the data Y_{ij} $(i=1, ..., k, j=1,...,n_k)$ are observed according to a model

$$Y_{ij} = \theta_i + \epsilon_{ij}, \quad i = 1,...,k; \; j = 1,...,n_k,$$

where the θ_i $(i=1,...,k)$ are unknown parameters that correspond to the common means of random samples and the ϵ_{ij} $(i=1,...,k; j=1,...,n_k)$ are error random variables. We further assume that each ϵ_{ij} $(i=1,...,k; j=1,...,n_k)$ is independent of the others and has a normal distribution $N(0, \sigma_i^2)$, and $\sigma_i^2 = \sigma^2$ for all i.
    In ANOVA, the hypothesis we like to test is then

$$H_0 : \theta_1 = \theta_2 = \cdots = \theta_k \text{ versus } H_1 : \theta_i \neq \theta_j \quad \text{for some } 1 \leq i, j \leq k.$$

An intuitive way to test the above hypothesis is to perform a series of two-sided t test, each of which as the form

$$H_0 : \theta_i = \theta_j \text{ versus } H_1 : \theta_i \neq \theta_j; \quad 1 \leq i, \; j \leq k.$$

If at least one of these null hypotheses can be rejected, the ANOVA hypothesis can be rejected. According to the Bonferroni correction, the resulting p-values should be multiplied by $k(k-1)/2$. This method for testing the ANOVA hypothesis is computationally intensive and has low power.

Let $\mathbf{a} = (a_1, \ldots, a_k)$ be a set of numbers that satisfy $\sum_{i=1}^{k} a_i = 0$. $\mathbf{a}$ is called a contrast. Let $\mathbf{A}$ be the set of all contrasts. The ANOVA hypothesis is then equivalent to

$$H_0 : \sum_{i=1}^{k} a_i \theta_i = 0 \quad \text{for all } (a_1, \ldots, a_k) \text{ such that } \sum_{i=1}^{k} a_i = 0$$

versus

$$H_1 : \sum_{i=1}^{k} a_i \theta_i \neq 0 \quad \text{for some } (a_1, \ldots, a_k) \text{ such that } \sum_{i=1}^{k} a_i = 0$$

or

$$H_0 : \sum_{i=1}^{k} a_i \theta_i = 0 \quad \text{for all } \mathbf{a} \in \mathbf{A} \text{ versus } H_1 : \sum_{i=1}^{k} a_i \theta_i \neq 0 \quad \text{for some } \mathbf{a} \in \mathbf{A}.$$

We first show how to test the hypothesis

$$H_0 : \sum_{i=1}^{k} a_i \theta_i = 0 \text{ versus } H_1 : \sum_{i=1}^{k} a_i \theta_i \neq 0$$

when a contrast $\mathbf{a} = (a_1, \ldots, a_k)$ is known.

Because Y_{ij} $(i = 1, \ldots, k; j = 1, \ldots, n_k)$ has a normal distribution with mean θ_i and variance σ^2, the average of each sample has also a normal distribution. That is,

$$\bar{Y}_{i \cdot} = \frac{1}{n_i} \sum_{j=1}^{n_i} Y_{ij} \sim N(\theta_i, \sigma^2/n_i), \quad i = 1, \ldots, k.$$

Furthermore, $\sum_{i=1}^{k} a_i \bar{Y}_{i \cdot}$ is also normally distributed, as

$$\sum_{i=1}^{k} a_i \bar{Y}_{i \cdot} \sim N \left(\sum_{i=1}^{k} a_i \theta_i, \sigma^2 \sum_{i=1}^{k} \frac{a_i^2}{n_i} \right).$$

With derivations similar to the t test, we have that

$$\frac{\sum_{i=1}^{k} a_i \bar{Y}_{i \cdot} - \sum_{i=1}^{k} a_i \theta_i}{\sqrt{S_p^2 \sum_{i=1}^{k} a_i^2/n_i}} \sim t_{N-k},$$

where $N = \sum_{i=1}^{k} n_i$ is the total number of data, and S_p^2 is the pooled variance estimate and calculated as

$$S_p^2 = \frac{1}{N-k} \sum_{i=1}^{k} (n_i - 1) S_i^2 = \frac{1}{N-k} \sum_{i=1}^{k} \sum_{j=1}^{n_i} (Y_{ij} - \bar{Y}_{i\cdot})^2.$$

The hypothesis

$$H_0 : \sum_{i=1}^{k} a_i \theta_i = 0 \text{ versus } H_1 : \sum_{i=1}^{k} a_i \theta_i \neq 0$$

can then be tested via a similar way as two-sided t test by reject the null hypothesis when

$$\left| \frac{\sum_{i=1}^{k} a_i \bar{Y}_{i\cdot}}{\sqrt{S_p^2 \sum_{i=1}^{k} a_i^2/n_i}} \right| > t_{N-k,\alpha/2}.$$

Now we come back to the ANOVA hypothesis

$$H_0 : \sum_{i=1}^{k} a_i \theta_i = 0 \quad \text{for all } \mathbf{a} \in \mathbf{A} \quad \text{versus} \quad H_1 : \sum_{i=1}^{k} a_i \theta_i \neq 0 \quad \text{for some } \mathbf{a} \in \mathbf{A},$$

$\mathbf{A}$ is the set of all contrasts, that is,

$$\mathbf{A} = \left\{ \mathbf{a} = (a_1, \ldots, a_k) : \sum_{i=1}^{k} a_i = 0 \right\}.$$

For a specific $\mathbf{a}$, there exists a set of $\boldsymbol{\theta} = (\theta_1, \ldots, \theta_k)$ such that $\sum_{i=1}^{k} a_i \theta_i = 0$. Define this set as

$$\Theta_{\mathbf{a}} = \left\{ \boldsymbol{\theta} = (\theta_1, \ldots, \theta_k) : \sum_{i=1}^{k} a_i \theta_i = 0 \right\}.$$

We have that

$$\sum_{i=1}^{k} a_i \theta_i = 0 \quad \text{for all } \mathbf{a} \in \mathbf{A} \iff \boldsymbol{\theta} \in \Theta_{\mathbf{a}} \quad \text{for all } \mathbf{a} \in \mathbf{A} \iff \boldsymbol{\theta} \in \bigcap_{\mathbf{a} \in \mathbf{A}} \Theta_{\mathbf{a}}$$

For each **a**, a two-sided t test as derived previously can be performed with the rejection region being

$$T_{\mathbf{a}} = \left| \frac{\sum_{i=1}^{k} a_i \bar{Y}_{i\cdot} - \sum_{i=1}^{k} a_i \theta_i}{\sqrt{S_p^2 \sum_{i=1}^{k} a_i^2/n_i}} \right| > t_{N-k,\alpha/2}.$$

The union of the rejection regions of all these t tests is therefore the rejection region of the ANOVA hypothesis. This is equivalent to reject the null if $\sup_{\mathbf{a}} T_{\mathbf{a}} > t$, where t is a constant to be determined. Nontrivial derivations show that

$$\sup_{\mathbf{a}:\Sigma a_i=0} T_{\mathbf{a}}^2 = \sup_{\mathbf{a}:\Sigma a_i=0} \frac{\left(\sum_{i=1}^{k} a_i \bar{Y}_{i\cdot} - \sum_{i=1}^{k} a_i \theta_i \right)^2}{S_p^2 \sum_{i=1}^{k} a_i^2/n_i}$$

$$= \frac{\sum_{i=1}^{k} n_i \left(\left(\bar{Y}_{i\cdot} - \bar{\bar{Y}} \right) - \left(\theta_i - \bar{\theta} \right) \right)^2}{S_p^2} \sim (k-1) F_{k-1,N-k}$$

where

$$\bar{\bar{Y}} = \frac{\sum_{i=1}^{k} n_i \bar{Y}_{i\bullet}}{\sum_{i=1}^{k} n_i} = \frac{1}{\sum_{i=1}^{k} n_i} \sum_{i=1}^{k} \sum_{j=1}^{n_i} Y_{ij} \quad \text{and} \quad \bar{\theta} = \frac{1}{\sum_{i=1}^{k} n_i} \sum_{i=1}^{k} n_i \theta_i.$$

In other words, $\sup_{\mathbf{a}:\Sigma a_i=0} T_{\mathbf{a}}^2/(k-1)$ has an F distribution with $k-1$ and $N-k$ degrees of freedom. With this result, the ANOVA hypothesis can be tested by reject the null if

$$F = \frac{\sum_{i=1}^{k} n_i \left(\bar{Y}_{i\cdot} - \bar{\bar{Y}} \right)^2 / (k-1)}{S_p^2} > F_{k-1,N-k,\alpha}$$

This is called the ANOVA F test. It can be shown that

$$\sum_{i=1}^{k} \sum_{j=1}^{n_i} \left(Y_{ij} - \bar{\bar{Y}} \right)^2 = \sum_{i=1}^{k} n_i \left(\bar{Y}_{i\cdot} - \bar{\bar{Y}} \right)^2 + \sum_{i=1}^{k} \sum_{j=1}^{n_i} \left(Y_{ij} - \bar{Y}_{i\cdot} \right)^2.$$

Furthermore, under the ANOVA null, $Y_{ij} \sim N\left(\theta_i, \sigma^2 \right)$, and

$$\frac{1}{\sigma^2} \sum_{i=1}^{k} n_i \left(\bar{Y}_{i\cdot} - \bar{\bar{Y}} \right)^2 \sim \chi_{k-1}^2, \quad \frac{1}{\sigma^2} \sum_{i=1}^{k} \sum_{j=1}^{n_i} \left(Y_{ij} - \bar{Y}_{i\cdot} \right)^2 \sim \chi_{N-k}^2, \quad \text{and}$$

$$\frac{1}{\sigma^2} \sum_{i=1}^{k} \sum_{j=1}^{n_i} \left(Y_{ij} - \bar{\bar{Y}} \right)^2 \sim \chi_{N-1}^2.$$

These observations suggest the following ANOVA table (Table 2.3).

Table 2.3 One-way ANOVA table

	Degrees of freedom	Sum of squares	Mean squares	F statistic
Between treatment groups	$k-1$	**SSB** $$\sum_{i=1}^{k} n_i \left(\bar{Y}_{i\cdot} - \bar{\bar{Y}}\right)^2$$	**MSB** $\dfrac{\text{SSB}}{k-1}$	$F = \dfrac{\text{MSB}}{\text{MSW}}$
Within treatment groups	$N-k$	**SSW** $$\sum_{i=1}^{k}\sum_{j=1}^{n_i} \left(Y_{ij} - \bar{Y}_{i\cdot}\right)^2$$	**MSW** $\dfrac{\text{SSW}}{N-k}$	
Total	$N-1$	**SST** $$\sum_{i=1}^{k}\sum_{j=1}^{n_i} \left(Y_{ij} - \bar{\bar{Y}}\right)^2$$		

Table 2.4 Two-way ANOVA data

		1		2		J		
	1	Y_{111} ... Y_{11n}	$\bar{Y}_{11\cdot}$	Y_{121} ... Y_{12n}	$\bar{Y}_{12\cdot}$	Y_{1J1} ... Y_{1Jn}	$\bar{Y}_{1J\cdot}$	$\bar{Y}_{1\cdot\cdot}$
	2	Y_{211} ... Y_{21n}	$\bar{Y}_{21\cdot}$	Y_{221} ... Y_{22n}	$\bar{Y}_{22\cdot}$	Y_{2J1} ... Y_{2Jn}	$\bar{Y}_{2J\cdot}$	$\bar{Y}_{2\cdot\cdot}$
	...	...	...	...	...	...	...	...
	I	Y_{I11} ... Y_{I1n}	$\bar{Y}_{I1\cdot}$	Y_{I21} ... Y_{I2n}	$\bar{Y}_{I2\cdot}$	Y_{IJ1} ... Y_{IJn}	$\bar{Y}_{IJ\cdot}$	$\bar{Y}_{I\cdot\cdot}$
		$\bar{Y}_{\cdot1\cdot}$		$\bar{Y}_{\cdot2\cdot}$		$\bar{Y}_{\cdot J\cdot}$		

2.6.2 Two-Way Analysis of Variance

Besides the one-way ANOVA, sometimes we also perform two-way ANOVA. The data of the two-way ANOVA typically look like as follows (Table 2.4).

There are $I \times J$ groups of data in the table, indexed by two subscripts i (row index) and j (column index). Each group has n data. The total number of data are therefore nIJ. The mean for the ijth group is denoted as $\bar{Y}_{ij\cdot} = \frac{1}{n}\sum_{k=1}^{n} Y_{ijk}$, the mean for the ith row $\bar{Y}_{i\cdot\cdot} = \frac{1}{nJ}\sum_{j=1}^{J}\sum_{k=1}^{n} Y_{ijk}$, the mean for the jth column $\bar{Y}_{\cdot j\cdot} = \frac{1}{nI}\sum_{i=1}^{I}\sum_{k=1}^{n} Y_{ijk}$, and the mean of all data $\bar{Y}_{\cdot\cdot\cdot} = \frac{1}{nIJ}\sum_{i=1}^{I}\sum_{j=1}^{J}\sum_{k=1}^{n} Y_{ijk}$.

We first introduce a non-interaction two-way ANOVA model. We assume that $Y_{ijk} = \mu + \alpha_i + \beta_j + \epsilon_{ijk}, i = 1,\ldots,I, j = 1,\ldots,J, k = 1,\ldots,n$, where the noise $\epsilon_{ijk} \sim N(0,\sigma^2)$, iid, and the means $\sum_{i=1}^{I}\alpha_i = \sum_{j=1}^{J}\beta_j = 0$. We like to test the hypothesis

$$H_0 : \alpha_1 = \cdots = \alpha_I = 0 \quad \text{and} \quad \beta_1 = \cdots = \beta_J = 0.$$

Table 2.5 Two-way ANOVA table (non-interaction model)

	Degrees of freedom	Sum of squares	Mean squares	F statistic
Between groups	$I+J-2$	**SSB** $$n \sum_{i=1}^{I} \sum_{j=1}^{J} \left[(\bar{Y}_{i..} - \bar{Y}_{...})^2 + (\bar{Y}_{.j.} - \bar{Y}_{...})^2 \right]$$	**MSB** $$\frac{SSB}{I+J-2}$$	$F = \dfrac{MSB}{MSW}$
Row factor	$I-1$	**SSR** $$nJ \sum_{i=1}^{I} \bar{Y}_{i..} - \bar{Y}_{...}$$	**MSR** $$\frac{SSR}{I-1}$$	$F = \dfrac{MSR}{MSW}$
Column factor	$J-1$	**SSC** $$nI \sum_{j=1}^{J} (\bar{Y}_{.j.} - \bar{Y}_{...})$$	**MSC** $$\frac{SSC}{J-1}$$	$F = \dfrac{MSC}{MSW}$
Within groups	$N-I-J+1$	**SSW** $$\sum_{i=1}^{I} \sum_{j=1}^{J} \sum_{k=1}^{n} (Y_{ijk} - \bar{Y}_{i..} - \bar{Y}_{.j.} + \bar{Y}_{...})$$	**MSW** $$\frac{SSW}{N-I-J+1}$$	
Total	$N-1$	**SST** $$\sum_{i=1}^{I} \sum_{j=1}^{J} \sum_{k=1}^{n} (Y_{ijk} - \bar{Y}_{...})^2$$		

versus

$$H_1 : \alpha_i \neq 0 \quad \text{or} \quad \beta_j \neq 0 \quad \text{for some } i, j$$

Similar to the one-way ANOVA F test table, for the non-interaction two-way ANOVA hypothesis, we have the following F test table (Table 2.5).

With this table, one can reject the null hypothesis if $F = MSB/MSW > F_{I+J-2,N-I-J+1,\alpha}$. Furthermore, one can test only the row factor using the statistic $F = MSR/MSW \sim F_{I-1,N-I-J+1}$ or the column factor using the statistic $F = MSC/MSW \sim F_{J-1,N-I-J+1}$.

We now introduce an interaction two-way ANOVA model, in which we assume $Y_{ijk} = \mu + \alpha_i + \beta_j + \gamma_{ij} + \epsilon_{ijk}$, $i = 1, \ldots, I; j = 1, \ldots, J, k = 1, \ldots, n$, where the noise $\epsilon_{ijk} \sim N(0, \sigma^2)$, iid, and the means $\sum_{i=1}^{I} \alpha_i = \sum_{j=1}^{J} \beta_j = \sum_{i=1}^{I} \gamma_{ij} = \sum_{j=1}^{J} \gamma_{ij} = 0$. We like to test the hypothesis

$$H_0 : \alpha_i = 0 \quad \text{and} \quad \beta_j = 0 \quad \text{and} \quad \gamma_{ij} = 0 \quad \text{for all } i, j.$$

versus

$$H_1 : \alpha_i \neq 0 \quad \text{or} \quad \beta_j \neq 0 \quad \text{or} \quad \gamma_{ij} \neq 0 \quad \text{for some } i, j$$

Table 2.6 Two-way ANOVA table (interaction model)

	Degrees of freedom	Sum of squares	Mean squares	F statistic
Between groups	$I \times J - 1$	**SSB** $n \sum\limits_{i=1}^{I} \sum\limits_{j=1}^{J} \left(\bar{Y}_{ij\cdot} - \bar{Y}_{...}\right)^2$	**MSB** $\dfrac{\text{SSB}}{I \times J - 1}$	$F = \dfrac{\text{MSB}}{\text{MSW}}$
Row factor	$I-1$	**SSR** $nJ \sum\limits_{i=1}^{I} \left(\bar{Y}_{i\cdot\cdot} - \bar{Y}_{...}\right)^2$	**MSR** $\dfrac{\text{SSR}}{I - 1}$	$F = \dfrac{\text{MSR}}{\text{MSW}}$
Columnfactor	$J-1$	**SSC** $nI \sum\limits_{j=1}^{J} \left(\bar{Y}_{\cdot j\cdot} - \bar{Y}_{...}\right)^2$	**MSC** $\dfrac{\text{SSC}}{J - 1}$	$F = \dfrac{\text{MSC}}{\text{MSW}}$
Interaction factors	$(I-1)(J-1)$	**SSI** $\sum\limits_{i=1}^{I} \sum\limits_{j=1}^{J} \sum\limits_{k=1}^{n} \left(\bar{Y}_{ij\cdot} - \bar{Y}_{i\cdot\cdot} \right.$ $\left. -\bar{Y}_{\cdot j\cdot} + \bar{Y}_{...}\right)^2$	**MSI** $\dfrac{\text{SSC}}{(I - 1)(J - 1)}$	$F = \dfrac{\text{MSI}}{\text{MSW}}$
Within groups	$N - I \times J$	**SSW** $\sum\limits_{i=1}^{I} \sum\limits_{j=1}^{J} \sum\limits_{k=1}^{n} \left(Y_{ijk} - \bar{Y}_{ij\cdot}\right)^2$	**MSW** $\dfrac{\text{SSW}}{N - IJ}$	
Total	$N-1$	**SST** $\sum\limits_{i=1}^{I} \sum\limits_{j=1}^{J} \sum\limits_{k=1}^{n} \left(Y_{ijk} - \bar{Y}_{...}\right)^2$		

Similar to the non-interaction model, we have the following F test table for the interaction two-way ANOVA hypothesis (Table 2.6).

With this table, one can reject the ANOVA null if $F = \text{MSB}/\text{MSW} > F_{I \times J - 1, N - I \times J, \alpha}$. Furthermore, we can test whether the interaction between the row factor and the column factor exists by rejecting the null hypothesis if $F = \text{MSI}/\text{MSW} > F_{(I-1)(J-1), N - I \times J, \alpha}$. Furthermore, one can test only the row factor using the statistic $F = \text{MSR}/\text{MSW} \sim F_{I-1, N - I \times J}$ or the column factor using the statistic $F = \text{MSC}/\text{MSW} \sim F_{J-1, N - I \times J}$.

2.7 Regression Models

In the one-way ANOVA, we are interested in analysis how one categorical variable (factor) influence the means of a response variable. This analysis is extended to account for two factors in the two-way ANOVA and can be further extended to multi-way ANOVA. In this section, we will see how one or more continuous (predictor) variables influence the response variable. We will first prove the concept via a detailed analysis of simple linear regression that contains only one predictor variable and then extend the derivation to multiple predictor variables.

2.7.1   Simple Linear Regression

In simple linear regression, we deal with a set of observations $\{(y_1, x_1), \ldots, (y_n, x_n)\}$, and we assume that the response variable Y and the predictor variable x have the following form:

$$Y_i = \alpha + \beta x_i + \epsilon_i.$$

In this formulation, Y_i is called the response variable, and x_i is called the predictor variable. ϵ_i is a random variable reflects the noise in this model. α is the intercept, and β is the slope. Note that in this simple linear regression model, Y_i and ϵ_i are random variables, while the others are not.

Define the sample means as

$$\bar{x} = \frac{1}{n} \sum_{i=1}^{n} x_i \quad \text{and} \quad \bar{y} = \frac{1}{n} \sum_{i=1}^{n} y_i$$

and sums of squares as

$$S_{xx} = \sum_{i=1}^{n} (x_i - \bar{x})^2, \quad S_{xy} = \sum_{i=1}^{n} (x_i - \bar{x})(y_i - \bar{y}), \quad \text{and} \quad S_{yy} = \sum_{i=1}^{n} (y_i - \bar{y})^2.$$

It is easy to show that

$$\hat{b} = \frac{S_{xy}}{S_{xx}} \quad \text{and} \quad \hat{a} = \bar{y} - \hat{b}\bar{x}$$

can minimize the following residual sum of squares (RSS)

$$\text{RSS} = \sum_{i=1}^{n} [y_i - (a + bx_i)]^2$$

The estimates $\hat{a}$ and $\hat{b}$ are called the least squares estimates of α and β, respectively. Now, assume that the noise ϵ_i has mean 0 and variance σ^2, that is,

$$E\epsilon_i = 0 \quad \text{and} \quad \text{Var}\epsilon_i = \sigma^2.$$

If we restrict the variables α and β as linear combinations of the response variable Y_i, say

$$\alpha = \sum_{i=1}^{n} c_i Y_i \quad \text{and} \quad \beta = \sum_{i=1}^{n} d_i Y_i.$$

We can get that the unbiased estimators of β and α are, respectively,

$$\hat{\beta} = \frac{S_{xY}}{S_{xx}} \quad \text{and} \quad \hat{\alpha} = \bar{Y} - \hat{\beta}\bar{x}.$$

They are called the best linear unbiased estimators (BLUE).

With an even stronger assumption that the noise ϵ_i has a normal distribution $N(0, \sigma^2)$, we have that the response variable Y_i has a normal distribution

$$Y_i \sim N\left(\alpha + \beta x_i, \sigma^2\right), \quad i = 1, \dots, n.$$

Applying the maximum likelihood methodology, we obtain that the MLE estimators for α and β are

$$\hat{\beta} = \frac{S_{xY}}{S_{xx}} \quad \text{and} \quad \hat{\alpha} = \bar{Y} - \hat{\beta}\bar{x}.$$

They have the same form as the best linear unbiased estimators. Furthermore, the MLE estimator for the variance σ^2 is

$$\hat{\sigma}^2 = \frac{1}{n} \sum_{i=1}^{n} \left(Y_i - \hat{\alpha} - \hat{\beta}x_i\right)^2,$$

and an unbiased estimator for the variance σ^2 is

$$S^2 = \frac{n}{n-2}\hat{\sigma}^2, \quad i = 1, \dots, n.$$

It can be shown that the estimators $\hat{\alpha}$ and $\hat{\beta}$ has normal distributions

$$\hat{\alpha} \sim N\left(\alpha, \frac{\sigma^2}{nS_{xx}} \sum_{i=1}^{n} x_i^2\right) \quad \text{and} \quad \hat{\beta} \sim N\left(\beta, \frac{\sigma^2}{S_{xx}}\right)$$

and $(n-2)S^2/\sigma^2$ has a χ^2 distribution with $n-2$ degrees of freedom, say

$$\frac{(n-2)S^2}{\sigma^2} \sim \chi^2_{n-2}$$

Therefore

$$\frac{\hat{\alpha} - \alpha}{S\sqrt{\left(\sum_{i=1}^{n} x_i^2\right)/nS_{xx}}} \sim t_{n-2} \quad \text{and} \quad \frac{\hat{\beta} - \beta}{S\sqrt{S_{xx}}} \sim t_{n-2}$$

Table 2.7 Regression ANOVA table

	Degrees of freedom	Sum of squares	Mean squares	F statistic
Regression slope	1	SSS	MSS	$F = \dfrac{\text{MSS}}{\text{MSR}}$
		$\dfrac{S_{xY}^2}{S_{xx}} = \sum\limits_{i=1}^{n} (\hat{Y}_i - \bar{Y})^2$	SSS	
Regression residual	$n-2$	SSR $=$ RSS	MSR	
		$S_{YY} = \sum\limits_{i=1}^{n} \left(Y_i - \hat{\alpha} - \hat{\beta} x_i\right)^2$	$\dfrac{\text{RSS}}{n-2}$	
Total	$n-1$	SST		
		$\sum\limits_{i=1}^{n} (Y_i - \bar{Y})^2$		

These results can be used to test the hypothesis

$$H_0 : \alpha = 0 \quad \text{versus } H_1 : \alpha \neq 0$$

and

$$H_0 : \beta = 0 \quad \text{versus } H_1 : \beta \neq 0$$

Since the square of a t distribution is an F distribution, we can also use F test to reject the null hypothesis that $\beta = 0$ if

$$\frac{\hat{\beta}^2}{S^2/S_{xx}} > F_{1,n-2,\alpha}.$$

These facts suggest the regression ANOVA table (Table 2.7).

The goodness of fit of the regression model is typically described by a statistic called the coefficient of determination, defined as

$$r^2 = \frac{\text{Regression sum of squares}}{\text{Total sum of squares}} = \frac{\sum_{i=1}^{n} (\hat{y}_i - \bar{y})^2}{\sum_{i=1}^{n} (y_i - \bar{y})^2} = \frac{S_{xy}^2}{S_{xx} S_{yy}}.$$

One of the purposes of building the regression model is to predict the future response, given the predictor variable. An intuitive prediction method is to plug the estimated parameters $\hat{\alpha}$ and $\hat{\beta}$ in the simple linear regression model and calculate the corresponding value of the response variable. It can be shown that at a certain point x_0, the quantity $\hat{\alpha} + \hat{\beta} x_0$ has a normal distribution

$$\hat{y}_0 = \hat{\alpha} + \hat{\beta} x_0 \sim N\left(\alpha + \beta x_0, \sigma^2 \left(\frac{1}{n} + \frac{(x_0 - \bar{x})^2}{S_{xx}}\right)\right),$$

while the response Y_0 has a normal distribution

$$Y_0 \sim N\left(\alpha + \beta x_0, \sigma^2 \left(1 + \frac{1}{n} + \frac{(x_0 - \bar{x})^2}{S_{xx}}\right)\right),$$

if we assume Y_0 follows the same regression model as the already observed data. Hence, a $100(1 - \rho)\%$ confidence interval of $\alpha + \beta x_0$ is given as

$$\hat{\alpha} + \hat{\beta} x_0 - t_{n-2,\rho/2} S \sqrt{\frac{1}{n} + \frac{(x_0 - \bar{x})^2}{S_{xx}}} \leq \alpha + \beta x_0 \leq \hat{\alpha} + \hat{\beta} x_0$$

$$+ t_{n-2,\rho/2} S \sqrt{\frac{1}{n} + \frac{(x_0 - \bar{x})^2}{S_{xx}}},$$

with the understanding that

$$\frac{(\hat{\alpha} + \hat{\beta} x_0) - (\alpha + \beta x_0)}{S \sqrt{1/n + (x_0 - \bar{x})^2 / S_{xx}}} \sim t_{n-2}.$$

And a $(1 - \rho) \times 100\%$ prediction interval of Y_0 is given as

$$\hat{\alpha} + \hat{\beta} x_0 - t_{n-2,\rho/2} S \sqrt{1 + \frac{1}{n} + \frac{(x_0 - \bar{x})^2}{S_{xx}}} \leq Y_0 \leq \hat{\alpha} + \hat{\beta} x_0$$

$$+ t_{n-2,\rho/2} S \sqrt{1 + \frac{1}{n} + \frac{(x_0 - \bar{x})^2}{S_{xx}}},$$

because $Y_0 - (\hat{\alpha} + \hat{\beta} x_0)$ has a t distribution

$$\frac{Y_0 - (\hat{\alpha} + \hat{\beta} x_0)}{S \sqrt{1 + 1/n + (x_0 - \bar{x})^2 / S_{xx}}} \sim t_{n-2}.$$

2.7.2 Logistic Regression

In simple linear regression model, we see how the predictor variable influences the response variable in a linear way. Generally, the relationship between the predictor and response can be extended via a nonlinear link function. Of particular interest is a binomial link, in which the response variable has a Bernoulli distribution, while the parameter in this distribution is influenced by the predictor variable. The logistic regression model deals with such situation.

Suppose that Y_i has a Bernoulli distribution with $P(Y_i = 1) = p_i$ and $P(Y_i = 0) = 1 - p_i$. The logistic regression model assumes that

$$\log\left(\frac{p_i}{1 - p_i}\right) = \alpha + \beta x_i.$$

An equivalent presentation is that

$$P(Y_i = 1) = p_i = \frac{e^{\alpha + \beta x_i}}{1 + e^{\alpha + \beta x_i}} = \pi(x_i)$$

and

$$P(Y_i = 0) = 1 - p_i = \frac{1}{1 + e^{\alpha + \beta x_i}} = 1 - \pi(x_i)$$

The likelihood function of the parameters α and β is then

$$L(\alpha, \beta | \mathbf{y}) = \prod_{i=1}^{n} \pi(x_i)^{y_i} (1 - \pi(x_i))^{1 - y_i}$$

Applying a numeric methodology such as the Newton–Raphson method to maximize the logarithm of this function yields the MLE estimator of α and β. With the parameters being estimated, we can further use this model to do prediction.

2.8 Statistical Computing Environments

Many of the statistical inference methods introduced above require labor insensitive computation of numeric data. To facilitate these calculations, quite a few statistical computing environments such as SAS, SPSS, S-Plus, Matlab, and R have been developed. Within these environments, R is free and open source and thus received more and more attention. In this section, we will briefly introduce the use of R. Students can refer to the book by Peter Dalgaard [2], the user manual of R, and other related textbooks for further studies.

2.8.1 Downloading and Installation

R can be freely downloaded via the home page of the R project (http://www.r-project.org/). R is available as binary executable for all Windows and MacOS platforms and as source code for the Unix and Linux platforms.

2.8.2   Storage, Input, and Output of Data

Data are stored in R as *objects*. The basic type of data is *vector*, which is simply a one-dimensional array of numeric numbers. A vector can be constructed using functions such as c(...), seq(...), and rep(...). The length of a vector is the elements in the vector and can be obtained using length(...). A two-dimensional array is called a *matrix*, which can be constructed using the function of matrix(...). The size of each dimension of a matrix can be obtained using dim(...). In the case of three and more dimensions, one can use the array(...) to construct an object. Besides numeric data, one can also use categorical data, which are called *factors* in R. The function of constructing a factor is factor(...). One can use as.factor(...) to convert a numeric vector to a factor.

Sometimes it is useful to combine a group of objects into a large object. The combination of such a collection of objects is called a *list*. A list can be constructed using list(...). The objects in a list may have different length and dimensions. In the simplest case, when a list contains a collection of vectors of the same length, the list is called a *data frame*. Data frame is by far the most widely used data object in R. One can treat a data frame as a data set, a two-dimensional table of data. A data frame can be constructed using data.frame(...).

Data are typically stored in plain text files as two-dimensional tables. For example, in the analysis of microarray, the expression values of multiple genes across many conditions are stored as a two-dimensional, in which each row being a gene and each column standing for a condition. In R, this kind of data can be read into a data frame using read.table(...).

As the complementary operator of reading data from a text file, a data frame can be written to a plain text file using write.table(...).

2.8.3   Distributions

R provides abundant functions to deal with most univariate distributions. For each distribution, there are usually four functions: probability density/mass function (pdf/pmf), cumulative distribution function, quantile function, and random number generator. R uses a single character prefix plus the name of the distribution to name these four functions. For example, for the normal distribution, the pdf is dnorm(...), the cdf is pnorm(...), the quantile function is qnorm(...), and the random number generator is rnorm(...). For the beta distribution, the pdf, cdf, quantile function, and random number generator are named as dbeta(...), pbeta(...), qbeta(...), and rbeta(...), respectively.

R also provides functions for random sampling. The function sample(...) is used to make a random sample from a finite population.

R provides functions to summarize data. One can use mean(...), median(...), var(...), and sd(...) to obtain the mean, median, variance, and standard derivation of a sample, respectively. For two samples, cov(...) and cor(...) can be used to obtain the covariance and correlation coefficient.

2.8.4 Hypothesis Testing

R provides functions for various hypothesis tests. For t tests, the function is t.test(...). For Wilcoxon signed rank test and rank sum (Mann–Whitney) test, the function is wilcox.test(...). Other functions include chisq.test(...), cor.test(...), fisher.test(...), ks.test(...), and many others. Typically, the names of these functions are given according to the name of the tests.

2.8.5 ANOVA and Linear Model

Linear models including ANOVA models and linear regression models can be performed in R using lm(...). summary.lm(...) can be used to obtain summaries of an ANOVA or regression result, and anova.lm(...) can be used to obtain the ANOVA table.

References

1. Casella G, Berger RL (2002) Statistical inference, 2nd edn. Duxbury/Thomson Learning, Pacific Grove
2. Dalggaard P (2002) Introductory statistics with R. Springer, New York

Chapter 3
Topics in Computational Genomics

Michael Q. Zhang and Andrew D. Smith

3.1 Overview: Genome Informatics

Genomics began with large-scale sequencing of the human and many model organism genomes around 1990; rapid accumulation of vast genomic data brings a great challenge on how to decipher such massive molecular information. As bioinformatics in general, genome informatics is also data driven; many computational tools developed can soon be obsolete when new technologies and data types become available. Keeping this in mind if a student wants to work in this fascinating new field, one must be able to adapt quickly and to "shoot the moving targets" with the "just-in-time ammunition."

In this chapter, we begin by reviewing the progress in genomics and its informatics in the first section. In the following sections, we will discuss a few selected computational problems in more detail. For each problem, we state clearly what the central question is, try to give a brief biological background, introduce current approaches, and raise open challenges. To understand genomics, a basic knowledge in molecular biology (a combination of genetics and biochemistry) is prerequisite. We would highly recommend the book "Recombinant DNA" by Watson, Gilman, Witkowski, and Zoller for an easy background reading. And a good place to read more on the history is the Landmarks in Genetics and Genomics website of NIH/NHGRI (http://www.genome.gov/25019887).

M.Q. Zhang (✉)
Department of Molecular and Cell Biology, The University of Texas at Dallas,
800 West Campbell Rd, RL11, Richardson, TX, USA 75080

Tsinghua National Laboratory for Information Science and Technology,
Tsinghua University, Beijing 100084, China
e-mail: mzhang@cshl.edu

A.D. Smith
Cold Spring Harbor Laboratory, Cold Spring Harbor, NY, USA

R. Jiang et al. (eds.), *Basics of Bioinformatics: Lecture Notes of the Graduate Summer School on Bioinformatics of China*, DOI 10.1007/978-3-642-38951-1_3,
© Tsinghua University Press, Beijing and Springer-Verlag Berlin Heidelberg 2013

The central subject of genetics, hence large-scale genetics – genomics – is genes, their structure and function – their regulation and evolution. The very concept of "gene" itself has also been evolved dramatically [66]. Classically, like the elementary particles in physics, genes are considered as the fundamental heritable units of life and are conserved from generation to generation. Like quantum quarks, found they also come with different "colors" (alleles), and they appeared to recombine and distribute randomly from the parents to the offspring. Later was able to show genes are not quite free particles; they live on finite number of strings and a distance between a pair may be defined by recombination frequency. Such statistical inference is so remarkable because genes were discovered before they could be observed under a microscope and before their physical identity – DNA – was worked out. All students should get familiar with this history and reflect on the power of logical thinking. From "one-gene one-protein model", "gene is a piece of DNA", to Watson-Crick double helix structure and the genetic code, gene evolved from an abstract concept to a physical entity. Students should also master the principles of these seminal experiments. It is the genetics and biochemistry that makes the molecular biology a quantitative and predictive science. It was the automatic DNA sequencing technology that made the large-scale genome sequencing become possible. By 2003, the human genome, together with several model organism genomes, was sequenced, as Gilbert once predicted at the beginning of the Human Genome Project:

> The new paradigm now emerging, is that all the genes will be known (in the sense of being resident in databases available electronically), and that the starting point of a biological investigation will be theoretical.

The emerging field of Bioinformatics has both a scientific aspect as the study of how information is represented and transmitted in biological systems and an engineering aspect as the development of technology for storage, retrieval, and display of complex biological data. The ultimate goal of genetics/genomics is to estimate the conditional probability $\Pr(P|G, E)$, where P is the phenotype (or traits), G is the genotype (or alleles), and E is the environment. Before this could be systematically studied, a "parts list" would have to be completed. This includes various functional (both structural and regulatory) elements in the genome, which would have to be identified and characterized. Here is a partial list of comments questions in genome informatics:

1. Where are the genes?
2. What are the regulatory regions?
3. What is the exon/intron organization of each gene?
4. How many different functional RNA transcripts can each gene produce? When and where are they expressed?
5. What are the gene products? Structure and function?
6. How do different genes and gene products interact?
7. How are they controlled and regulated?

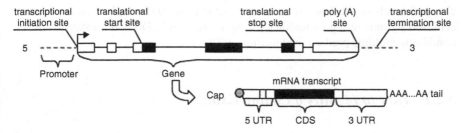

Fig. 3.1 A typical vertebrate PolII gene contains multiple exons

8. How is genome evolved? What is the relationship among different species?
9. What are the mutations, polymorphism, and selection?

In the following, we will describe a few typical problems in detail.

3.2   Finding Protein-Coding Genes

According to the Central Dogma, genetic information flows from

$$DNA \rightarrow RNA \rightarrow Protein$$

Eukaryotic gene expression starts from transcription, in which a pre-mRNA copy of the gene is made; through RNA processing and transport, in which the pre-mRNA transcript is capped at the 5'-end, the introns are spliced and the exons are ligated, poly(A)-tail is synthesized at the 3'-end, and the matured mRNA is transported from the nucleus out into the cytoplasm and ends with protein translation. A typical vertebrate protein-coding gene structure is depicted in Fig. 3.1, together with its mRNA transcript. It contains six exons, including three coding exons (in black). Given a genomic DNA sequence, finding a protein-coding gene consists of (a) identification of the gene boundaries and (b) delineation of the exon-intron organization. Computational prediction of gene boundaries and noncoding exons is extremely difficult (see the next section); most predictions have been focusing on coding regions (CDS in Fig. 3.1). Experimental methods include cDNA/EST/CAGE-tag sequencing, exon trapping, and tiling microarrays. Since a gene may only express in certain cell types and under specific conditions, not every transcript can be identified experimentally. Two common strategies have been used in ab initio gene prediction algorithms: (a) detecting individual exon candidates and connecting them by, e.g., Dynamic Programming (DP) and (b) segmenting DNA sequence into exon/intron/splice-site states by (standard or generalized) hidden Markov models (HMMs). Basic ab initio gene prediction algorithms have not changed much in the last 10 years (see review by [93]); high accuracy can be achieved by integrating evolutionary conservation and cDNA information and by combining multiple prediction algorithms [40].

3.2.1 How to Identify a Coding Exon?

It would be very simple to treat this problem as a discrimination problem. Given labeled training samples (X_i, Y_i), $i = 1, \ldots, N$: when $Y = 1$, the corresponding sample X is a true exon sample (i.e., $X = 3'\text{ss} - \text{CDS} - 5'\text{ss}$, $3'\text{ss}$ may contain 50nt $3'$ splice-site sequence ending with AG and $5'\text{ss}$ may contain 10nt $5'$splice-site sequence, and the coding sequence CDS does not have a STOP codon at least in one of the three reading frames), and when $Y = 0$, X is a matching pseudo-exon. One can train any standard classifier (LDA, QDA, SVM, etc., see [34]) as the predictor by feature selection and cost minimization (e.g., minimizing classification errors with cross-validation). Many discriminant methods have thus developed; the key is to choose good discriminant feature variables and appropriate pseudo-exon samples. Good discriminant feature variables often include $5'\text{ss}$ score, $3'\text{ss}$ score (including branch-site score), in-frame hexamer coding potential scores, and exon-size. HEXON [80] and [92] are exon finder methods based on LDA and QDA, respectively.

For the first coding exon, one would use Translational Initiation Site (TIS or Kozak) score to replace the $3'\text{ss}$ core; and for the last coding exon, the exon must end with a STOP codon instead of a $5'\text{ss}$.

3.2.2 How to Identify a Gene with Multiple Exons?

Although assembling multiple exons into a gene was straightforward by using DP, fully probabilistic state models (such as HMMs) have become favorable because weighting problem becomes simple counting of relative observed state frequencies. Under an HMM, a DNA sequence is segmented (partitioned) into disjoint fragments or states (because in the duality of the regions and boundaries, we refer region as state and boundary as transition between states). There are only finite number of states (or labels: TIS, exon, $5'\text{ss}$, intron, $3'\text{ss}$, STOP), and one needs to train an HMM model to get the parameters: $P(s|q)$ is probability of emitting a base s in a state q and the transition probabilities $T(q_1|q_2)$ from a state q_1 to a state q_2. For a given partition (assignment of labels, or parse) Z, the joint probability is simply given by the product

$$P(Z, S) = P(s_1|q_1)T(q_1|q_2)P(s_2|q_2) \ldots T(q_{N-1}|q_N)P(s_N|q_N)$$

for a sequence of length N. A recursive algorithm (similar to DP) called the Viterbi algorithm may be used to find the most probable parse [71] corresponding to the optimal transcript (exon/intron) prediction. The advantage of HMMs is that one can easily add more states (such as intergenic region, promoter, UTRs, poly(A), and frame- or strand-dependent exons/introns) as well as flexible transitions between the states (to allow partial transcript, intronless genes, or even multiple genes).

HMMgene [49] is based on HMM which can be optimized easily for maximum prediction accuracy. Genie introduced generalized HMM (GHMM, [50]) and used the neural networks as individual sensors for splice signals as well as for coding content. Genscan [20], fgenesh [75], and TigrScan [56] are also based on GHMM, and it allows exon-specific length distribution whereas the intrinsic length distribution for a HMM is geometric (i.e., decaying exponentially). Augustus [83] added a new intron submodel. But these generative models often ignore complex statistical dependences. CRAIG [13], a recently discriminative method for ab initio gene prediction, appears to be promising, which is based on a conditional random field (CRF) model with semi-Markov structure that is trained with an online large-margin algorithm related to multiclass SVMs.

The future challenges are to predict alternative spliced exons and short intronless genes.

3.3 Identifying Promoters

Computational prediction of the transcriptional start site (TSS) is the most difficult problem in gene finding as most of the first exons are untranslated regions (UTRs) and there was a lack of high-quality training data. Fortunately, recent experimental advance (5′RACE, CAGE-tag sequencing, PolII (or PIC, or H3K4me3) ChIP-chip or ChIP-Seq, etc.) (e.g., [76]) has produced genome-wide mapping of mammalian core-promoter/TSS data for a few cell lines. These genomic methods are not as accurate as the traditional methods, such as nuclease S1 protection or primer extension assays (Fig. 3.2), but the latter methods cannot be scaled up for genome-wide studies.

At the molecular level, promoter activation and transcription initiation (beginning 5′-end pre-mRNA synthesis by PolII) is a complex process (in addition to Cell vol. 10, 2002, interested readers are strongly recommended to read the book "Transcriptional Regulation in Eukaryotes: Concepts, Strategies, and Techniques". After the chromatin remodeling, the key step is the pre-initiation complex (PIC) binding of the core-promoter (100 bp around TSS) and initiation is mainly regulated by transcription factors bound in the proximal promoter (1 kb upstream) and in the first intron region. Although several core-promoter elements have been identified (Fig. 3.3), with each element being short and degenerate and not every element occurring in a given core-promoter, the combinatorial regulatory code within

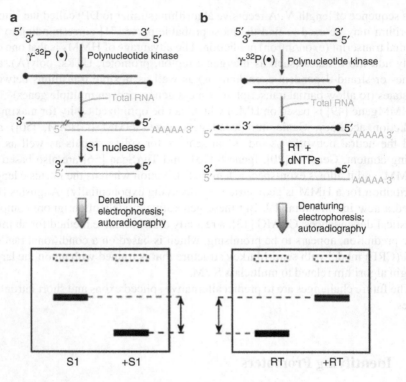

Fig. 3.2 Nuclease S1 protection or primer extension assays

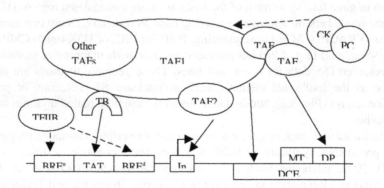

Fig. 3.3 Some core-promoter elements have been identified

core-promoters remains elusive [85]. Their predictive value has also been very limited, despite some weak statistical correlations among certain subsets of the elements which were uncovered recently. Further biochemical characterization of core-promoter binding factors under various functional conditions is necessary before a reliable computational classification of core-promoters becomes possible.

An example of the type of question that must be answered is how CK2 phosphory-lation of TAF1 may switch TFIID binding specificity from a DCE to DPE function Fig. 3.3.

A number of statistical and machine learning approaches that can discrim-inate between the known promoter and some pseudo-promoter sequences have been applied to TSS prediction. In a large-scale comparison [6], eight prediction algorithms were compared. Among the most successful algorithms were Eponine [32] (which trains Relevant Vector Machines to recognize a TATA-box motif in a G+C rich domain and uses Monte Carlo sampling), McPromoter [65] (based on neural networks, interpolated Markov models, and physical properties of promoter regions), FirstEF [28] (based on quadratic discriminant analysis of promoters, first exons, and the first donor site), and DragonGSF [5] (based on artificial neural networks). However, DragonGSF is not publicly available and uses additional binding site information based on the TRANSFAC database, exploiting specific information that is typically not available for unknown promoters. Two new de novo promoter prediction algorithms have emerged that further improve in accuracy. One is ARTS [81], which is based on Support Vector Machines with multiple sophisticated sequence kernels. It claims to find about 35 % true positives at a false-positive rate of 1/1,000, where the abovementioned methods find only about half as many true positives (18 %). ARTS uses only downstream genic sequences as the negative set (non-promoters), and therefore it may get more false-positives from upstream nongenic regions. Furthermore, ARTS does not distinguish if a promoter is CpG-island related or not and it is not clear how ARTS may perform on non-CpG-island-related promoters. Another novel TSS prediction algorithm is CoreBoost [94] which is based on simple LogitBoosting with stumps. It has a false-positive rate of 1/5,000 at the same sensitivity level. CoreBoost uses both immediate upstream and downstream fragments as negative sets and trains separate classifiers for each before combining the two. The training sample is 300 bp fragments (-250, $+50$); hence it is more localized than ARTS which has training sample of 2 kb fragments (-1 kb, $+1$ kb). The ideal application of TSS prediction algorithms is to combine them with gene prediction algorithms [6, 93] or with the CAGE-tag and ChIP-chip PIC mapping data [23, 46]. The future challenges are how to integrate CAGE-tag [76] and epigenomic [46] data to identify tissue- or developmental specific (alternative) promoters and TSSs.

3.4   Genomic Arrays and aCGH/CNP Analysis

Genomic (BAC/oligo or tiling) microarrays were developed to answer new impor-tant questions, such as what is the total human transcriptome [24], especially the locations of alternative transcripts (including non-poly(A) transcripts) and noncod-ing transcripts (including antisense and, pseudo-gene/retro-elements)? Where are all the in vivo binding sites for a given transcription factor ([46])? And how the genome is rearranged (duplicated, deleted, inverted, translocated, etc.) in disease, development, or evolution [78]? Using aCGH to detect CNVs has become a very powerful method for mapping copy number polymorphisms (CNPs) and cancer genes (oncogenes are often found in the amplified regions and tumor-suppressor genes are found in the deleted region).

Since typical aCGH data is very noisy, smoothing techniques may be used to find the number of breakpoints and the corresponding levels. Simple models assume that data are a realization of a Gaussian noise process and use the maximum likelihood criterion adjusted with a penalization term for taking into account model complexity: suppose normalized CGH values $x_1, \ldots, x_N$; breakpoints $0 < y_1 < \ldots < y_N < x_N$; and levels $\mu_1, \ldots, \mu_n$ and error variances $\sigma_1, \ldots, \sigma_N$, the likelihood is

$$\prod_{i=1}^{y_1} \frac{1}{\sigma_1 \sqrt{2\pi}} e^{-\frac{1}{2}\left(\frac{x_i-\mu_1}{\sigma_1}\right)^2} \cdots \prod_{i=y_N+1}^{n} \frac{1}{\sigma_{N+1} \sqrt{2\pi}} e^{-\frac{1}{2}\left(\frac{x_i-\mu_{N+1}}{\sigma_{N+1}}\right)^2}.$$

Maximum likelihood estimators of the means and the variances can be found explicitly by maximizing the log likelihood with a penalty term to control the number N of the breakpoints

$$f(y_1, \ldots, y_N) = \sum_{i=1}^{N+1} (y_{i+1} - y_i) \log \widehat{\sigma_i} + \lambda N.$$

More sophisticated methods are available: SW-ARRAY [70] based on DP, a variance-based automatic pipeline CNVfinder [38], and BreakPtr based on HMM [48].

A comprehensive view of CNVs among 270 HapMap samples was reported using high-density SNP genotyping arrays and BAC array CGH [73]. A novel algorithm, which combines GIM for intensity pre-processing, SW-ARRAY for pairwise CNV detection, and a maximum clique algorithm for CNV extraction, is applied to the Affymetrix GeneChip Human Mapping 500K Early Access (500K EA) arrays data to identify 1,203 CNVs ranging in size from 960 to 3.4 Mb [47]. Recently, a new HMM-based algorithm PennCNV has also been developed for CNV diction in whole-genome high-density SNP genotyping data [88].

3.5 Introduction on Computational Analysis of Transcriptional Genomics Data

Transcriptional regulatory data includes data about gene expression (most commonly from gene expression microarrays), data about binding sites of TFs (such as ChIP-chip data or ChIP-seq data), and several other kinds of data compiled in databases, often manually curated from individual research projects (such as the TRANSFAC database).

3.6 Modeling Regulatory Elements

Transcription factor binding sites are often referred to as *cis*-regulatory elements. The word "element" is used because these sites are the elemental units that are combined inside regulatory regions, such as promoters, to encode the information of the transcriptional regulatory programs. These binding sites are the actual nucleotides in the genome that are recognized by the DNA binding domains of transcription factors and are usually thought of as being contiguous in the sequence. When modeling binding sites, we seek to incorporate as much useful information as possible. However, the most sophisticated models have two main problems: (1) they do not lend themselves to algorithmic manipulation, and (2) they are overly complex and lead to over fitting current data. So currently models that attempt to incorporate too much information are inappropriate for general analysis. In this section we describe commonly used methods of modeling regulatory elements. It is important throughout to distinguish *motifs* from *regulatory elements* or binding sites. In a general setting, the term "motif" describes a recurring property in a data or a statistical summary for a data sample or a recurring component in the data. Here we use this term to describe sets of genomic sites; in our application these can be viewed as samples from genomic sequences. We will describe different representations for motifs, but it is important to remember that binding sites are DNA segments that may match a motif (*motif occurrences*) but are not the motif itself.

3.6.1 Word-Based Representations

The simplest way to describe a set of binding sites is with words over the DNA alphabet $\{A, C, G, T\}$. A *consensus sequence* for a set of binding sites is the word containing, at each position, the base appearing most frequently at that position across the binding sites. This assumes that the binding sites all have the same length, as do most representations for binding sites. Consensus sequences are useful in many contexts, primarily because their simplicity allows them to be easily manipulated, and statistics related to them are often easily derived. Consensus sequences are easily remembered and communicated and are very easily manipulated in a computer.

However, many TFs bind to sites with significant *degeneracy*, meaning that any two binding sites for the same TF may have quite different sequences. The approach of considering a sequence as *similar* to the consensus by counting the number of positions where the sequences mismatch the consensus is still simple but ignores the fact that different positions in a site will be more or less important to the binding affinity of the TF for that site. In general, a consensus sequence representation for TF binding sites is not adequate for use in computational analysis. Representations like regular expressions allow a great deal of flexibility, for example, wildcard characters can be used to indicate that a particular position may be occupied by either of a pair of bases. Regular expressions have been used successfully to model protein domains [4] but have seen relatively limited use in describing regulatory elements. The IUPAC has developed a nomenclature for nucleic acids that contains special symbols to represent subsets of the nucleotides, for example, the symbol "R" is used for purines (see [43] for the full nomenclature and further details). However, relaxations like this still do not result in one of the most useful characteristics in binding-site models: a statistical foundation.

3.6.2 The Matrix-Based Representation

The most popular way to represent motifs is the matrix-based representation, which has been validated repeatedly through use in successfully large-scale analysis projects. This representation has gone by many names: profiles, alignment matrices, position-frequency matrices, and weight matrices; the terminology can be confusing. In addition, there are a few different (but related) kinds of matrices that people use to represent motifs, and certain names have been used by different researchers to describe different kinds of matrices. In this section we will describe what is referred to as a *count matrix* and a *position-weight matrix* through the rest of this tutorial.

We will first describe the *count matrix*. Let M be a matrix with four rows (one for each DNA base) and w columns (with w being the *width* of M). When the binding sites for the TF associated with M are aligned (without gaps), column i of M contains the counts of bases appearing in column i of the alignment. We use $M_i(j)$ to denote the jth row in the ith column of M, and this value is the number of times base j appears in column i of the alignment of binding sites. So each entry in M must be nonnegative, and the sum of the entries in any column of M must equal the number of binding sites in the alignment. We can visualize such a matrix as follows:

$$M = M_1 M_2 \ldots M_w = \begin{bmatrix} M_1(A) & M_2(A) & \ldots & M_w(A) \\ M_1(C) & M_2(C) & \ldots & M_w(C) \\ M_1(G) & M_2(G) & \ldots & M_w(G) \\ M_1(T) & M_2(T) & \ldots & M_w(T) \end{bmatrix}.$$

A *position-weight matrix*, or PWM, is very similar to a count matrix, except that the columns of a PWM are normalized. To construct a PWM, first take the count matrix obtained from an alignment of sites and divide the entry in each column by the sum of the entries in that column. We remark that the term PWM is frequently used in the literature to describe other kinds of matrices, including the *scoring matrices* described in Sect. 3.7.

Count matrices and PWMs contain almost equivalent information, and many databases and programs can treat them as equivalent (e.g., the TRANSFAC matrix table contains both count matrices and PWMs). Through this tutorial, unless otherwise stated or made clear by context, when we refer to motifs, we assume they are represented as PWMs.

Criticisms of matrix-based representations for binding sites are usually focused on the simplifying assumptions of this representation. The two major assumptions made by the matrix-based representations are that:

- Binding sites for a TF all have the same width.
- Contributions of different positions in the binding site to the site's function (usually directly related to the binding affinity of the TF and site) are independent.

While these assumptions are generally not true, they can describe the binding specificity of most TFs with high accuracy. Over time the matrix-based representation has become regarded as the most useful in general, and while the sites for any particular TF may be better modeled using some other representation, the matrix-based representation can usually provide a sufficiently accurate description.

3.6.3   Other Representations

As we just stated, the assumptions of independent positions and fixed-width binding sites have been very useful, but we know they are not strictly valid. A conceptually naive model that eliminates these assumptions would list probabilities of each sequence being a binding site for the given TF. Such a model is impractical due to the number of parameters required to describe each motif. Profile hidden Markov models (HMMs) were used by [57] to represent binding site motifs in a way that allows occurrences to have indels (gaps). This technique is taken directly from work on proteins, where profile HMMs have seen great successes. Durbin et al. [33] give details on using profile HMMs to model motifs (focusing on proteins). A slightly different approach was taken by Gupta and Liu [41], who incorporated parameters for insertions and deletions into a "stochastic dictionary model" for motifs, which is a more general Bayesian model to describe entire sequences based on their pattern content. Models that describe dependencies between positions in binding sites have also been designed. Barash et al. [8] described models based on Bayesian networks, where the probability of a base at a given position may depend on the identity of bases at other positions. The structure of these dependencies can be arbitrary: the probability of the first position might depend on the base appearing at the second position, the third position, both the second and third positions, or neither. Zhou and Liu [95] defined a "generalized weight matrix" that allows correlations between arbitrary pairs of positions but is restricted in complexity by requiring that pairs be independent.

3.7   Predicting Transcription Factor Binding Sites

If we have a motif that describes the binding specificity of some TF, the most common first task is to find sites in sequences that appear similar to the pattern described by the motif. Those sites are called *occurrences*, and we will refer to them as such even when we have not stated an exact criteria for which sites are *similar* to a motif. In this section we describe some fundamental ideas about occurrences of matrix-based motifs and their occurrences in sequences.

3.7.1 The Multinomial Model for Describing Sequences

Consider a sequence S over the DNA alphabet $\{A, C, G, T\}$ as having been generated by sampling w symbols according to a multinomial model over the alphabet. For example, the probabilities could be $f = (f(A), f(C), f(G), f(T)) = (0.2, 0.3, 0.3, 0.2)$. Thinking of sequences in this way allows us to calculate the probability of observing a particular sequence. If $S = ACG$, and therefore $w = 3$, then the likelihood of S being generated by the multinomial f would be

$$L(S = ACG) = f(A)f(C)f(G) = 0.2 \times 0.3 \times 0.3 = 0.018. \quad (3.1)$$

Sequences can be described by a variety of simple models, and some kinds of sequences are described very well using simple models. The multinomial model is highly useful in many contexts, but we know that sequences often exhibit dependencies between positions. For example, using a multinomial model as above would overestimate the probability of observing the sequence $CGCG$ in the human genome, because of general CpG depletion. More complex models (e.g., Markov models) allow dependence between positions in sequences to be described. When estimating the parameters of a multinomial model, we usually just use the frequencies of nucleotides in some set of sequences (e.g., promoters) or the entire genome.

Motifs as Probabilistic Models

Matrix-based motifs can be interpreted in several ways. One useful interpretation is as a statistical model that generates sequences, similar to the multinomial model described in the above paragraph. Under this interpretation, we can assign a probability to each sequences of width equal to the width of the motif. Let M be a matrix, just like the position-weight matrix described in Sect. 3.6, where each column in M has been normalized to have unit sum. So entry $M_i(j)$ gives the probability that a sequence generated from M will have the jth base at position ith. The probability for a given sequence is the product of the matrix entries that correspond to the bases at each position:

$$\Pr(S \text{ was generated from } M) = \prod_{i=1}^{w} M_i(s_i).$$

These probability values become useful when compared with the probability of the same sequence being generated by sampling bases independently from a multinomial distribution. The ratio of the two probabilities indicates how much more likely is it that the sequence was generated from the motif model:

$$\frac{\Pr(S \text{ was generated from } M)}{\Pr(S \text{ was generated from } f)}.$$

Taking the log of this ratio (resulting in a log-likelihood ratio), positive values indicate that the sequences fits more closely with the motif, and negative values indicate that the sequence fits more closely to the distribution described by the simple multinomial. We refer to the log of this ratio as the *match score* of the sequence S with respect to the motif M.

Pseudocounts

The matrix M will often have 0 entries, and these are usually corrected by using a *pseudocount*, which adds a value to each entry of the matrix [33]. The pseudo-count values can be calculated in a few different ways. The most simple method is to add a small constant positive value ε to each entry:

$$M_i(j)' = M_i(j) + \varepsilon,$$

prior to taking the ratio with respect to f. The value of ε is usually much smaller than any value already in M. When the entries in the matrix are counts (nonnegative integers instead of rationals), adding a count of 1 to each entry is called "Laplace's method":

$$M_i'(j) = M_i(j) + 1,$$

and this method can be adapted to take a base composition into account:

$$M_i'(j) = M_i(j) + w f_i,$$

where w can be some arbitrary weight. Additional details on mathematical properties of these matrices can be found in Rahmann et al. [72].

3.7.2 Scoring Matrices and Searching Sequences

To facilitate computation of match scores, we can use the motif and multinomial models to construct *scoring matrices*. Let M be a position-weight matrix with nonnegative entries such that each column has unit sum. Let f be the parameters of a multinomial distribution describing the expected frequency of bases (e.g., in the genome; in promoters). For entry i in column j,

$$score_i(j) = \log \frac{M_i(j)}{f_j}.$$

Notice that this requires entries in M and f to be nonzero, which is usually ensured by using some pseudocount method (see Sect. 3.7). In practice, base frequencies in f are never 0.

3.7.3 Algorithmic Techniques for Identifying High-Scoring Sites

The naive method of identifying motif occurrences is to align the scoring matrix with each possible position in a sequence and calculating the associated match score. This method works well in many applications, producing perfectly accurate scores with a time complexity of $O(wn)$ for motif width w and sequence length n. However, there are applications where very large amounts of sequence must be searched for occurrences of hundreds or thousands of motifs. Finding motif occurrences is often a subproblem involved in motif discovery (see Sect. 3.10), and in such cases identifying occurrences of some candidate motif can be a major bottleneck in the computation.

Here we describe three programs that implement three different approaches for finding motif occurrences in sequences. The MATCH program [45] is a tool developed in close association with the TRANSFAC database [59]. The scoring function used in MATCH is based on scoring matrices, as described above. The search used in MATCH is very simple but incorporates a heuristic speedup based on the idea of a matrix "core." The core is a set of consecutive matrix columns with high information content relative to the rest of the matrix. Cores of size 5 are used by MATCH, which pre-computes the match score between each 5-mer and the core scoring matrix. The search proceeds by identifying the locations in the sequences where a subset of the 5-mers scoring above some cutoff occurs. Those 5-mers are then extended to the full width of the original matrix to obtain the final match score. This method is heuristic because it could miss occurrences that score highly with respect to the full motif but score below the cutoff with respect to the core.

The POSSUMSEARCH program of Beckstette et al. [9] is based on the enhanced suffix array data structure [1]. The storm program STORM [77] implements a search similar to that of POSSUMSEARCH, with suffix trees used in the place of suffix arrays. The technique of using suffix trees to increase the speed of identifying matrix occurrences was also used by Dorohonceanu and Nevill-Manning [31]. These methods all incorporate two important ideas: heavily preprocessing the sequences in which to search and using "look-ahead" scoring. The preprocessing in the form of enhanced suffix arrays or suffix trees allows the search to only match identical substrings of the sequence once against a prefix of the matrix. This helps the search a great deal because the small DNA alphabet means that relatively short segments may appear many times in a sequence and would otherwise require matching many times. The "look-ahead" scoring is similar to a branch and bound search. While matching the motif against a segment of the sequence, if the score for matching with the initial positions is sufficiently low, it can suggest that regardless of the identity of the remaining positions in the segment, a high score cannot be attained. This knowledge allows the algorithms to make an early decision that the segment cannot possibly be an occurrence. Versions of MATCH, STORM, and POSSUMSEARCH are freely available for academic use.

3.7.4   Measuring Statistical Significance of Matches

There are several ways to determine if match scores for a motif are statistically significant. One of the first methods was proposed by Staden [82]. For a motif of width w, this method assumes sequences are generated by selecting w bases, independently at random according to a multinomial model describing base frequencies. The p-value for a match score is the proportion of sequences generated according to that procedure that would have a match score at least as high. Staden [82] also described a simple dynamic programming algorithm for calculating those p-values. An alternative, incorporated with MATCH [45], identifies putative sites based on their expected frequency. Given input sequences, scoring thresholds are set based on p-value selected from the distribution of all potential site scores. Additional methods are described by Schones et al. [77]. When selecting a method for measuring statistical significance of matches, the most important consideration is the hypothesis underlying each method – each method makes a different set of assumptions, and different assumptions might be appropriate in different contexts. Scoring thresholds can be described using the *functional depth* of the score. The functional depth of a score cutoff is a normalization of the score to the [0, 1] interval. The normalization takes the cutoff score and subtracts the minimum possible score for the motif. Then this value is divided by the difference between the maximum and minimum possible scores for the motif.

Even with a highly accurate motif model, and using the most stringent matching criteria, scanning genomic sequences as described in this section is not by itself an effective means of identifying functional transcription factor binding sites. Using such a simple procedure will result in high rates of both false-positive and false-negative predictions. The main problems are not the assumptions of independent positions in the motif or the assumptions of fixed binding-site width. Most of the difficulty stems from the complexity of transcription factor function, including genome organization, chromatin structure, and protein–protein interactions. Still, when used in conjunction with other information, the methods just described can become highly effective and represent a fundamental technique in regulatory sequence analysis. In later sections we will describe how additional information can be incorporated into the process.

3.8 Modeling Motif Enrichment in Sequences

The following problem is commonly encountered in projects to reverse engi-
neer transcriptional regulatory circuits. Analysis of microarray gene expression
experiments has identified a set of genes that appear to be co-regulated. A set
of candidate TFs responsible for the observed co-regulation has been compiled,
each with characterized binding specificity. Reasonable hypotheses state that motif
corresponding to the correct candidate will:

- Occur more often in promoters of the co-regulated genes
- Have stronger occurrences in those promoters
- Occur with a particular strength or frequency in a significantly high proportion
  of those promoters

Motifs with various combinations of these three properties for a particular set of
sequences are said to be *enriched* or *overrepresented* in the sequences. These properties are always evaluated relative to what we expect to observe in the sequences.

3.8.1 Motif Enrichment Based on Likelihood Models

Many important measures of motif enrichment in sequences are based on using
a mixture model to describe the sequences, where one component of the mixture
describes the motif occurrences in the sequences, and the other component describes
the remaining parts of the sequences. The importance of this general formulation
comes from its use in a variety of influential algorithms, and the flexibility that
makes this method able to incorporate different kinds of information. An early
use of mixture models to describe sequences containing motif occurrences was by
Lawrence and Reilly [52], and the model was later extended by Bailey and Elkan [3],
Liu et al. [55] and others. In Sect. 3.10.2 we will explain how these likelihood
formulations of enrichment are used in motif discovery.

We saw in Sect. 3.6 how matrices can be used as probabilistic models for their
occurrences. Suppose we are given a sequence S, a multinomial distribution f
over the DNA bases, a motif M, and a location z where M occurs in S. Then by
multiplying the likelihoods for the corresponding parts of S, the likelihood that S
was generated according to the two models is

$$L(S|f, M, z) = \prod_{i=1}^{z-1} f(S_i) \prod_{j=1}^{w} M_i(S_{z+j-1}) \prod_{i=z+w}^{n} f(S_i),$$

where $w = |M|$ and $n = |S|$.

This calculation can easily be extended to the case where M occurs in multiple
(non-overlapping) locations in S. Let $Z = \{z_1, \ldots, z_k\}$ be a set of start positions in

S for occurrences of M, where the difference $|z_i - z_j| \geq w$ for any $z_i, z_j \in Z$, and let $Z^c = \{i : 1 \leq i \leq n, i < z, z + w \leq i, \forall z \in Z\}$. Then we define

$$L(S|f, M, Z) = \prod_{i \in Z^c} f(S_i) \prod_{z \in Z} \prod_{j=1}^{w} M_j(S_{z+j-1}). \qquad (3.2)$$

It is also easy to see how this likelihood formulation can be adapted to describe multiple distinct motifs, each with a set of occurrences.

When measuring enrichment of a motif, we are usually concerned with enrichment in a set of sequences, rather than a single sequence. The likelihood expressed in Eq. 3.2 can be easily adapted for the case of a set $\mathcal{F}$ of sequences as

$$L(\mathcal{F}|f, M, Z) = \prod_{S \in \mathcal{F}} L(S|f, M, Z), \qquad (3.3)$$

where in this case Z describes occurrence locations for all $S \in \mathcal{F}$. As with the case of a single sequence, the occurrence indicators may be constrained to indicate some specific number of locations in each sequence.

At this point we comment that the base composition f used in these likelihood calculations will often be determined beforehand, either from the sequence S or some larger set of sequences of which S is a member, or f may be determined using, for example, the frequencies of bases in a genome. Sometimes f is calculated using S but only at positions in Z^c as defined above. In those cases f depends on Z. Unless otherwise indicated, we will assume f is fixed in our calculations and will be implicit in our formulas.

For a specified set $\mathcal{F}$ of sequences and motif M, we can use Eq. 3.3 to calculate the likelihood of S given M, over all possible occurrence locations Z as

$$L(\mathcal{F}|M) = \sum_{Z} \Pr(Z) L(\mathcal{F}|M, Z),$$

where the sum is over Z that are valid according to the constraints we wish to model about for numbers of occurrences per sequence and disallowing overlapping occurrences. Often the Z are assumed to be uniformly distributed, which simplifies the above formula and makes it feasible to evaluate. Similarly, we can calculate the likelihood of $\mathcal{F}$ given M, when maximized over valid Z, as

$$L_{\max}(\mathcal{F}|M) = \max_{Z} L(\mathcal{F}|M, Z).$$

The way these likelihood formulas are commonly used in practice is straightforward: for two motifs M_1 and M_2, if $L(\mathcal{F}|M_1) > L(\mathcal{F}|M_2)$, then we say that M_1 is more enriched than M_2 in $\mathcal{F}$. Comparing enrichment of motifs is an important part of many motif discovery algorithms. When we are only concerned with a single motif, we can ask if the likelihood of $\mathcal{F}$ is greater under the mixture model than if we assume $\mathcal{F}$ has no motif occurrences, using only the base composition (as illustrated in Eq. 3.1). If the base composition f is held fixed, then the logarithm of the likelihood ratio

$$L(\mathcal{F}|f, M, Z)/L(\mathcal{F}|f)$$

In the *one occurrence per sequence* (OOPS) model, we assume that each sequence was generated from a mixture model and exactly one location per sequences is the start of a motif occurrence. The *zero or one occurrence per sequence* (ZOOPS) allows slightly more flexibility: when we attempt to identify the motif occurrences whose locations maximize the likelihood, we may assume any particular sequence contains no occurrence of the motif. By relaxing our assumptions further, we arrive at the *two-component mixture* (TCM) model. In this model each sequence may have any number of occurrences (as long as they do not overlap).

3.8.2 Relative Enrichment Between Two Sequence Sets

The methods described above consider enrichment relative to what we expect if the sequences had been generated randomly from some (usually simple) statistical model. Biological sequences are rarely described well using simple statistical models, and currently no known models provide adequate descriptions of transcriptional regulatory sequences like promoters. It is usually more appropriate to measure enrichment of motifs in a given set of sequences relative to some other set of sequences. The set of sequences in which we wish to test motif enrichment is called the *foreground* set, and enrichment of a motif is then measured relative to what we observe in some *background* set of sequences.

At the beginning of Sect. 3.8, we gave three characteristics that we intuitively associate with motif enrichment, and those three can easily be understood when a background sequence set is used. Motifs that are enriched in the foreground relative to the background should occur more frequently in the foreground than the background, the foreground occurrences should be stronger than those in the background, and more sequences in the foreground than in the background should contain an occurrence.

The exact measures of enrichment could be adapted from the likelihood-based measures of Sect. 3.8.1, perhaps by examining the difference or ratio of the enrichment calculated for a motif in the foreground sequences and that calculated in the background sequences. However, a more general and extremely powerful method is available when measuring relative enrichment: using properties of motif occurrences in the sequences to classify the foreground and background sequences. For example, if we fix some criteria for which sites in a sequence are occurrences of a motif, and under that criteria 90 % of our foreground sequences contain at least one occurrence, but only 20 % of our background sequences contain an occurrence, then (1) we could use the property of "containing an occurrence" to predict the foreground sequences with high accuracy, and (2) that motif is clearly enriched in the foreground relative to the background. This classification-based approach also benefits from a great deal of theory and algorithms from the machine learning community. The idea of explicitly using a background sequence set in this way is due originally to [7].

The purpose of using a background set is to provide some way of expressing properties of sequences that are not easily described using some simple statistical distribution. For that reason selection of the background set is often critical. The background set should be selected to control extraneous variables and should often be as similar as possible to the foreground in all properties except the defining property of the foreground. While it is not always possible to find an ideal background set, multiple background sets can be used to control different characteristics.

When the foreground is a set of proximal promoters for co-regulated genes, a natural background to use might be a set of random promoters. In a case like this, simply taking random sequences would not control for the properties common to many promoters, such as CpG islands or TATA-box motifs. Using random promoters can control for these kinds of patterns and help reveal motifs that are specific to the foreground sequence set, and not promoters in general. When the foreground has been derived from microarray expression data, and the promoters of interest correspond to upregulated genes, reasonable background may be promoters of house-keeping genes or promoters of downregulated genes in the same experiment. Using house-keeping genes for control is akin to selecting a set of promoters that are not likely to be regulated specifically in any given condition. Using downregulated genes may illuminate the interesting differences related to the particular experiment. Similarly, if the foreground contains sequences showing high binding intensity in a ChIP-on-chip experiment, we might consider sequences showing low affinity in the same experiment as a background. Alternatively, we might suspect that certain motifs are enriched in the general regions where binding has been observed. Using a background set of sequences also taken from those regions might be able to control such effects (possibly related to chromatin structure in those regions).

Often the size of the foreground sequence set is dictated by the experiment that identified the sequences, but the size of the background set can be selected. There is no correct size for a background sequence set, and each individual program or statistical method applied to the sequences may either limit the size of the background set or require a minimum size. In general, however, it is advisable to have a large background set and to have a background set that is similar in size (count and length) to the foreground set.

A related means of measuring enrichment, similar in spirit to the classification method, has been applied when it is difficult to assign sequences to discrete classes, such as a foreground and background. This technique uses regression instead of classification and attempts to use properties of sequences to fit some experimentally derived function, such as binding intensity in a ChIP-on-chip experiment. This strategy has been used by Das et al. [26, 27], Bussemaker et al. [21], and Conlon et al. [25].

Before concluding this section, we remark that a great deal of work has been done on measures of enrichment for motifs that are represented as words. These measures of enrichment usually involve some statistical evaluations of the number of occurrences of words in sequences, since in general word-based representations do not consider the strength of an occurrence. The central questions concern the probabilities of observing words, either as exact or approximate matches (see Sect. 3.10.1), with particular frequencies in individual sequences or sets of sequences with high frequency in a set of sequences or a single sequence. These and many related questions have been answered very well in the literature [2, 15, 63, 89].

3.9 Phylogenetic Conservation of Regulatory Elements

In Sect. 3.7 we explained why scanning genomic sequences with scoring matrices is likely to result in a very high rate of false-positive predictions. We describe how comparative genomics can help resolve this problem. Cross-species comparison is the most popular and generally useful means of adding confidence to computational predictions of binding sites, and in this context is often called *phylogenetic footprinting*, because comparison between species allows us to examine the evolutionary "footprint" resulting from constrained evolution at important genomic sites [16] and as an analog to the biochemical footprinting technique [69]. In this section we describe the most common general methods for using phylogenetic conservation to identify regulatory elements and discuss some related issues.

3.9.1 Three Strategies for Identifying Conserved Binding Sites

Here we will assume that we already have multispecies alignments for the genomic regions of interest, and that we have high confidence in the accuracy of those alignments.

3.9.1.1 Searching Inside Genomic Regions Identified as Conserved

The most basic method is to use the multispecies alignments to define regions that are conserved and, then use techniques such as those described in Sect. 3.7 to predict binding sites within those regions. It is often hypothesized, and in some cases demonstrated, that noncoding genomic regions with high conservation across several species play important regulatory roles [10, 67, 91]. Although this method is very simple, it is also extremely crude. Such highly conserved regions are the "low-lying fruit" and are not necessarily relevant to any particular regulatory context. In cases where extremely stringent criteria for conservation is not useful, the proper definition for conserved regions (in terms of size and degree of conservation) may be difficult to determine.

3.9.1.2 Using a Continuous Measure of Conservation

The strategy of defining conserved regions will only identify individual sites that exist within much larger regions that also appear to be under selection. Functional regulatory elements that show strong conservation across a number of species are also frequently identified as very small islands of conservation. For these reasons it is often useful to describe conservation by assigning a conservation value to each base in the genome, recognizing that one base might be under selection while the adjacent bases may be undergoing neutral evolution.

One popular measure of conservation that assigns a score to each base in a genome is the *conservation* score presented in the UCSC Genome Browser. This score is calculated using the phastCons algorithm of Siepel et al. [79], which is based on a *phylogenetic hidden Markov model* (phylo-HMM) trained on multi-species genome alignments (e.g. the 17-species vertebrate alignments available through the UCSC Genome Browser). Phylo-HMMs originated in the work of Felsenstein and Churchill [37] and as used in phastCons essentially describe the genomic alignment columns as either evolving according a constrained evolutionary model or a neutral model. The phastCons scores can be thought of as equivalent to the likelihoods that individual alignment columns are under negative selection. The phastCons scores have become important tools in genomics, but they do have some problems for regulatory sequence analysis, including disproportionate weight given to positions that align with very distant species (which are the regions most difficult to align accurately), and a smoothing parameter that introduced dependence between the scores at adjacent positions. Smoothing is desirable for coding sequences, in which the regions under selection are relatively large, but for regulatory regions, abrupt transitions might be more appropriate. Parameters in the phastCons algorithm can be adjusted to alleviate these problems, and in general the existing pre-computed phastCons scores remain highly useful.

3.9.1.3 Evaluating Sites Using the Full Alignment

Both of the two strategies just described ultimately ignore information in the alignment columns. It is desirable to also consider the pattern of substitutions in the alignment columns within a site when we predict whether or not that site is under selection. Recent work by Moses et al. [60, 61] has resulted in much more sophisticated techniques that show great promise in helping us exploit cross-species conservation in identifying regulatory elements. This work is based on models initially developed to account for site-specific evolution in protein-coding sequences [42]. A given motif is used to construct a set of evolutionary models, one model for each column of the motif. A likelihood is calculated for each alignment column using the evolutionary model constructed from the corresponding motif column, and the product of these likelihoods is taken. The special property of each column-specific evolutionary model is that the substitution matrix is structured to favor substitutions resulting in bases that have greater frequency in the corresponding motif column. The final likelihood for a candidate binding site under such a model can be compared with the likelihood under a neutral model, indicating whether the site appears more likely to have evolved under selective pressure associated with the motif.

In addition to considering the patterns of substitutions in sites, we might also want to consider that evolutionary forces likely work to constraining properties of entire binding sites, rather than individual positions within those sites. Other functional elements in the genome, particularly RNAs that fold into important structures, have been found to have compensatory substitutions. This likely also happens in binding sites, for which the functional requirement is the overall affinity of the site for the binding domain of its cognate TF. Measuring conservation at a site by independently evaluating the aligned sites in each species could account for such events, but currently it is not clear how best to combine such individual species scores.

3.9.2   Considerations When Using Phylogenetic Footprinting

3.9.2.1   Which Alignments to Use

The question of which alignments to use is extremely difficult to answer. Often the easiest choice will be to use pre-computed full-genome alignments, such as those produced using TBA/MULTIZ [14] or MLAGAN [18]. The alternative of actually computing the multispecies alignments allows one to select the set of species used and to tune the alignment parameters to something more appropriate for regulatory regions. Most global multiple-sequence alignment algorithms currently in use were originally designed to align coding sequences, including the popular CLUSTAL algorithm [86]. The CONREAL algorithm [11] is designed for pairwise alignment of promoters but must assume while computing the alignment that functional regulatory elements exist in the sequences. The PROMOTERWISE algorithm [35] is also designed specifically for promoters and allows inversion and translocation events, the utility of which is debatable. Neither of these algorithms are designed for more than two sequences.

3.9.2.2   What Set of Species to Use

Although it is well known that conservation of putative binding sites suggests function, it is not known what proportion of functional binding sites are conserved. Lack of binding site conservation could be due to the loss of that particular regulatory function: a particular TF can regulate a particular gene in one species, but not in another species even if that other species has strong orthologs for both the TF and target. Another possibility is that the TF-target relationship is conserved between two species, but the binding sites through which the regulatory relationship is implemented are not orthologous.

Much of the diversity seen between closely related species such as mammals is attributed to differences in gene regulation, and therefore it makes sense that many regulatory elements will not be conserved. More closely related species will have a greater number of conserved binding sites, and those sites will have a greater degree of conservation. When selecting the species to use, one should consider how deeply conserved is the particular biological phenomenon that initiated the search for regulatory elements. We also might want to check that the DNA binding domain of the TF in question has a high degree of conservation in the species selected. If there is significant adaptation of the DNA binding domain, the aligned sites might show large differences in order to retain their function. If there is reason to believe that the binding specificity of the orthologous TF has changed in a particular species, that species may best be left out of the analysis.

3.9.2.3 Binding Site Turnover

Another issue that must be considered is the possibility of binding site "turnover" [30]. Turnover refers to the process by which orthologous regulatory regions, with equivalent function in two species, containing equivalent binding sites do not align. It is assumed that those binding sites do not share a common ancestral site but are both evolving under pressure to preserve a similar function. The commonly proposed mechanism is for a single site to exist in the ancestral sequence, but an additional site capable of similar function emerges by random mutation along one lineage. Because only one site is needed, the original site can mutate in the lineage that acquired the new site. Such a scenario would require many conditions to be satisfied, and the spontaneous emergence of new sites that can perform the function of existing sites seems to require that the precise location of the site (relative to other sites or the TSS) does not matter. Additionally, it is reasonable to assume that short binding sites are more likely to be able to emerge by chance. However, there is increasing evidence that such turnover events are important [44, 64], and efforts are being made to model these events to help understand regulatory sequences [12, 61, 62].

3.10 Motif Discovery

The problem of motif discovery is to identify motifs that optimize some measure of enrichment without relying on any given set of motifs; this is often referred to as de novo motif identification. Regardless of the motif representation or measure of motif enrichment being optimized, motif discovery is computationally difficult. The algorithmic strategies that have been applied to motif discovery are highly diverse, but several techniques have emerged as useful after 20 years of research and are critical components of the most powerful motif discovery programs. Tompa et al. [87] provide a review and comparison of many available motif discovery programs. In this section we will describe some of these techniques, grouping them into the broad categories of either word-based and enumerative or based on general statistical algorithms.

3.10.1 Word-Based and Enumerative Methods

When motifs are represented using words, as described in Sect. 3.6, the motif discovery problem can be solved by simply generating (i.e., enumerating) each possible word of some specified width and then evaluating the enrichment of each word in the sequences. Such an algorithm is an exact algorithm in that the most enriched word will be identified. When the measure of enrichment for each word can be evaluated rapidly, and the width of the words is sufficiently short, this technique works very well. This technique was used very early by Waterman et al. [90], and with advanced data structures (e.g., suffix trees, suffix arrays, and hashing) can be feasible for widths much greater than 10 bases, depending on the size of the sequence data and measure of enrichment used.

We also explained in Sect. 3.6 that word-based representations are not the most appropriate for TF binding sites, because they cannot adequately describe the degeneracy observed in binding sites for most TFs. However, because words are more easily manipulated and much more amenable to algorithmic optimization than matrix-based motifs, methods have been developed to increase their expressiveness.

One way to increase the expressiveness of word-based motifs is to relax our definition of what it means to be an occurrence of a word. A common relaxation is to fix some number of mismatches and define a match to a word as any subsequence of our input sequences that has at most that many mismatches when compared with the word. The occurrences are said to be approximate matches to the word-based motif, and such a representation can be highly effective when the likelihood of degeneracies in binding sites for a TF is roughly uniform across the positions in the sites. Several algorithms have been developed to discover words that have a surprisingly high number of such approximate occurrences in a set of sequences [19, 36, 53, 58, 68, 84].

Another way to increase the expressiveness of word-based representations is to actually expand the representation to include wildcard characters and even restricted regular expressions. PRATT [17] and SPLASH [22] are among the first regular expression-based motif discovery algorithms. Given a set of sequences on a fixed alphabet and a substitution matrix, these algorithms discover motifs composed of tokens (care positions), no-care positions, and gaps (PRATT can find motifs with flexible gaps). These algorithms proceed by enumerating regular expressions of a specified length and extending significant motifs. Motif sites are identified deterministically through matching, where each substring in the input sequence either does or does not match a regular expression. Motif significance is evaluated analytically, by evaluating the likelihoods of motifs (and associated sites) based on motif structure and the size and composition of the input sequence set.

While the most popular motif discovery programs in use today produce motifs that are represented as matrices, almost without exceptions these methods employ some sort of word-based procedure at some point in the algorithm. This should be seen as analogous to the way in which sequence database search algorithms, such as BLAST, initially screen for short exact matches but subsequently apply a more rigorous procedure to evaluate those initial matches.

3.10.2   General Statistical Algorithms Applied to Motif Discovery

Here we describe how two related and general algorithms, expectation maximization (EM) and Gibbs sampling, are applied to the motif discovery problem. Both of these algorithms can be used to identify maximum likelihood estimates of parameters for mixture models and are easily applied to estimate parameters of the enrichment models described in Sect. 3.8.1.

For our explanation of EM and Gibbs sampling, we use notation from Sect. 3.8.1 with slight modifications. The occurrence indicators Z, which were used as sets of locations for the beginnings of motif occurrences, will now have a value for every possible location in the sequences that could be the start of a motif occurrence. So given a motif M of width w and a sequence S of length n, $Z(S_k)$ indicates the event that an occurrence of M starts at position k in S, for $1 \leq k \leq n - w + 1$. In this way the occurrence indicators Z can take values of exactly 0 or 1, corresponding to complete knowledge about the occurrences of M in S. But now we introduce the variables $\bar{Z}$, which lie *between* 0 and 1, and $\bar{Z}(S_k)$ is interpreted as the probability or the expectation that an occurrence of M begins at position k in S: $\Pr(Z(S_k) = 1) = E(Z(S_k)) = \bar{Z}(S_k)$. In Sect. 3.8.1 we placed restrictions on the values taken by Z to require, for example, that each sequence have exactly one motif occurrence (recall the OOPS model). Analogous restrictions can be used when fractional values in Z are interpreted as expectations: the OOPS model would require that the *sum* of values of $\bar{Z}$ for a particular sequence be exactly one:

$$\sum_{k=1}^{n-w+1} \bar{Z}(S_k) = 1,$$

where $n = |S|$ and $w = |M|$. Different restrictions on $\bar{Z}$ can be made to correspond to the requirements of the ZOOPS or TCM models.

3.10.3 Expectation Maximization

The expectation maximization algorithm is a general method for finding maximum likelihood estimates of parameters to a mixture model [29] and was originally applied to motif discovery by Lawrence and Reilly [52]. Starting with some initial value for the motif M (possibly a guess), the idea is to calculate the expected values of the occurrence indicators Z and then update M as the maximum likelihood estimate given the expected values $\bar{Z}$. These two steps, called the *expectation step* and *maximization step*, are iterated until the likelihood corresponding to the maximum likelihood estimate of M converges.

Expectation Step. In the expectation step, for the given M, we calculate the expected value $\bar{Z}$ using the formula

$$\bar{Z}(S_k) = \frac{L(S_k|M)}{L(S_k|M) + L(S_k|f)}, \tag{3.4}$$

where $L(S_k|M)$ and $L(S_k|f)$ are the likelihoods for the width w subsequence of S, starting at position k, according to the motif M and base composition f, respectively (refer back to Sect. 3.8.1 for how these individual likelihoods are calculated).

Maximization Step. In the maximization step, we update the estimate of the motif M using the values of $\bar{Z}$ from the expectation step. Define the function $I(X)$ to take a value 1 when X is true and 0 otherwise. Then the maximum likelihood estimate of the value for base j in the ith column of M is

$$M_i(j) = \sum_{S \in \mathcal{F}} \sum_{k=1}^{n-w+1} \frac{\bar{Z}(S_k) I(S_{k+i-1} = j)}{N}, \tag{3.5}$$

where N is a normalizing factor to ensure that the columns of M sum to exactly 1. The value of N can be obtained directly from $\bar{Z}$, and is actually the sum of the values in $\bar{Z}$ over all sequences, and all positions in the sequences. When f is not specified as fixed, but instead estimated from the sequences, it can be updated during the maximization step in a manner similar to that of $M_i(j)$.

The above description assumes that we have uniform prior probabilities on the values of Z during the expectation step. The formula for $\bar{Z}(S_k)$ can be easily modified to incorporate prior probabilities on Z that account for additional information that might be available about which positions in the sequences are more likely to be the start of a motif occurrence. Although there is no rule for how the initial value of M is obtained, the most common approach is to use a word-based algorithm, and construct the initial matrix-based motif from word-based motifs. Also, motif discovery programs based on EM typically try many different initial motifs, because the algorithm tends to converge quickly. MEME [3] is the most widely known motif discovery program-based EM, but several other programs use

EM at some point, often just prior to returning the motifs, because EM does a nice job of optimizing a motif that is nearly optimal. The major early criticism of EM in motif discovery is that it provides no guarantee that the motifs produced are globally optimal, and instead depends critically on the starting point (i.e., the initial motif) selected. It is now recognized that in practice, most matrix-based motif discovery algorithms have this problem, which illustrates the value of word-based methods in identifying good starting points.

3.10.4   Gibbs Sampling

The Gibbs sampling algorithm is a general method for sampling from distributions involving multiple random variables [39]. It was first applied to motif discovery in the context of protein motifs by Lawrence et al. [51]. As with the EM algorithm, Gibbs proceeds in two alternating steps: one using an estimate of the motif to update information about its occurrences and the other step using information about those occurrences to refine the estimate of the motif. Applied to motif discovery, the main difference between Gibbs sampling and EM is that instead of using the expected values of Z to update the motif M, Gibbs randomly samples exact occurrence locations for M (i.e., 0 or 1 values for variables in Z). We denote the indicators for the sampled occurrence location with $\hat{Z}$.

Sampling Step. In the sampling step, for the current estimate of M, we first calculate the expected values of variables in $\bar{Z}$ using the Eq. 3.4. Then we update the values of $\hat{Z}$ as

$$\hat{Z}(S_k) = \begin{cases} 1 \text{ with probability } \bar{Z}(S_k), \\ 0 \text{ with probability } 1 - \bar{Z}(S_k), \end{cases} \tag{3.6}$$

though the sampling procedure might require some consideration for maintaining any conditions on Z. For example, the OOPS condition would require that exactly one k satisfy $\hat{Z}(S_k) = 1$ for any $S \in \mathcal{F}$.

Predictive Update Step. In the predictive update step, the parameters of the motif M are estimated from the sampled $\hat{Z}$ variables as follows:

$$M_i(j) = \sum_{S \in \mathcal{F}} \sum_{k=1}^{n-w+1} \frac{\hat{Z}(S_k) I(S_{k+i-1} = j)}{N}, \tag{3.7}$$

where N is the same normalizing factor used in Eq. 3.5.

In the original application of Gibbs sampling to motif discovery, each predictive update step updated the values of $\hat{Z}$ for only one sequence S, and the values corresponding to other sequences retained their values from previous iterations [51]. While the original application was concerned with the OOPS model, the Gibbs sampling strategy has seen many generalizations and extensions. Liu et al. [55]

provide a detailed and rigorous treatment. Note that, in contrast with EM, Gibbs sampling does not converge absolutely, and implementations use different criteria to determine how many iterations to execute. Programs implementing the Gibbs sampling strategy for motif discovery include GIBBSSAMPLER [55], ALIGNACE [74], and MDSCAN [54], which are widely used. ALIGNACE is designed specifically for intergenic DNA sequences and masks occurrences of earlier-discovered motifs to facilitate discovery of multiple independent motifs. The MDSCAN program is designed to work on a ranked sequence set and uses overrepresented words in the highest ranking sequences to obtain starting points for the Gibbs sampling procedure.

References

1. Abouelhoda MI, Kurtz S, Ohlebusch E (2004) Replacing suffix trees with enhanced suffix arrays. J Discret Algorithms 2(1):53–86
2. Apostolico A, Bock ME, Lonardi S (2002) Monotony of surprise and large-scale quest for unusual words. In: Proceedings of the sixth annual international conference on computational biology. ACM Press, New York, pp 22–31
3. Bailey TL, Elkan C (1995) Unsupervised learning of multiple motifs in biopolymers using expectation maximization. Mach Learn 21(1–2):51–80
4. Bairoch A (1992) PROSITE: a dictionary of site and patterns in proteins. Nucl Acids Res 20:2013–2018
5. Bajic V, Seah S (2003) Dragon gene start finder identifies approximate locations of the 5′ ends of genes. Nucleic Acids Res 31:3560–3563
6. Bajic V, Tan S, Suzuki Y, Sugano S (2004) Promoter prediction analysis on the whole human genome. Nat Biotechnol 22:1467–1473
7. Barash Y, Bejerano G, Friedman N (2001) A simple hyper-geometric approach for discovering putative transcription factor binding sites. Lect Notes Comput Sci 2149:278–293
8. Barash Y, Elidan G, Friedman N, Kaplan T (2003) Modeling dependencies in protein-DNA binding sites. In: Miller W, Vingron M, Istrail S, Pevzner P, Waterman M (eds) Proceedings of the seventh annual international conference on computational molecular biology, ACM Press, New York, pp 28–37. doi http://doi.acm.org/10.1145/640075.640079
9. Beckstette M, Stothmann D, Homann R, Giegerich R, Kurtz S (2004) Possumsearch: fast and sensitive matching of position specific scoring matrices using enhanced suffix arrays. In: Proceedings of the German conference in bioinformatics. German Informatics Society, Bielefeld, Germany, pp 53–64
10. Bejerano G, Pheasant M, Makunin I, Stephen S, Kent WJ, Mattick JS, Haussler D (2004) Ultraconserved elements in the human genome. Science 304(5675):1321–1325
11. Berezikov E, Guryev V, Plasterk RH, Cuppen E (2004) CONREAL: conserved regulatory elements anchored alignment algorithm for identification of transcription factor binding sites by phylogenetic footprinting. Genome Res 14(1):170–178. doi:10.1101/gr.1642804
12. Berg J, Willmann S, Lassig M (2004) Adaptive evolution of transcription factor binding sites. BMC Evol Biol 4(1):42. doi:10.1186/1471-2148-4-42. URL http://www.biomedcentral.com/1471-2148/4/42
13. Bernal A, Crammer K, Hatzigeorgiou A, Pereira F (2007) Global discriminative learning for high-accuracy computational gene prediction. PLoS Comput Biol 3:e54
14. Blanchette M, Kent WJ, Riemer C, Elnitski L, Smit AF, Roskin KM, Baertsch R, Rosenbloom K, Clawson H, Green ED, Haussler D, Miller W (2004) Aligning multiple genomic sequences with the threaded blockset aligner. Genome Res 14(4):708–715

15. Blanchette M, Sinha S (2001) Separating real motifs from their artifacts. In: Brunak S, Krogh A (eds) Proceedings of the annual international symposium on intelligent systems for molecular biology. Bioinformatics 17 (Suppl 1):S30–S38
16. Blanchette M, Tompa M (2002) Discovery of regulatory elements by a computational method for phylogenetic footprinting. Genome Res 12(5):739–748
17. Brazma A, Jonassen I, Ukkonen E, Vilo J (1996) Discovering patterns and subfamilies in biosequences. In: Proceedings of the annual international symposium on intelligent systems for molecular biology, St. Louis, Missouri, USA, pp 34–43
18. Brudno M, Do CB, Cooper GM, Kim MF, Davydov E, Green ED, Sidow A, Batzoglou S (2003) LAGAN and Multi-LAGAN: efficient tools for large-scale multiple alignment of genomic DNA. Genome Res 13(4):721–731
19. Buhler J, Tompa M (2002) Finding motifs using random projections. J Comput Biol 9(2):225–242
20. Burge C, Karlin S (1997) Prediction of complete gene structure in human genomic DNA. J Mol Biol 268:78–94
21. Bussemaker HJ, Li H, Siggia ED (2001) Regulatory element detection using correlation with expression. Nat Genet 27(2):167–171
22. Califano A (2000) SPLASH: structural pattern localization analysis by sequential histograms. Bioinformatics 16(4):341–357
23. Carninci P, et al (2006) Genomewide analysis of mammalian promoter architecture and evolution. Nat Genet 38:626–635
24. Cheng J, Kapranov P, Drenkow J, Dike S, Brubaker S, Patel S, Long J, Stern D, Tammana H, Helt G, Sementchenko V, Piccolboni A, Bekiranov S, Bailey DK, Ganesh M, Ghosh S, Bell I, Gerhard DS, Gingeras TR (2005) Transcriptional maps of 10 human chromosomes at 5-nucleotide resolution. Science 308(5725):1149–1154
25. Conlon EM, Liu XS, Lieb JD, Liu JS (2003) Integrating regulatory motif discovery and genome-wide expression analysis. Proc Natl Acad Sci USA 100(6):3339–3344
26. Das D, Banerjee N, Zhang MQ (2004) Interacting models of cooperative gene regulation. Proc Natl Acad Sci USA 101(46):16234–16239
27. Das D, Nahle Z, Zhang M (2006) Adaptively inferring human transcriptional subnetworks. Mol Syst Biol 2:2006.0029
28. Davuluri R, Grosse I, Zhang M (2002) Computational identification of promoters and first exons in the human genome. Nat Genet 229:412–417; Erratum: Nat Genet 32:459
29. Dempster AP, Laird NM, Rubin DB (1977) Maximum likelihood from incomplete data via the EM algorithm. J R Stat Soc B 39:1–38
30. Dermitzakis ET, Clark AG (2002) Evolution of transcription factor binding sites in mammalian gene regulatory regions: conservation and turnover. Mol Biol Evol 19(7):1114–1121
31. Dorohonceanu B, Nevill-Manning C (2000) Accelerating protein classification using suffix trees. In: Proceedings of the 8th international conference on intelligent systems for molecular biology (ISMB). La Jolla, California, USA, pp 128–133
32. Down T, Hubbard T (2002) Computational detection and location of transcription start sites in mammalian genomic DNA. Genome Res 12:458–461
33. Durbin R, Eddy SR, Krogh A, Mitchison G (1999) Biological sequence analysis: probabilistic models of proteins and nucleic acids. Cambridge: Cambridge University Press
34. Duta R, Hart P, Stock D (2000) Pattern classification, 2 edn. Wiley, New York
35. Ettwiller L, Paten B, Souren M, Loosli F, Wittbrodt J, Birney E (2005) The discovery, positioning and verification of a set of transcription-associated motifs in vertebrates. Genome Biol 6(12):R104
36. Evans PA, Smith AD (2003) Toward optimal motif enumeration. In: Dehne FKHA, Ortiz AL, Sack JR (eds) Workshop on algorithms and data structures. Lecture notes in computer science, vol 2748, Springer Berlin Heidelberg, pp 47–58
37. Felsenstein J, Churchill G (1996) A Hidden Markov Model approach to variation among sites in rate of evolution. Mol Biol Evol 13(1):93–104

38. Fiegler H, et al (2006) Accurate and reliable high-throughput detection of copy number variation in the human genome. Genome Res 16:1566–1574
39. Gelfand AE, Smith AFM (1990) Sampling-based approaches to calculating marginal densities. J Am Stat Assoc 85:398–409
40. Guigó R, et al (2006) EGASP: the human ENCODE Genome Annotation Assessment Project. Genome Biol 7(Suppl 1):S2.1–S2.3
41. Gupta M, Liu J (2003) Discovery of conserved sequence patterns using a stochastic dictionary model. J Am Stat Assoc 98(461):55–66
42. Halpern A, Bruno W (1998) Evolutionary distances for protein-coding sequences: modeling site-specific residue frequencies. Mol Biol Evol 15(7):910–917
43. IUPAC-IUB Commission on Biochemical Nomenclature (1970) Abbreviations and symbols for nucleic acids, polynucleotides and their constituents: recommendations 1970. J Biol Chem 245(20):5171–5176. URL http://www.jbc.org
44. Javier Costas FC, Vieira J (2003) Turnover of binding sites for transcription factors involved in early drosophila development. Gene 310:215–220
45. Kel A, Gossling E, Reuter I, Cheremushkin E, Kel-Margoulis O, Wingender E (2003) MATCHTM: a tool for searching transcription factor binding sites in DNA sequences. Nucl Acids Res 31(13):3576–3579
46. Kim TH, Barrera LO, Zheng M, Qu C, Singer MA, Richmond TA, Wu Y, Green RD, Ren B (2005) A high-resolution map of active promoters in the human genome. Nature 436:876–880
47. Komura D, et al (2006) Genome-wide detection of human copy number variations using high-density DNA oligonucleotide arrays. Genome Res 16:1575–1584
48. Korbel JO, et al (2007) Systematic prediction and validation of breakpoints associated with copy-number variants in the human genome. Proc Natl Acad Sci USA 104:10110–10115
49. Krogh A (1997) Two methods for improving performance of an HMM and their application for gene finding. Proc Int Conf Intell Syst Mol Biol 5:179–186
50. Kulp D, Haussler D, Reese M, Eeckman F (1996) A generalized hidden Markov model for the recognition of human genes in DNA. Proc Int Conf Intell Syst Mol Biol 4:134–142
51. Lawrence C, Altschul S, Boguski M, Liu J, Neuwald A, Wootton J (1993) Detecting subtle sequence signals: a Gibbs sampling strategy for multiple alignment. Science 262:208–214
52. Lawrence C, Reilly AA (1990) An expectation maximization (EM) algorithm for the identification and characterization of common sites in unaligned biopolymer sequences. Proteins Struct Funct Genet 7:41–51
53. Li M, Ma B, Wang L (2002) On the closest string and substring problems. J ACM 49(2):157–171
54. Liu XS, Brutlag DL, Liu JS (2002) An algorithm for finding protein-DNA binding sites with applications to chromatin immunoprecipitation microarray experiments. Nat Biotechnol 20(8):835–839
55. Liu JS, Lawrence CE, Neuwald A (1995) Bayesian models for multiple local sequence alignment and its Gibbs sampling strategies. J Am Stat Assoc 90:1156–1170
56. Majoros W, Pertea M, Salzberg S (2004) TigrScan and GlimmerHMM: two open source ab initio eukaryotic genefinders. Bioinformatics 20:2878–2879
57. Marinescu VD, Kohane IS, Riva A (2005) The MAPPER database: a multi-genome catalog of putative transcription factor binding sites. Nucl Acids Res 33(Suppl 1):D91–D97
58. Marsan L, Sagot MF (2000) Extracting structured motifs using a suffix tree – algorithms and application to promoter consensus identification. In: Minoru S, Shamir R (eds) Proceedings of the annual international conference on computational molecular biology. ACM Press, New York, pp 210–219
59. Matys V, Fricke E, Geffers R, Gossling E, Haubrock M, Hehl R, Hornischer K, Karas D, Kel AE, Kel-Margoulis OV, Kloos DU, Land S, Lewicki-Potapov B, Michael H, Munch R, Reuter I, Rotert S, Saxel H, Scheer M, Thiele S, Wingender E (2003) TRANSFAC(R): transcriptional regulation, from patterns to profiles. Nucl Acids Res 31(1):374–378

60. Moses AM, Chiang DY, Pollard DA, Iyer VN, Eisen MB (2004) MONKEY: identifying conserved transcription-factor binding sites in multiple alignments using a binding site-specific evolutionary model. Genome Biol 5(12):R98

61. Moses AM, Pollard DA, Nix DA, Iyer VN, Li XY, Biggin MD, Eisen MB (2006) Large-scale turnover of functional transcription factor binding sites in drosophila. PLoS Comput Biol 2(10):e130

62. Mustonen V, Lassig M (2005) Evolutionary population genetics of promoters: predicting binding sites and functional phylogenies. Proc. Natl. Acad. Sci. USA 102(44):15936–15941. doi:10.1073/pnas.0505537102. URL http://www.pnas.org/cgi/content/abstract/102/44/15936

63. Nicodeme P, Salvy B, Flajolet P (2002) Motif statistics. Theor Comput Sci 287:593–617

64. Odom DT, Dowell RD, Jacobsen ES, Gordon W, Danford TW, MacIsaac KD, Rolfe PA, Conboy CM, Gifford DK, Fraenkel E (2007) Tissue-specific transcriptional regulation has diverged significantly between human and mouse. Nat Genet 39(6):730–732; Published online: 21 May 2007

65. Ohler U, Liao G, Niemann H, Rubin G (2002) Computational analysis of core promoters in the drosophila genome. Genome Biol 3(12):RESEARCH0087

66. Pearson H (2006) What is a gene?. Nat Genet 441:398–340

67. Pennacchio LA, Ahituv N, Moses AM, Prabhakar S, Nobrega MA, Shoukry M, Minovitsky S, Dubchak I, Holt A, Lewis KD, Plajzer-Fick I, Akiyama J, Val SD, Afzal V, Black BL, Couronne O, Eisen MB, Visel A, Rubin EM (2006) In vivo enhancer analysis of human conserved non-coding sequences. Nature 444(7118):499–502

68. Pevzner P, Sze S (2000) Combinatorial approaches to finding subtle signals in DNA sequences. In: Bourne P, et al (eds) Proceedings of the annual international symposium on intelligent systems for molecular biology. Menlo Park, AAAI Press, pp 269–278

69. Portugal J (1989) Footprinting analysis of sequence-specific DNA-drug interactions. Chem Biol Interact 71(4):311–324

70. Price TS, Regan R, Mott R, Hedman A, Honcy B, Daniels RJ, Smith L, Greenfield A, Tiganescu A, Buckle V, Ventress N, Ayyub H, Salhan A, Pedraza-Diaz S, Broxholme J, Ragoussis J, Higgs DR, Flint J, Knight SJL (2005) SW-ARRAY: a dynamic programming solution for the identification of copy-number changes in genomic DNA using array comparative genome hybridization data. Nucl Acids Res 33(11):3455–3464

71. Rabiner L (1989) A tutorial on hidden markov models and selected applications in speech recognition. Proc IEEE 77:257–286

72. Rahmann S, Muller T, Vingron M (2003) On the power of profiles for transcription factor binding site detection. Stat Appl Genet Mol Biol 2(1):7

73. Redon R, Ishikawa S, Fitch KR, Feuk L, Perry GH, Andrews TD, Fiegler H, Shapero MH, Carson AR, Chen W, Cho EK, Dallaire S, Freeman JL, Gonzalez JR, Gratacos M, Huang J, Kalaitzopoulos D, Komura D, MacDonald JR, Marshall CR, Mei R, Montgomery L, Nishimura K, Okamura K, Shen F, Somerville MJ, Tchinda J, Valsesia A, Woodwark C, Yang F, Zhang J, Zerjal T, Zhang J, Armengol L, Conrad DF, Estivill X, Tyler-Smith C, Carter NP, Aburatani H, Lee C, Jones KW, Scherer SW, Hurles ME (2006) Global variation in copy number in the human genome. Nature 444:444–454

74. Roth F, Hughes J, Estep P, Church G (1998) Finding DNA regulatory motifs within un-aligned noncoding sequences clustered by whole-genome mRNA quantitation. Nat Biotechnol 16(10):939–945

75. Salamov A, Solovyev V (2000) Ab initio gene finding in Drosophila genomic DNA. Genome Res 10:516–522

76. Sandelin A, et al (2007) Mammalian RNA polymerase II core promoters: insights from genome-wide studies. Nat Rev Genet 8.424–436

77. Schones D, Smith A, Zhang M (2007) Statistical significance of cis-regulatory modules. BMC Bioinform 8:19

78. Sebat J, et al (2004) Large-scale copy number polymorphism in the human genome. Science 305:525–528

79. Siepel A, Bejerano G, Pedersen JS, Hinrichs AS, Hou M, Rosenbloom K, Clawson H, Spieth J, Hillier LW, Richards S, Weinstock GM, Wilson RK, Gibbs RA, Kent WJ, Miller W, Haussler D (2005) Evolutionarily conserved elements in vertebrate, insect, worm, and yeast genomes. Genome Res 15(8):1034–1050

80. Solovyev VV, et al (1994) Predicting internal exons by oligonucleotide composition and discriminant analysis of spliceable open reading frames. Nucl Acids Res 22:5156–5163

81. Sonnenburg S, Zien A, Ratsch G (2006) ARTS: accurate recognition of transcription starts in human. Bioinformatics 22:e472–e480

82. Staden R (1989) Methods for calculating the probabilities of finding patterns in sequences. Comput Appl Biosci 5(2):89–96

83. Stanke M, Waack S (2003) Gene prediction with a hidden markov model and a new intron submodel. Bioinformatics 19(Suppl 2):II215–II225

84. Sumazin P, Chen G, Hata N, Smith AD, Zhang T, Zhang MQ (2005) DWE: discriminating word enumerator. Bioinformatics 21(1):31–38

85. Thomas M, Chiang C (2006) The general transcription machinery and general cofactors. Crit Rev Biochem Mol Biol 41:105–178

86. Thompson JD, Higgins DG, Gibson TJ (1994) CLUSTAL W: improving the sensitivity of progressive multiple sequence alignment through sequence weighting, position-specific gap penalties and weight matrix choice. Nucl Acids Res 22(22):4673–4680

87. Tompa M, Li N, Bailey TL, Church GM, De Moor B, Eskin E, Favorov AV, Frith MC, Fu Y, Kent WJ, Makeev VJ, Mironov AA, Noble WS, Pavesi G, Pesole G, Regnier M, Simonis N, Sinha S, Thijs G, van Helden J, Vandenbogaert M, Weng Z, Workman C, Ye C, Zhu Z (2005) Assessing computational tools for the discovery of transcription factor binding sites. Nat Biotechnol 23(1):137–144

88. Wang K, Li M, Hadley D, Liu R, Glessner J, Grant SF, Hakonarson H, Bucan M (2007) PennCNV: an integrated hidden Markov model designed for high-resolution copy number variation detection in whole-genome SNP genotyping data. Genome Res 17(11):1665–1674

89. Waterman MS (1995) Introduction to computational biology: maps, sequences and genomes. Chapman and Hall, London

90. Waterman MS, Arratia R, Galas DJ (1984) Pattern recognition in several sequences: consensus and alignment. Bull Math Biol 46:515–527

91. Woolfe A, Goodson M, Goode DK, Snell P, McEwen GK, Vavouri T, Smith SF, North P, Callaway H, Kelly K, Walter K, Abnizova I, Gilks W, Edwards YJK, Cooke JE, Elgar G (2005) Highly conserved non-coding sequences are associated with vertebrate development. PLoS Biol 3(1):e7

92. Zhang M (1997) Identification of protein coding regions in the human genome by quadratic discriminant analysis. Proc Natl Acad Sci USA 94:565–568

93. Zhang M (2002) Computational prediction of eukaryotic protein-coding genes. Nat Rev Genet 3:698–709

94. Zhao X, Xuan Z, Zhang MQ (2006) Boosting with stumps for predicting transcription start sites. Genome Biol 8:R17

95. Zhou Q, Liu JS (2004) Modeling within-motif dependence for transcription factor binding site predictions. Bioinformatics 20(6):909–916

Chapter 4
Statistical Methods in Bioinformatics

Jun S. Liu and Bo Jiang

4.1 Introduction

The linear biopolymers, DNA, RNA, and proteins, are the three central molecular building blocks of life. DNA is an information storage molecule. All of the hereditary information of an individual organism is contained in its genome, which consists of sequences of the four DNA bases (nucleotides), A, T, C, and G. RNA has a wide variety of roles, including a small but important set of functions. Proteins, which are chains of 20 different amino acid residues, are the action molecules of life, being responsible for nearly all the functions of all living beings and forming many of life's structures. All protein sequences are coded by segments of the genome called genes. The universal genetic doe is used to translate triplets of DNA bases, called codons, to the 20-letter alphabet of proteins. How genetic information flows from DNA to RNA and then to protein is regarded as the central dogma of molecular biology. Genome sequencing projects with emergences of microarray techniques have resulted in rapidly growing and publicly available databases of DNA and protein sequences, structures, and genome-wide expression. One of the most interesting questions scientists are concerned with is how to get any useful biological information from "mining" these databases.

Statistics studies how to best extract information from noisy data, how to infer the underlying structure from uncertain observations, and how to quantify uncertainties in both natural phenomena and inference procedures. The role of statistics played in the study of biology aims to develop and apply new and powerful methods for

J.S. Liu (✉)
Department of Biostatistics, Harvard University, Cambridge, MA 02138, USA
e-mail: jliu@stat.harvard.edu

B. Jiang
MOE Key Laboratory of Bioinformatics and Bioinformatics Division, TNLIST/Department of Automation, Tsinghua University, Beijing 100084, China

R. Jiang et al. (eds.), *Basics of Bioinformatics: Lecture Notes of the Graduate Summer School on Bioinformatics of China*, DOI 10.1007/978-3-642-38951-1_4,
© Tsinghua University Press, Beijing and Springer-Verlag Berlin Heidelberg 2013

"mining" biological data and to evaluate results by calculating significance level, deriving asymptotic distributions, and designing simulation study for comparing methods.

In this chapter, we will describe some applications of statistics in bioinformatics, including basics of statistical modeling, microarray data analysis, pair-wise and multiple sequence alignments, sequence pattern discovery, transcriptional regulation analysis, and protein structure predictions.

4.2 Basics of Statistical Modeling and Bayesian Inference

People routinely make inferences about unknown or unseen events based on what they observed, what they experienced, and what they believed. To infer means "to conclude based on facts and/or premises." In statistical inference, the facts are the data, the premises are carried by a probability model, and the conclusions are made on unknowns. The probability model, which is abstracted from the scientific problem under study, quantifies uncertainties using a probabilistic law to connect data and parameters of interests. A typical parametric statistical model takes the form $p(\mathbf{y}|\theta)$, where $\mathbf{y}$ is the observed data, θ is the parameter of interest, and $p()$ denotes a specific probability distribution that relates θ to $\mathbf{y}$. The goals of statistical inference are to optimally estimate θ and to quantify the uncertain of this estimation.

4.2.1 Bayesian Method with Examples

Statisticians differ in their interpretations of probability and can be generally categorized as Frequentist or Bayesian. A comparison of essences of Frequentist and Bayesian approaches is made in Table 4.1.

Science needs not only objective evidence, criterion, and proofs but also subjective endeavor, such as a good judgment in "guesswork." Bayesian statistics rides between subjective and objective worlds, by using probability theory as the

Table 4.1 *Frequentist* versus *Bayesian*: a comparison

Differences	Frequentist	Bayesian
Interpretation of probability	Long-run frequency	Measure of uncertainty
Common methods	*Criterion based*: asymptotic, minimax, or simulation methods for evaluation	*Use rules of probability*: decision-theoretic, inverse probability, and simulation methods for evaluation
Inference	*Pre-data analysis*: confidence regions, *p*-values	*Post-data analysis*: posterior distribution
Modeling effort	Can be nonparametric	Mostly model-based

Fig. 4.1 Simple case:
flipping a thumbtack

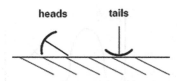

rule of combining and propagating evidences and information and as the language
to convey uncertainty. Generally speaking, it is a process of fitting a probability
model to a data set and summarizing the results by a probability distribution
of the unknowns. A Bayesian procedure involves three steps: (1) setting up a
full probability model, (2) deriving the posterior distribution for the parameter of
interest, and (3) evaluating the fit and conducting model critique. The advantages
of using Bayesian statistics lies in many folds. First, it explicitly quantifies the
uncertainty in parameter estimation and uses rule of probability to process all
the evidence jointly. Second, because of its principled way of organizing data
and information and the availability of Monte Carlo-based computational tools, a
greater emphasis can be given to the scientific problem instead of mathematical
convenience. Moreover, according to admissibility theorem, sensible answers tend
to be Bayesian ones. Bayesian method has been successfully applied to solve
many typical problems, such as Behrens-Fisher problem, missing data problem in
hierarchical model, and nuisance parameter problem.

The Bayesian method is based on the Bayes theorem, which is a simple
mathematical formula to invert the probability $p(A|B) = p(B|A) p(A) / p(B)$,
where A and B are two "events." If we change them to θ and $\mathbf{y}$, we have $p(\theta|\mathbf{y}) =
p(\mathbf{y}|\theta) p(\theta) / p(\mathbf{y})$. Here is a simple example. After flipping a thumbtack onto
floor, there may be two cases: head or tail (Fig. 4.1). Suppose we have observed
h heads and t tails after $(h + t)$ flips. To learn about the probability of getting a
head, denoted as θ, from these observations using the Bayesian method, we start by
assuming a prior distribution $p(\theta)$ on θ, which may be viewed as a description of
our ignorance or prior knowledge about it. We can write the posterior probability of
θ conditioned on observations as

$$p(\theta|h \text{ heads}, t \text{ tails}) \propto p(\theta)p(h \text{ heads}, t \text{ tails}|\theta) \propto p(\theta)\theta^h(1 - \theta)^t.$$

A graphical illustration is given by Fig. 4.2. A very convenient choice of $p(\theta)$ is
a member of the two parameter beta family,

$$p(\theta|\alpha, \beta) \propto \theta^{\alpha-1}(1 - \theta)^{\beta-1},$$

with $\alpha, \beta > 0$. This class of family has the nice property (called *conjugate*)
that the resulting posterior distributions are again beta distributions. That is, the
posterior distribution becomes Beta($h + t$). The above procedure can be generalized
to more than two states by replacing the binomial distribution with the multinomial
distribution and the Beta prior by a Dirichlet distribution prior.

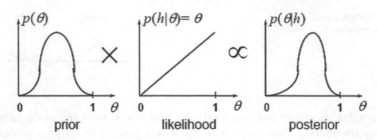

Fig. 4.2 Learning about probability θ

4.2.2 *Dynamic Programming and Hidden Markov Model*

4.2.2.1 Dynamic Programming

Dynamic programming is a method for solving problems exhibiting the property of overlapping subproblems and or a chain structure. The following example is known as the "secretary problem," which can be solved by dynamic programming.

Example 4.1 Suppose m candidates are applying to one secretary job. We only get to observe a candidate's quality, summarized by a real number X_j, during the interview and we have to decide whether or not to offer the candidate the job on the spot. Suppose these candidates are arranged in a random sequential order for interviews. Our goal is to maximize the probability of finding the best secretary with *no looking back*. Now, we try to solve this problem by reasoning backward. If we delay our decision until the last candidate, the probability of hiring the best secretary is $P_m = 1/m$. If we wait until having two persons left, we can choose among them by the following strategy: if the $(m-1)$th person is better than all previous ones, take her/him; otherwise, wait till the last one. Thus, the probability of hiring the best secretary of this strategy is

$$P_{m-1} = \frac{1}{m} + \left(1 - \frac{1}{m-1}\right) P_m.$$

Following the same logic, if we let go the first $(j-1)$ persons and take the *best one so far* starting from the jth person, the probability of hiring the best secretary is given by

$$P_j = \frac{1}{m} + \left(1 - \frac{1}{j}\right) P_{j+1}.$$

By solving this recursion, we have

$$P_j = \frac{j-1}{m} \left(\frac{1}{j-1} + \cdots + \frac{1}{m-1}\right) \approx \frac{j-1}{m} \log\left(\frac{m}{j-1}\right),$$

Fig. 4.3 Illustration of
hidden Markov model

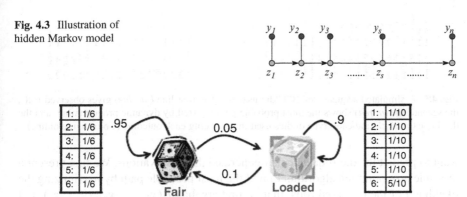

Fig. 4.4 Occasionally dishonest casino

which is maximized at $j \approx m/e = m/2.718\ldots$. Therefore, the optimal strategy is to reject the first 37 % of the candidates and recruit the first person who is better than all that have been previously interviewed. The chance to get the best one is approximately 37 %!

4.2.2.2   Markov Chain and Hidden Markov Model

Markov chain is a stochastic process in which the probability of the current state is only dependent on the nearest past. Let $X_t, t = 1, 2, \ldots$ denote the Markov chain, of which each take values in $\{1, 2, \ldots, b\}$. The 1-*step transition probability* is defined as $P_{i,j}^{t,t+1} = P(X_{t+1} = j | X_t = i)$, which can also be written in a matrix form. For *time-homogeneous* Markov chain, the transition probability is independent of time t, that is $P_{i,j}^{t,t+1} = P_{i,j}$. *Time-homogeneous* Markov chain satisfies *Chapman–Kolmogorov* equation: $P_{i,j}^{n+m} = \sum_{k=1}^{b} P_{i,k}^{n} P_{k,j}^{m}$ or, in the matrix form, $\mathbf{P}^{n+m} = \mathbf{P}^n \times \mathbf{P}^m$.

A hidden Markov model (HMM) assumes that the underlying mechanism can be modeled by a Markov chain, but one only gets to have an indirect observation on the chain, such as a noisy or convoluted version of it. Figure 4.3 gives an illustration of the hidden Markov model, in which the y's are observed and the z's are governed by a Markov transition rule $p_{ij} = P(z_{t+1} = j | z_t = i)$. In HMM, y can be viewed as a "noisy copy" of the z, that is, $y_t | z_t \sim f_t(\cdot | z_t)$. We explain some of the key concepts and methods for dealing with HMM through the following simple example.

Example 4.2 An occasionally dishonest casino uses two kinds of dice (Fig. 4.4). A fair die has equal probabilities of rolling out any number and a loaded die has probability 1/2 to roll a 6, and 1/10 to roll any other number. The probability that the casino switches from the fair to loaded die is 0.05 and the probability of switching back from loaded to fair is 0.1. Any number of rolls can thus be described by a hidden Markov model with $y_1, \ldots, y_n$ denoting observed rolls and $z_1, \ldots, z_n$ denoting types of dies generating each outcome. Given an observed sequence of rolls, we

```
2 1 6 2 1 6 6 5 6 6 6 3 5 2 3 2 1 2 6 4 6 2 2 5 3 3 3 1 4 3 1 5 1 3 6 1 6 3 5 1 6 3 1 2 3 1 4 6 3 6
2 2 2 2 2 2 2 2 2 2 2 1 1 1 1 1 1 1 1 1 1 1 1 1 1 1 1 1 1 1 1 1 1 1 1 1 1 1 1 1 1 1 1 1 1 1 1 1 1 1
1 1 1 1 2 2 2 2 2 2 2 1 1 1 1 1 1 1 1 1 1 1 1 1 1 1 1 1 1 1 1 1 1 1 1 1 1 1 1 1 1 1 1 1 1 1 1 1 1 1
5 1 3 3 5 6 1 3 5 5 4 6 3 2 4 1 6 2 5 4 2 4 4 2 1 2 3 2 6 3 6 6 6 4 5 6 2 2 4 6 6 1 4 6 3 4 2 6 4 6
1 1 1 1 1 1 1 1 1 1 1 1 1 1 1 1 1 1 1 1 1 1 1 1 1 1 2 2 2 2 2 2 2 2 2 2 2 2 2 2 2 2 2 2 2 2 2 2 2 2
1 1 1 1 1 1 1 1 1 1 1 1 1 1 1 1 1 1 1 1 1 1 1 1 1 1 2 2 2 2 2 2 2 2 2 2 2 2 2 2 2 2 2 2 2 1 1 2 2 2
```

Fig. 4.5 A simulated sequence of 100 "die usages." The first line (*in blue*) gives observed rolls, the second line (*in red*) shows the most probable path inferred by dynamic programming, and the third line (*in black*) indicates type of dies used in generating each outcome (Color figure online)

want to predict the die sequence that generated the observations. We first present a dynamic programming algorithm to find the most probable path by maximizing the likelihood of the observed rolls, that is, find $\arg\max\limits_{z_1,\dots,z_n} P(y_1,\dots,y_n|z_1,\dots,z_n)$. Let $f_i(z)$ be the optimal probability value for the first i observations given that $z_i = z$. Then we have the recursion

$$f_i(z) = P(y_i|z_i = z) \times \max_{z'}\{f_{i-1}(z')\,P(z_i = z|z_{i-1} = z')\}. \tag{4.1}$$

The optimal path can be found by tracing backward:

$$\hat{z}_n = \arg\max_z f_n(z) \text{ and } \hat{z}_i = \arg\max_z\{f_i(z)P(z_{i+1}=\hat{z}_{i+1}|z_i=z)\},\ i=1,\dots,n-1.$$

This procedure is only of a linear order with respect to n. A simulated example and the inferred most probable path is shown in Fig. 4.5.

The problem can also be solved by sampling (hidden) types of dies $z_1,\dots,z_n$ from its posterior distribution $P(z_1,\dots,z_n|y_1,\dots,y_n)$. Let $g_i(z)$ be the marginal probability for $z_i = z$ conditioned on the first i observations. By substituting the sign of *max* in (4.1) with a sign of *sum*, $g_i(z)$ can be calculated by recursive forward summation:

$$g_i(z) = P(y_i|z_i = z) \times \sum_{z'}\{g_{i-1}(z')\,P(z_i = z|z_{i-1} = z')\}.$$

Then, we can draw z's by backward sampling as follows:

$$z_n \sim P(z_n = z|y_1,\dots,y_n) = \frac{g_n(z)}{\sum_{z'} g_i(z')}$$

and

$$z_i \sim P(z_i = z|y_1,\dots,y_n,z_{i+1},\dots,z_n) = \frac{g_i(z)\,P(z_{i+1}|z_i = z)}{\sum_{z'}\{g_i(z')\,P(z_{i+1}|z_{i-1} = z')\}}.$$

The unknown z's can be estimated by their posterior means computed by averaging over the multiple simulated samples from the above procedure. It can also be estimated by the marginal posterior mode.

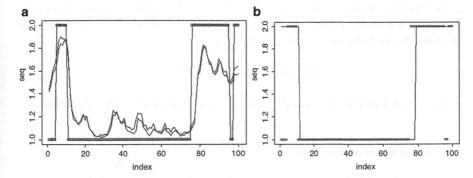

Fig. 4.6 Solving the dishonest casino problem using the forward–backward algorithms. The *dotted line* indicates the true type of the die used in generating each outcome, and the *solid line* shows the inferred type by using (**a**) the posterior mean (*curly lines*) and marginal posterior mode and (**b**) the joint posterior mode. The *overlapping lines* in (**a**) are the exact mean and its Monte Carlo approximation, respectively

An additional benefit of using samples from the posterior distribution of the z's to do the estimation is that one can obtain an explicit measure of estimation uncertainty. As shown in Fig. 4.6, when the posterior probability for $z_i = 1$ is neither close to 1 nor close to 0, it reflects our uncertainty of its value and our non-commitment to its estimate.

4.2.3   Metropolis–Hastings Algorithm and Gibbs Sampling

4.2.3.1   Metropolis–Hastings Algorithm

The basic idea of Metropolis–Hastings algorithm or any other Markov chain Monte Carlo (MCMC) algorithm is to simulate a Markov chain in the state space of **x** so that the limiting/stationary/equilibrium distribution of this chain is the target distribution π. Starting with any configuration $\mathbf{x}^{(0)}$ and a *proposal function* $T(\mathbf{x}, \mathbf{x}')$, the algorithm proceeds by iterating the following steps: (1) a small but random perturbation of the current configuration is made, (2) the "goodness ratio"

$$r\left(\mathbf{x}, \mathbf{x}'\right) = \frac{\pi\left(\mathbf{x}'\right)}{\pi\left(\mathbf{x}\right)}$$

resulting from this perturbation is computed, (3) a uniform random variable is generated independently, (4) and the new configuration is accepted if the random number is smaller than or equal to the "goodness ratio" and rejected otherwise. Metropolis et al. [1] required their proposal function to be symmetric, i.e., $T(\mathbf{x}, \mathbf{x}') = T(\mathbf{x}', \mathbf{x})$. Hasting [2] generalized Metropolis et al.'s recipe and suggested using

$$r\left(\mathbf{x}, \mathbf{x}'\right) = \min\left\{1, \pi\left(\mathbf{x}'\right) T\left(\mathbf{x}', \mathbf{x}\right) \middle/ \pi\left(\mathbf{x}\right) T\left(\mathbf{x}, \mathbf{x}'\right)\right\}$$

as the "goodness ratio." One can verify that the Metropolis–Hastings algorithm prescribes a transition rule with respect to which the target distribution $\pi(\mathbf{x})$ is invariant by checking the *detailed balance* condition:

$$\pi(\mathbf{x}) A(\mathbf{x}, \mathbf{x}') = \pi(\mathbf{x}') A(\mathbf{x}', \mathbf{x}), \tag{4.2}$$

where $A(\mathbf{x}, \mathbf{x}') = T(\mathbf{x}, \mathbf{x}') r(\mathbf{x}, \mathbf{x}')$ is the actual transition rule of the algorithm.

4.2.3.2 Gibbs Sampling Algorithm

Gibbs sampler [3] is a special Markov Chain Monte Carlo (MCMC) scheme. Its most prominent feature is that the underlying Markov chain is constructed by composing a sequence of conditional distributions along a set of directions (often along the coordinate axis). Suppose we can decompose the random variable into d components, i.e., $\mathbf{x} = (x_1, \ldots, x_d)$. In the Gibbs sampler, one randomly or systematically chooses a coordinate, say x_1, and then updates it with a new sample x_1' drawn from the conditional distribution $\pi(\cdot|\mathbf{x}_{[-1]})$, where $\mathbf{x}_{[-1]}$ refers to $\{x_j; j = 2, \ldots, d\}$. It is easy to check that every single conditional update step in Gibbs sampler satisfies the *detailed balance* condition in Eq. (4.2) and thus leaves π invariant.

Here is an illustrative example from [4]. Consider the simulation from a bivariate Gaussian distribution. Let $\mathbf{x} = (x_1, x_2)$ and let the target distribution be

$$N\left(\begin{pmatrix} 0 \\ 0 \end{pmatrix}, \begin{pmatrix} 1 & \rho \\ \rho & 1 \end{pmatrix}\right).$$

The Markov chain $\mathbf{x}^{(t)} = (x_1^{(t)}, x_2^{(t)})$ corresponding to a systematic-scan Gibbs sampler is generated as

$$x_1^{(t+1)}|x_2^{(t)} \sim N\left\{\rho x_2^{(t)}, (1 - \rho^2)\right\},$$

$$x_2^{(t+1)}|x_1^{(t+1)} \sim N\left\{\rho x_1^{(t+1)}, (1 - \rho^2)\right\}.$$

The sampling trajectory for a chain starting from, say, $\mathbf{x}^{(0)} = (10, 10)$ is shown in Fig. 4.7. One can see that the chain quickly settled in the "normal range" of the target distribution.

Gibbs sampler can be used to solve the missing data problem efficiently. Let $\mathbf{x} = (\mathbf{y}, \mathbf{z})$, where $\mathbf{y}$ is observed but $\mathbf{z}$ is missing. The observed-data posterior distribution, $p(\theta|\mathbf{y}) = \int p(\theta|\mathbf{y}, \mathbf{z}) p(\mathbf{z}|\mathbf{y}) d\mathbf{z}$, can be simulated via a two-component Gibbs sampler, which can be heuristically illustrated by the diagram in Fig. 4.8, in which we iteratively draw $\mathbf{z} \sim p(\bullet|\mathbf{y}, \theta)$ and $\theta \sim p(\bullet|\mathbf{z}, \mathbf{y})$. In the example of the dishonest casino in Sect. 4.2.2, if both the parameters and the hidden states in the hidden Markov model are unknown, we can use the two-component Gibbs sampler

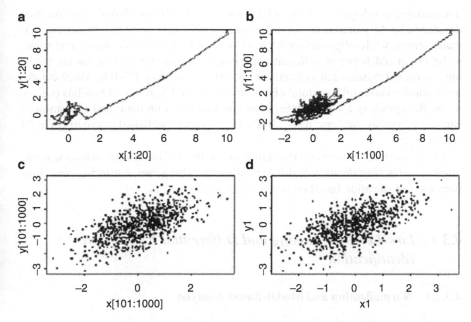

Fig. 4.7 Trajectories of (**a**) the first 20 iterations and (**b**) the first 100 iterations of a Gibbs sampler iterations. Scatter plots of (**c**) 101–1,000 iterations of the Gibbs sampler and (**d**) 900 random variables directly sampled from the bivariate Gaussian distribution

Fig. 4.8 A graphical
illustration of the
two-component Gibbs
sampler

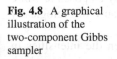 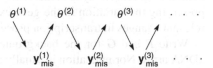

to simulate from the posterior distribution. More precisely, one can first draw the hidden path conditioned on parameters and then draw the parameters conditioned on the previously obtained hidden path, as illustrated in Fig. 4.8.

4.3   Gene Expression and Microarray Analysis

DNA microarray technology is a promising high-throughput method for measuring the gene expression levels of thousands of genes at the whole-genome scale. In this technique, RNA isolated from samples are labeled with fluorochromes or biotin before being hybridized to a microarray consisting of large numbers of cDNA/oligonucleotide orderly arranged onto a microscope slide. After hybridization under stringent conditions, a scanner records the intensity of the emission signals that is proportional to transcript levels in the biological samples. Two somewhat different microarray technologies are competing in the market: cDNA microarrays, developed by Stanford University, and oligonucleotide microarrays, developed by Affymetrix. cDNA arrays are cheaper, and more flexible as custom-made arrays, while oligonucleotide arrays are more automated, stable, and easier to be compared between different experiments. There are several useful public microarray databases, such as Stanford Microarray Database (SMD), which curates most Stanford and collaborators' cDNA arrays; Gene Expression Omnibus (GEO), an NCBI repository for gene expression and hybridization data; and Oncomine, a cancer microarray database with $\sim$9 K cancer-related published microarrays in 31 cancer types.

In this section, we focus on the statistical analysis of microarray data, including normalization, identification of differentially expressed genes, clustering, visualization, and classification based on gene expression.

4.3.1 Low-Level Processing and Differential Expression Identification

4.3.1.1 Normalization and Model-Based Analysis

Normalization corrects for overall chip brightness and other factors that may influence numerical values of expression intensities and reduces unwanted variation across chips by using information from multiple chips. The basic idea of normalization is to remove systematic biases in the data as completely as possible while preserving the variation in the gene expression that occurs because of biologically relevant changes in transcription processes.

Write R and G for the background-corrected red and green die intensities in cDNA array. Normalization is usually applied to the log-ratios of expression, which can be written as

$$M = \log_2 R - \log_2 G.$$

The log-intensity of each spot is given as

$$A = \frac{(\log_2 R - \log_2 G)}{2},$$

which is a measure of the overall brightness of the spot. The relationship between the dye bias and the expression intensity can be seen best in an MA plot, a scatterplot of the M-values against the A-values for an array [5]. Lowess normalization [6] is based on MA plot. Each M-value is normalized by subtracting from it the corresponding value of fitted lowess (local weighted regression) curve as function of A-values.

An alternative method is to select an invariant set of genes as described for oligonucleotide arrays by Tseng et al. [7]. A set of genes are said to be invariant if their ranks are the same for both red and green intensities. When there are sufficient invariant genes, the use of invariant genes is similar to lowess normalization as described above. In addition, if there were substantial scale differences between microarrays, it is usually useful to normalize between arrays by a simple scaling of the M-values from a series of arrays so that each array has the same median absolute deviation.

Some issues other than the simple normalization in "low-level" analysis might arise in experiments, such as probe-specific biases, corner effects, and outliers or scratches in experiment. Li and Wong [8] proposed a statistical model for the probe-level analysis and developed model-based estimates for gene expression. Let θ_i denote an expression value for the gene in the ith sample. Since each gene may be assigned about 20 probes, their probe-level model is given as:

$$\text{MM}_{ij} = v_j + \theta_i \alpha_j + \varepsilon \quad \text{and} \quad \text{PM}_{ij} = v_j + \theta_i \alpha_j + \theta_i \phi_j + \varepsilon. \quad (4.3)$$

Here, MM_{ij} and PM_{ij} denote the *Perfect Match* and *Mismatch* signals for the ith array and the jth probe pair for the gene, v_j is the baseline response of the jth probe pair due to nonspecific hybridization, α_j is the rate of increase of the MM response of the jth probe pair, ϕ_j is the additional rate of increase in the corresponding PM response, and ε is a random error. Model (4.3) can be reduced to a simpler form $y_{ij} = \text{PM}_{ij} - \text{MM}_{ij} = \theta_i \phi_j + \varepsilon_{ij}$, $\varepsilon_{ij} \sim N(0, \sigma^2)$, with the identifiability constraint $\sum_j \phi_j^2 = J$.

4.3.1.2   Identifying Differentially Expressed Genes

In the analysis of gene expression data, it is of great biological and clinical interest to decide which genes are differentially expressed between different groups of experiments. As a hypothesis testing problem, the decision is often made using the p-value, the probability of the test statistics being as or more extreme under the null distribution. The p-value has a uniform distribution under the null if the test statistics is continuous and a small p-value represents evidence against the null. For example, suppose $\{X_1, \ldots, X_{n_1}\}$ and $\{Y_1, \ldots, Y_{n_2}\}$ are expression values of a particular gene measured for individuals of two distinct groups (e.g., brain tissues vs. muscle tissues), satisfying $X_i \sim N(\mu_1, \sigma^2)$ and $Y_j \sim N(\mu_2, \sigma^2)$. To test the null hypothesis of $\mu_1 = \mu_2$, one can use the test statistics

$$T = \frac{(\bar{X} - \bar{Y})}{\sqrt{s_p^2 / n_1 + s_p^2 / n_2}},$$

where $s_p^2 = \{(n_1 - 1) s_1^2 + (n_2 - 1) s_2^2\} / (n_1 + n_2 - 2)$. Under the null, T has a t-distribution with $(n_1 + n_2 - 2)$ degrees of freedom, and the p-value of the observed value t^* of the test statistics is equal to $2 (1 - F_{t,\text{df}} (|t^*|))$, where $F_{t,\text{df}} (\cdot)$ is the cdf of the t-distribution with "df" degrees of freedom. If the two samples have unequal variances, i.e., $\sigma_1 \neq \sigma_2$, one should use $T' = (\bar{X} - \bar{Y}) / \sqrt{s_1^2 / n_1 + s_2^2 / n_2}$. In general, the distribution of T' can be approximated by the Welch approximation.

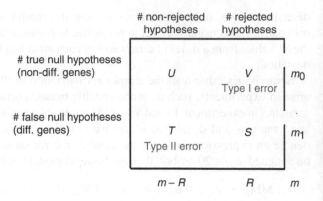

Fig. 4.9 Multiple hypotheses testing

There are also nonparametric ways of computing p-values for various test statistics (e.g., T or T'), such as the permutation test. For example, if the numbers of samples in two groups are 5 and 9, respectively, for each gene there are 2002 different T values obtainable from permuting the group labels. Under the null hypothesis that the distributions for the two groups are identical, all these T values are equally probable. We can thereby estimate the probability of getting a T as extreme as the observed one from the pool of permutated T values.

There is a serious consequence of performing statistical tests on many genes simultaneously, which is known as multiple testing, or "fishing expedition." That is, if one tests many hypotheses simultaneously, say 1,000, each at the significance level of 5 %, then one should expect to reject $\sim$50 of them even if there is nothing interesting in these 1,000 cases. A table of multiple hypothesis testing is given in Fig. 4.9. To deal with the multiple testing problem, an immediate approach is to control family-wise error rate (FWER), which is defined as the probability of accumulating at least one type I error, i.e., $P(V \geq 1)$. To control the FWER for testing m hypotheses at level α, we need to control the false rejection rate for each individual test at α/m, which is known as the Bonferroni correction. This may be appropriate for some applications (e.g., testing a new drug vs. several existing ones) but tends to be too conservative for our task of gene selection. A compromised approach was proposed by Benjamini and Hochberg [9], who advocate for controlling the false discovery rate (FDR), which is defined as the expected proportion of Type I errors among the rejected hypotheses, i.e., $E(Q)$, where

$$Q = \begin{cases} V/R & \text{if } R > 0 \\ 0 & \text{if } R = 0 \end{cases}.$$

A nonparametric empirical Bayes model, which is closely related to the frequentist FDR criterion, is introduced by Efron et al. [10], in which one can estimate both the "empirical null distribution" and a bound for the proportion of "alternatives" among the multiple tests.

4.3.2   Unsupervised Learning

Tasks of assigning objects to classes based on their "features" (e.g., measurements) can be broadly categorized as supervised learning, also known as classification or pattern recognition, and unsupervised learning, also known as clustering analysis. In supervised learning, one first has a set of "training samples" in which the classes are predefined and each individual's class label is known. The goal is to understand the basis for the classification from the training data. This information is then used to classify future observations. In unsupervised learning, the classes are unknown and need to be "discovered" from the data; and individuals are unlabeled. We focus on unsupervised learning here and will discuss supervised learning methods later.

A clustering analysis usually involves several distinct steps. First, a suitable distance (or similarity) measure between objects must be defined either explicitly or implicitly based on their observed features. Then, a clustering algorithm must be selected and applied. There are two general classes of clustering methods: criteria based or model-based, with the former including hierarchical clustering and K-means clustering and the latter mostly based on statistical mixture modeling.

4.3.2.1   Hierarchical Clustering

Hierarchical clustering provides a hierarchy of clusters, from the smallest, which has only one cluster including all objects, to the largest set, in which each observation is in its own cluster. Bottom-up methods, which proceed by series of fusions of the n objects into groups, are commonly used. The resulting dendrogram can be useful for assessing cluster strengths. Given a data table of the expression of m genes under n conditions, hierarchical clustering can be performed based on genes with similar expression profiles across various conditions.

4.3.2.2 K-Means Clustering

K-means clustering requires a pre-specification of the number of clusters. It then attempts to minimize the sum of squared within-cluster distances by iteratively assigning each object to the "closest" cluster. The distance between an object and a cluster is defined as that between the centroid of the cluster and the object. The results of K-means clustering depend on the specified number of clusters and initial assignments of clusters. People usually use several starting points and choose the "best" solution. Several criteria have been proposed for choosing the number of clusters. Given the n samples and k clusters, we let $W(k)$ denote the total sum of squares within clusters and $B(k)$ denote the sum of squares between cluster means. Calinski and Harabasz [11] proposed to maximize

$$\text{CH}(k) = \frac{B(k)\,(k-1)}{W(k)\,(n-k)}.$$

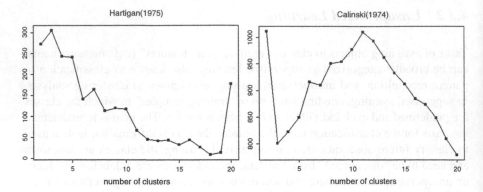

Fig. 4.10 Comparison of Hartigan's criterion and Calinski and Harabasz's criterion in choosing number of clusters

Hartigan [12] suggested to find the smallest k such that

$$H(k) = \left[\frac{W(k)}{W(k+1)} - 1 \right] (n - k - 1) < 10.$$

A comparison of the above two criteria is given in Fig. 4.10. In addition, Tibshirani et al. [13] proposed to maximize a gap statistics:

$$\text{Gap}_n(k) = E_n^* \left(\log \left(W(k) \right) \right) - \log \left(W(k) \right),$$

where $E_n^* (\cdot)$ denotes expectation under a sample of size n from an appropriate null reference distribution of the data.

A variant of K-means clustering, named K-medoid clustering, used medoid, the "representative object" within a cluster, instead of centroid that is the average of the samples within a cluster. Tseng and Wong [14] proposed a clustering method based on K-means clustering that produces tight and stable clusters without forcing all points into clusters. This method, named tight clustering, proceeds by repeatedly performing the K-means algorithm on data sets produced by resampling from the original data and then selecting clusters and cluster members that tend to be stable in majority of the resampled data sets. More precisely, candidates of tight clusters are selected according to the average comemebership matrix computed from the clustering results of all the resampled data sets.

4.3.2.3 Self-Organizing Maps

In contrast to the rigid structure of hierarchical clustering and the nonstructure of K-means clustering, self-organizing map (SOP) [15] imposes a partial structure on the clusters and is easily visualized and interpreted. In SOP, each cluster is mapped to a node of a two-dimensional graph, usually in the form of a size of $k = r \times s$ grid. Each node on the grid represents a cluster center or prototype vector. This arrangement

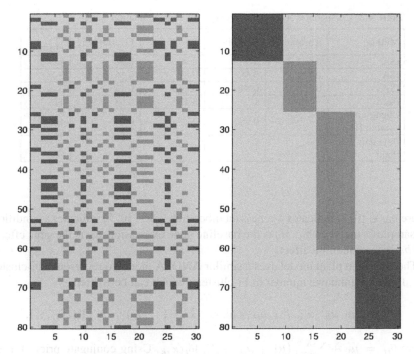

Fig. 4.11 Bi-clustering of gene expression

allows one to exploit and explore similarities among the clusters. The nodes (i.e., cluster centers) are first initialized at random positions. Then, SOP iteratively picks one data point (gene) x at random and moves nodes toward x in a two-dimensional space, the closest node most, remote nodes (in term of grid) less. The amount of movement is decreased with the number of iterations.

4.3.2.4   Two-Way Clustering

Genes do not necessarily belong to the same cluster under all conditions or for all samples. Two-way clustering (or bi-clustering) is a method for finding simultaneously subsets of genes and conditions/samples that show strong similarities (Fig. 4.11). These methods may help discover unexpected subclasses from expression profiles [16]. A simplest two-way clustering approach is to cluster genes and samples independently and display them together. More sophisticated approaches introduce certain criterion and use iterative algorithms to optimize such criterion. Lazzeroni and Owen [17] proposed a two-way additive model, called the plaid model, for bi-clustering. Suppose there are K bi-clusters and the expression value of the ith gene under the jth condition is Y_{ij}. The plaid model can be written as

$$Y_{ij} = \mu_0 + \sum_{k=1}^{K} \rho_{ik}\kappa_{jk}\left(\mu_k + a_{ik} + b_{jk}\right),$$

Table 4.2 Comparison of bi-clustering algorithms in simulation study

Method	Threshold	Sensitivity	Specificity	Overlapping	# clusters
ISA	0.6, 1.2	0.53	0.90	0.08	8
ISA	0.7, 1.1	0.68	0.84	0.16	8
ISA	0.6, 1	0.84	0.84	0.12	3
Plaid		1	0.73	0.63	11
Cheng and Church [18]		0.98	0	0	10
Bayesian Plaid		1	1	0	3

where $\rho_{ik} \in \{0, 1\}$ indicates a gene's membership, $\kappa_{jk} \in \{0, 1\}$ indicates a condition (or sample)'s membership, μ_0 is the baseline gene expression, a_{ik} is the gene effect, and b_{jk} is the condition effect.

The Bayesian plaid model uses a similar ANOVA model to represent a bi-cluster and allows an unknown number of bi-clusters, which can be written as

$$P\left(Y_{ij} \mid \mu_0, \mu_k, \alpha_{ik}, \beta_{jk}, \delta_{ik}, \kappa_{jk}, \tau_{\varepsilon}, \ \ k = 1 \dots K\right) = N\left(Y_{ij}; \theta_{ij}, \tau_{\varepsilon}\right),$$

where $\theta_{ij} = \mu_0 + \sum_{k=1}^{K} \left(\mu_k + \alpha_{ik} + \beta_{jk}\right)\delta_{ik}\kappa_{jk}$. Using conjugate priors for all the parameters, we can integrate out all the continuous variables conditional on the indicator variables. Thus, we can design a rather efficient MCMC algorithm to sample from the posterior distribution of the indicator variables, i.e.,

$$P(\delta_{ik}, \kappa_{jk}, \ \ k = 1 \dots K \mid \text{all the } Y_{ij}\text{'s}) \propto$$

$$\int P(Y's \mid \mu_0, \mu_k, \alpha_{ik}, \beta_{jk}, \delta_{ik}, \kappa_{jk}, \tau_{\varepsilon}, \ \ k = 1 \dots K) \mathrm{d}P(\mu_0, \mu_k, \alpha_{ik}, \beta_{jk}, \tau_{\varepsilon})$$

Table 4.2 compares a few published bi-clustering algorithms, including the algorithm proposed by Cheng and Church [18] and iterative signature algorithm (ISA; [19]), based on their operation characteristics, such as specificities, sensitivities, and overlapping rates. The comparisons are made on simulation data containing two clusters with additive overlapping effect.

4.3.2.5 GO on Clustering

The Gene Ontology (GO) database (http://www.geneontology.org/) provides a controlled vocabulary to describe gene and gene product attributes in any organism. GO can be used to evaluate and refine clustering by checking if GO terms are significantly enriched for members in the cluster and annotate unknown genes by inferring their functions based on the GO terms of other cluster members.

4.3.3 Dimension Reduction Techniques

In microarray data, one sample contains measurements of thousands of genes, and one gene often has measurements in tens of samples, making the visualization of data quite difficult. If we could plot the data in two-dimensional space, it will be easier to detect or confirm the relationship among data points and give us a better insight of the data. There are many methods to reduce dimensions, such as multidimensional scaling, principle component analysis, and singular value decomposition.

4.3.3.1 Multidimensional Scaling

Given distances between data points in a high dimensional space (e.g., correlations), multidimensional scaling (MDS) aims to find a representation of the data in a low dimensional space (such as 2-D) that approximates the original pair-wise distance relationship as much as possible. For example, given all pair-wise distances between major US cities, we can use MDS to construct a "best" two-dimensional map (representation). Technically, one can minimize the "stress" function defined as

$$\Phi = \sum \left[d_{ij} - f(\delta_{ij}) \right]^2,$$

where d_{ij} refers to the distance between points i and j on the map and $f(\delta_{ij})$ is a monotone transformation of the observed input distance δ_{ij}.

4.3.3.2 Principal Component Analysis and Singular Value Decomposition

The goal of the principal component analysis (PCA) is to recursively find a direction along which the data appear to be most variable. More precisely, it first finds the largest eigenvalue of the sample covariance matrix

$$\hat{\Sigma} = \sum_{i=1}^{n} (x_i - \bar{x}) (x_i - \bar{x})^{\mathrm{T}}$$

and its corresponding eigenvector, called the first principal component (PC). Then, it finds the second PC, which is the eigenvector for the second largest eigenvalue of $\hat{\Sigma}$. Figure 4.12 shows the projection of 387 cell-cycle-related genes, of which mRNA expressions were measured at 167 time points, onto the space spanned by its first and second PCs, a reduction of dimensions from 167 to 2.

Singular value decomposition (SVD) is closely related to PCA. The equation for the singular value decomposition of X is as following:

$$X = UDV^{\mathrm{T}},$$

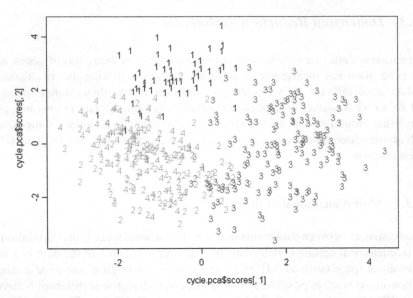

Fig. 4.12 387 cell-cycle-related genes visualized by PCA (*1* G1 phase, *2* G3 phase, *3* M phase, *4* S phase)

where X is the original $n \times p$ data (e.g., n genes, p experiments), U is a $n \times p$ orthogonal matrix, D is an $p \times p$ diagonal matrix, and V is also a $p \times p$ matrix. The columns of U are called the *left singular vectors* and form an orthogonal basis for the assay expression profiles. The rows of V contain elements of the *right singular vectors* and form an orthogonal basis for the gene transcriptional responses.

4.3.3.3 Gene Shaving

Results from dimension reduction may not be of direct interest to biologists and may be difficult to interpret. It is thus of interest to directly select genes that can explain a significant portion of the observed variation. In this sense, we want to find genes that have a large variation across samples. Gene shaving [20] is a method to find several groups of genes (possibly overlapping) so that in each group the between-gene variation is small, and the between-sample variation is large. More precisely, gene shaving seeks to find a sequence of subsets S_k of size k, so that $\text{var}_J \left(\sum_{i \in S_k} x_{iJ} / k \right)$ is maximized among all subsets of size k, where

$$\text{var}_J \left(\frac{1}{k} \sum_{i \in S_k} x_{iJ} \right) = \frac{1}{p} \sum_{j=1}^{p} \left(\frac{1}{k} \sum_{i \in S_k} \left(x_{ij} - \bar{x}_{i\bullet} \right) \right)^2 .$$

One needs to search all subsets to get the exact optimum. An approximate solution can be obtained, however, by reformulating the problem as in [20]. Let

$$w_i = \begin{cases} k^{-1} & \text{if } i \in S_k \\ 0 & \text{otherwise} \end{cases}, \quad \text{and} \quad \mathbf{w}^{\mathsf{T}}\mathbf{x}_j = (w_1, \ldots, w_G) \begin{pmatrix} x_{1j} \\ \vdots \\ x_{Gj} \end{pmatrix} = \sum_{g=1}^{G} w_g x_{gj} = \bar{x}_j.$$

Thus, our task is to find the "best" weight vector to maximize $\mathrm{var}_J(\mathbf{w}^{\mathsf{T}}\mathbf{x}_J)$, where the variance is taken with respect to random variable J. If we do not restrict w_i as 0 or k^{-1}, but only assume that $\sum_i w_i^2 = 1$, the "best" weight vector is just the first principal component of the sample covariance matrix. This leads to the PC shaving algorithm given below:

Start with the entire expression table $\mathbf{X}$; each row is centered to have zero mean. Compute the first principal component of rows of $\mathbf{X}$:

$$\mathbf{w} = \underset{\|\mathbf{w}\|_2 = 1}{\arg\max} \left(\|\mathbf{X}\mathbf{w}\|_2 \right).$$

Shave off 10 % of rows $\mathbf{x}_i$ with smallest $|\mathbf{x}_i^{\mathsf{T}}\mathbf{w}|$.
Repeat steps 2 and 3 until only one gene remains.
This produces a sequence of nested gene clusters. Select one of these by minimizing the gap statistics introduced in "K-means clustering."
Orthogonalize each row of $\mathbf{X}$ with respect to the average of all genes in the chosen cluster.
Repeat steps 1–5 to find the second cluster, the third cluster, ...

4.3.4   Supervised Learning

Supervised learning is among the oldest statistical tasks, of which the classification problem is a typical example: given Y as the class label (e.g., normal/disease) and X_i's as the covariates (expression values of certain genes), we learn the relationship between Y and X_i's from the training data so as to predict Y for future data where only X_i's are known. For example, gene expression data have been used to distinguish and predict different tumor types or long-survival patients versus short-survival patients. There are many popular classification methods, including discriminant analysis (linear/nonlinear, e.g., Fisher discriminant), logistic regression, classification and regression trees (CART), nearest neighbor methods, neural networks, support vector machines, Bayesian networks, boosting, and Bayesian additive regression trees (BART).

Before introducing these methods, we first discuss the problem of performance assessment (i.e., evaluation of the error rate) of classification algorithms on a given data set. If the goodness of a classifier is judged by comparing the observed and the fitted Y values using all the data (i.e., the training error), there is a strong tendency for the trained classifier to overfit the data and to underestimate the true prediction error rate. To avoid this problem, we need to divide the observations into two independent data sets L_1 and L_2. Classifiers are built using L_1, and error rates are estimated for L_2. However, if the number of samples in the data set is relatively small with respect to the number of features used in training, splitting data set may lead to insufficient number of training samples or inaccurate estimation of error. One remedy is to use N-fold cross-validation method by dividing data into N subsets with equal size. Classifiers are built using $(N-1)$ subsets, and error rates are computing for letting out subset. This step is repeated N times with each subset serving as testing set. The final cross-validation error is the average of N error rates.

4.3.4.1 Bayes Classification Rule

Suppose the probability distribution function (pdf) for G is $p_G(g)$, $g = 1$, $2, \ldots, K$. The conditional distribution of X given $G = g$ is $f_{X|G}(x|G = g)$. The training data (x_i, g_i), $i = 1, 2, \ldots, N$ are independent samples from the joint distribution of X and G. We have

$$f_{X,G}(x, g) = p_G(g) f_{X|G}(x|G = g).$$

Assume further that the loss of predicting G as $\hat{G} = g(X)$ is $L(G, \hat{G})$. The goal of the Bayes classification rule is to minimize

$$E_{X,G} L(G, g(X)) = E_X \left(E_{G|X} L(G, g(X)) \right).$$

Thus, it suffices to find $g(X)$ to minimize $E_{G|X} L(G, g(X))$ for each X, which leads to the optimal predictor

$$g(x) = \arg\min_g E_{G|X=x} L(G, g),$$

which is known as Bayes classification rule. For 0–1 loss, i.e.,

$$L(g, g') = \begin{cases} 1 & g \neq g' \\ 0 & g = g' \end{cases},$$

we have $E_{G|X=x} L(G, g) = 1 - \Pr(G = g|X = x)$, and the Bayes rule becomes the rule of maximum a posteriori (MAP) probability:

$$g(x) = \arg\min_g E_{G|X=x} L(G, g) = \arg\max_g \Pr(G = g|X = x).$$

Some classification algorithms use the classification rule by directly approximating the class density $\Pr(X|G = g)$. The Naïve Bayes method assumes that $\Pr(X|G = g)$ is a product of marginal densities, that is, all the variables are independent conditional on the label. Linear and quadratic discriminant analyses assume Gaussian densities. Mixtures of Gaussians, or general nonparametric density estimates, such as Parzen windows, have also been widely used.

4.3.4.2   Linear Discriminant Analysis (LDA)

In one-dimensional space, the LDA method simply finds two group means and cuts at some point, say, the middle point. In two-dimensional space, linear discriminant analysis connects two group means with a line and sets the classification boundary as a line in parallel to the "main direction" of the data, i.e., projecting to

$$\hat{\Sigma}^{-1} (\mu_1 - \mu_2).$$

In general, LDA is related to finding the decision boundary for samples from two Gaussian distribution with the same covariance matrix. Given estimated prior probabilities $\hat{\pi}_k = N_k/N$, class centers $\hat{\mu}_k = \sum_{g_i=k} x_i/N$, and the covariance matrix

$$\hat{\Sigma} = \frac{\sum_{k-1}^{K} \sum_{g_i=k} (x_i - \hat{\mu}_k)(x_i - \hat{\mu}_k)^{\mathrm{T}}}{(N-K)},$$

the classification decision of LDA for observation $\mathbf{x}$ is to find k that maximizes

$$\delta_k(\mathbf{x}) = \mathbf{x}\hat{\Sigma}^{-1}\hat{\mu}_k - \frac{1}{2}\hat{\mu}_k^{\mathrm{T}}\hat{\Sigma}^{-1}\hat{\mu}_k + \log \hat{\pi}_k.$$

LDA has some limitations: it does not consider nonlinear relationship and assumes that class means capture most of the information. A variant of LDA uses weighted voting by weighting genes based on how informative it is at classifying samples (e.g., t-statistics).

4.3.4.3   Logistic Regression

Given data $(y_i, \mathbf{x}_i)$, $i = 1, \ldots, n$, where y_i's are binary responses, the logistic regression model assumes that

$$\log \frac{P(y_i = 1|\mathbf{x}_i)}{P(y_i = 0|\mathbf{x}_i)} = \beta_0 + \beta_1 x_{i1} + \cdots + \beta_p x_{ip}.$$

In practice, one often estimates the β's from the training data using the maximum likelihood estimation (MLE) procedure. The decision boundary is determined by the regression equation, i.e., classify $y_i = 1$ if $\beta_0 + \beta_1 x_{i1} + \cdots + \beta_p x_{ip} > 0$.

Figure 4.13 shows the performances of logistic regression and LDA on a diabetes data set. In practice, logistic regression and LDA often give very similar results. If the additional assumption made by the LDA (i.e., the Gaussian assumption)

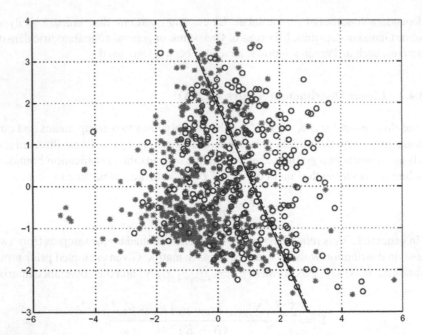

Fig. 4.13 Comparison of the logistic regression and LDA. *Stars*: samples without diabetes; *circles*: samples with diabetes. *Solid line*: classification boundary obtained by LDA. *Dash dot line*: boundary obtained by the logistic regression

is appropriate, LDA tends to estimate the parameters more efficiently by using more information about the data. Besides, samples without class labels can be used under the model of LDA. On the other hand, logistic regression relies on fewer assumptions and is more robust to gross outliers.

4.3.4.4 Nearest Neighbor Classifier

The k-nearest neighbor method, k-NN, due to Fix and Hodges [21], classifies an observation x as follows: first, it finds the k-observations in the learning set that are closest to x; then it predicts the class membership of x by the majority vote, i.e., choose the class that is most common among these k neighbors. As a relatively simple method, the k-NN classifier is shown to be competitive with more complex approaches such as aggregated classification trees and support vector machines.

4.3.4.5 Artificial Neural Networks

The name "Artificial Neural Networks" (ANNs) reflects an effort to mathematically model natural nervous systems and an attempt to understand human intelligence.

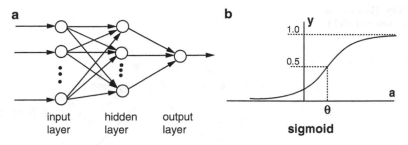

Fig. 4.14 (a) 3-layer multilayer perceptron, (b) Sigmoid function

There are three types of ANNs: feedforward, feedback, and competitive learning (self-organizing). Multilayer perceptron (MLP) is a broadly used feedforward neural network. Figure 4.14a shows a general picture of a 3-layer MLP. Given input **x**, the kth node of the output layer will compute the output as

$$g_k(\mathbf{x}) \equiv y_k = f\left(\sum_j w_{kj} f\left(\sum_i w_{ji} x_i + w_{j0}\right) + w_{k0}\right),$$

where f is the activation function and w's are weights of input neurons of each layer.

The sigmoid function shown in Fig. 4.14b is often used as activation functions, and weights are computed using the backpropagation (BP) algorithm with random starting values around zero. The structure of MLP, including numbers of hidden units and layers, is usually determined heuristically. Early stopping or penalty for large weights is used to prevent over-fitting. ANN algorithms have been used for sequence analysis, such as splicing site prediction [22, 23], and the classification and diagnostic prediction of cancers based on gene expression profiling [24]. But in general ANNs are nontrivial to tune and their performances are often sensitive to parameter tunings. There is also very little theoretical guidance on how to optimally use ANNs for learning.

4.3.4.6   Classification and Regression Tree

A classification and regression tree (CART) model represents the data by a tree structure. The algorithm starts with the root node, which contains all the data, and recursively splits the data at each node into two (or multiple) different nodes so as to reduce the "impurity" of data. There are several measures of impurities, such as the entropy

$$-\sum_{class} P(class) \log_2 (P(class))$$

Fig. 4.15 Geometrical
interpretation of SVM

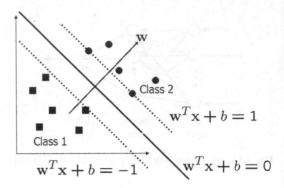

or the Gini index impurity

$$1 - \sum_{\text{class}} (P \text{ (class)})^2.$$

The CART algorithm assumes the independence of partitions and may split on different genes in the same level. The process of splitting stops when the total impurity is small enough or the number of samples in each node is too small. To avoid over-fitting, each split will automatically incur a cost. After splitting with the training set, the test set is used for pruning the obtained tree to reduce over-fit.

4.3.4.7 Support Vector Machines

Support vector machine (SVM) is a classifier derived from statistical learning theory (VC theory) by Vapnik and Chervonenkis in the 1970s. SVM is closely related to kernel methods and large-margin classifiers, reproducing kernel Hilbert space and Gaussian process, and has been studied by researchers from diverse communities from machine learning, optimization, and statistics. Intuitively, SVM seeks to find the "best" classification hyperplane in the sense that the decision boundary should be as far away from the data of both classes as possible. Let $\{x_1, \ldots, x_n\}$ be our data set, and let $y_i \in \{+1, -1\}$ be the class label of x_i. Suppose that the two classes can be perfectly separated by a hyperplane (linear decision boundary). Then, SVM maximizes the margin between the boundary and its nearest data points (Fig. 4.15), which is $m = 2 / \|w\|$. Thus, SVM solves a constrained optimization problem:

$$\min_{w} \frac{1}{2} \|w\|^2,$$

subjected to $y_i \left(w^T x_i + b\right) \geq 1$.

If the points of two classes are not separable by a linear classification boundary, a slack variable can be introduced as a trade-off between the classification of the data and maximization of the "margin." To extend to nonlinear decision boundary, the original feature data $\mathbf{x}_i$ is transformed to the feature space H of $\Phi(\mathbf{x}_i)$. The transformation is implicitly performed by using a key kernel idea: for certain classes of transformations, the inner product in the transformed space can be computed in the original space through a kernel function, i.e.,

$$\langle \Phi(\mathbf{x}_i), \Phi(\mathbf{x}_j) \rangle_H = K(\mathbf{x}_i, \mathbf{x}_j).$$

Some commonly used kernel functions are polynomial kernel with degree d

$$K(\mathbf{x}, \mathbf{y}) = (\mathbf{x}^\mathsf{T}\mathbf{y} + 1)^d,$$

radial basis function (RBF) kernel with width σ

$$K(\mathbf{x}, \mathbf{y}) = \exp\left(-\|\mathbf{x} - \mathbf{y}\| / 2\sigma^2\right),$$

and sigmoid kernel with parameter κ and θ

$$K(\mathbf{x}, \mathbf{y}) = \tanh\left(\kappa \mathbf{x}^\mathsf{T}\mathbf{y} + \theta\right).$$

Research on different kernel functions in different applications is very active.

The ideas of kernel and large margin have resulted in a family of new methods known as kernel machines, e.g., kernel PCA (KPCA), kernel Fisher discriminant analysis (KFD), and support vector clustering (SVC). A list of SVM implementations can be found at http://www.kernel-machine.org/software.html.

4.3.4.8 Boosting

Boosting is a method of improving the effectiveness of predictors, which relies on the existence of *weak learners*. A *weak learner* is a "rough and moderately inaccurate" predictor but one that can predict better than chance. The rules for boosting proceed by iterating the following steps: (1) set all weights of training samples as equal; (2) train a weak learner on the weighted samples; and (3) re-weight samples based on the performance of weak learner. At last, the prediction is made by weighted voting.

Here are some remarks on the application of supervised learning methods to classify samples based on gene expression profiles. First, screening of number of genes from 10 to 100 is advisable. Second, prediction models may include other predictor variables, such as age and sex of an individual. Moreover, outcomes may be continuous in clinical cases, e.g., blood pressure and cholesterol level.

4.4 Sequence Alignment

DNA and protein play central roles in the storage and manifestation of genetic information. Far from decoding the myth of life, the Human Genome Project (HGP) leads to more puzzles and more challenges, which attract many computational scientists. It is well known that the function of a protein is closely related to its three-dimensional structure, which is, to a large extent, determined by its one-dimensional amino acid sequence. However, the problem of predicting a protein's structure and function from its sequence still eludes researchers. Fortunately, since all genes and proteins evolve from one common ancestor, similar sequences often imply similar function and structure. Thus, detecting and quantifying sequence similarities via sequence alignments have traditionally been an important topic in computational biology. In this section, we will review some algorithms in pair-wise and multiple sequence alignments.

4.4.1 Pair-Wise Sequence Analysis

A dynamic programming method similar to those explained in Sect. 4.2.2 was first introduced for the pair-wise sequence alignment by Needleman and Wunsch [46] and Smith and Waterman [47]. The essence of their dynamic programming idea is to store the subproblem solutions for later use. More precisely, let the score for matching between nucleotides (or amino acids) x and y be $s(x, y)$ and let the gap penalty be γ. We first initialize an $N \times M$ matrix, where N and M are the lengths of the two sequences, respectively. We fill in the matrix from its upper-left corner to the lower-right corner in a recursive fashion with the (i, j)th entry filled with element $F(i, j)$, which records the best alignment score for the subsequence $[1:i]$ of the first sequence and subsequence $[1:j]$ of the second sequence and a pointer to one of the previous best alignments. The computation of $F(i, j)$ can be achieved by the following recursion:

$$F(i, j) = \max \begin{cases} F(i - 1, j - 1) + s(x_i, y_j) \\ F(i - 1, j) - \gamma \\ F(i, j - 1) - \gamma \end{cases} . \qquad (4.4)$$

In other words, the optimal sub-score for Problem (i, j) can only be one of the three possibilities: either align elements i and j or let one of them "hang out." The pointer is used to indicate which of the three options has been chosen (could be multiple). To obtain the final optimal solution, we trace back along the pointers. Figures 4.16 and 4.17 show an example for finding the best alignment of "AATCG" and "AGGC" with $s(x, y) = I_{\{x=y\}}$ and $\gamma = 1$.

The result of a sequence alignment depends on the scoring system and gap penalties employed. So far, we have used a simple linear function for penalizing

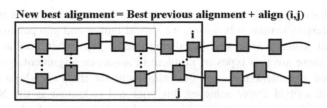

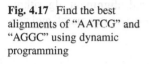

New best alignment = Best previous alignment + align (i,j)

Fig. 4.16 Illustration of dynamic programming for pair-wise alignment

Fig. 4.17 Find the best alignments of "AATCG" and "AGGC" using dynamic programming

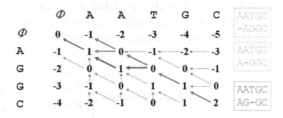

gaps in sequences. Under this model, the penalty increases linearly with the length of the gap. However, in nature, gaps tend to occur by groups. Thus, a more realistic approach is to penalize gaps less and less as we extend an existing one. We can model this by using a convex penalty function. The drawback of this approach is that the computation cost under this general convex penalty model increases to $O(N^3)$ operations and $O(N^2)$ space with N being the average length of the sequences. This is impractical for aligning protein sequences, which are in the range of hundreds, and impossible for aligning DNA sequences, which tend to be much longer. Therefore, a linear approximation of the convex gap function known as the affine gap penalty is preferred:

$$\gamma(g) = a + (g-1)b,$$

where g is the number of gaps and typically $a > b$ (e.g., $a = 12$ and $b = 2$). A computational trick to deal with the affine gap penalty is to "remember" at position (i, j) whether we are within a gap, or initiating a new gap, or in a match state. We accomplish this by keeping three functions, each recording a directional optimum in dynamic programming:

$$F_h(i, j) = \max \{F(i, j-1) - a; F_h(i, j-1) - b\},$$

$$F_v(i, j) = \max \{F(i-1, j) - a; F_v(i-1, j) - b\},$$

$$F(i, j) = \max \{F(i-1, j-1) + s(x_i, y_j); F_h(i, j); F_v(i, j)\}.$$

This algorithm allows for a time complexity of $O(N^2)$.

Another problem arises as the penalization of gaps at the ends. If two sequences are of similar length and we are interested in their global similarities, entire sequences need to be compared and the end gaps should be penalized. If two sequences have very different lengths and/or we are only interested in seeing if they share certain similar subregions, we should ignore end gap penalties most of the time and probably even stop penalizing middle gaps when they are too long. In general, there are two types of problem in sequence alignment: global versus local. Global alignment finds the best alignment for the two whole sequences. Algorithm described above achieves this goal and is known as the Needleman-Wunsch algorithm in computational biology. Local alignment finds high scoring subsequences (e.g., two proteins only share one similar functional domain) and can be achieved by modifying global alignment method. Local alignment algorithm, which is known as the Smith-Waterman algorithm, uses negative mismatch and gap penalties and seeks best score anywhere in the matrix:

$$F(i, j) = \max \begin{cases} F(i-1, j-1) + s(x_i, y_j) \\ F(i-1, j) - \gamma \\ F(i, j-1) - \gamma \\ 0 \end{cases}.$$

The only change from the Needleman-Wunsch algorithm is to include a 0 in the maximization procedure to compute $F(i, j)$. Intuitively, the 0 allows for "resetting" the alignment in a region (i.e., stop penalizing) when the other three options does not help us find an optimal sequence.

The scoring system for aligning DNA sequences usually assigns +5 for match and −4 for mismatch. For proteins, different amino acid residue pairs receive different scores reflecting their functional and structural similarities. The scores are generally affected by residue sizes, electric charges, van der Waals interactions, abilities to form salt bridges, etc. Schwartz and Dayhoff [25] proposed the percent accepted mutations (PAM) matrices from a database of 1,572 changes in 71 groups of closely related proteins (>85 % similar) by constructing a phylogenetic tree of each group and tabulating the probability of amino acid changes between every pair of amino acids. Henikoff and Henikoff [26] constructed BLOcks amino acid SUbstitution Matrices (BLOSUM) by checking >500 protein families in the PROSITE database [27] and finding ~2,000 blocks of aligned segments. BLOCKS database characterizes ungapped patterns with 3–60 amino acids long and can be used to evaluate the observed substitutions. BLOSUM N is constructed according to the rule that if sequences >$N\%$ are identical, their contributions are weighted to sum to 1. PAM250 and BLOSUM62 are most widely used.

With the completion of whole-genome sequencing of several organisms, there are growing needs for rapid and accurate alignments of sequences of thousands of or even millions of base pairs. The BLAST (basic local alignment search tool) algorithm [28] is a heuristic search algorithm that seeks words of length W (default 3 for protein and 12 for DNA) with a score of at least T when aligned with the query (scored with a substitution matrix). Words in the database satisfying this initial screening are extended in both directions in an attempt to find a locally optimal ungapped alignment or HSP (high-scoring pair) with a score of at least S or an

Table 4.3 BLAST programs

Name	Query	Database
Blastn	Nucleotide	Nucleotide
Blastp	Protein	Protein
Blastx	Nucleotide	Protein
Tblastn	Protein	Translated
Tblastx	Nucleotide	Translated

E value lower than the specified threshold. HSPs that meet these criteria will be reported by BLAST, provided they do not exceed the cutoff value specified for the number of descriptions and/or alignments to report. There are several versions of BLAST programs for various types of queries in different databases, as shown in Table 4.3.

Recently developed BLAT (BLAST-Like Alignment Tool, http://genome.ucs. ed/cgi-bin/hgBlat) can align a much longer region very fast with little resources. BLAT's speed stems from an index of all non-overlapping K-mers in the genome. It uses the index to find regions in the genome likely to be homologous to the query sequence. It performs an alignment between homologous regions and stitches together these aligned regions (often exons) into larger alignments (typically genes). Finally, BLAT revisits small internal exons possibly missed at the first stage and adjusts large gap boundaries that have canonical splice sites where feasible.

4.4.2 Multiple Sequence Alignment

It is important for understanding functions and structures or revealing evolutionary relationships to find common patterns shared by multiple protein or DNA sequences. The challenges in multiple sequence alignment (MSA) are mainly due to the choice of sequences, the choice of scoring function, and the need of optimization strategy. In general, MSA algorithms can be categorized as criteria-based approaches (extension of pair-wise alignment) or model-based approaches (EM, Gibbs, HMM). Table 4.4 summarizes the different types of MSA algorithms.

ClustalW [29] is the most commonly used MSA method. It consists of three main steps: (1) all pairs of sequences are aligned separately in order to calculate a distance matrix giving the divergence of each pair of sequences; (2) a guide tree is calculated from the distance matrix; and (3) the sequences are progressively aligned according to the branching order in the guide tree. ClustalW is a fast algorithm for MSA. However, as a progressive method, the algorithm of ClustalW is heuristic driven, and no explicit model is used to guide for the alignment. Overall, ClustalW is not sensitive enough for finding and aligning remotely related sequences.

Hidden Markov model (HMM) coupled with the Gibbs sampling algorithm introduced in Sects. 4.2.2 and 4.2.3 provides a model-based approach for sequence alignment. Figure 4.18 shows a generative model for DNA sequence alignment, where the squares are main states (M_i), diamonds are insert states (I_i), and circles

Table 4.4 Types of MSA algorithms

Type	Example	Strategy	Pros	Cons
Exact	MSA, DCA, OMA	High-dimensional dynamic programming	Close to "optimal"	Computationally very expensive, limitation on the number of sequences
Progressive	ClustalW, Pileup, MultAlign	Sequences added to the alignment one by one according to pre-computed dendrogram	Fast, little time and memory requirement. Perform well most of the time	Alignment cannot be refined once sequences are added. Bad on long insertions or deletions
Iterative	Stochastic (Gibbs sampler, HMM, DA, Genetic Algorithm) and non-stochastic (Praline, Probe)	Alignment is iteratively refined	Overcome the drawback of progressive method	Require more computational time
Consistency based	SAGA, COFFEE, T-Coffee	Using consistency-based scoring function	Do not depend on any substitution matrix. Position-based scheme	Require more computational time. Limited number of aligned sequences

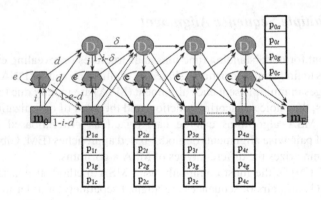

Fig. 4.18 Hidden Markov model for multiple sequence alignment

are delete states (D_i). Each main state except beginning (M_0) and ending states (M_E) is associated with an emission probability for four nucleotides:

$$\mathbf{p}_i = (p_{ia}, p_{ic}, p_{ig}, p_{it}).$$

Insertion states are associated with background probability:

$$\mathbf{p}_0 = (p_{0a}, p_{0c}, p_{0g}, p_{0t}),$$

Φ X1 X2 X3 X4 X5 X6
Φ
m1
m2
m3
m4

Fig. 4.19 Transition path (*red*) in HMM (Color figure online)

Motif 1
Motif 2
Motif 3
Motif 4

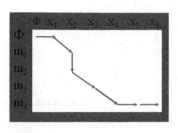

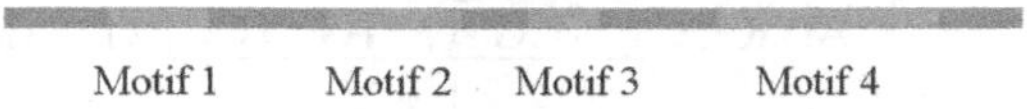

Fig. 4.20 An illustration of propagation model

and deletion states indicate gaps in the alignment. Different states are connected with Markov transition probabilities (e.g., e, d, i, and δ). The red path in Figs. 4.18 and 4.19, which is "hidden" in the model, shows the transition between states in generating a sequence with six nucleotides "ATCTTG." In the path, a horizontal line indicates an insertion, a vertical line indicates a deletion, and a diagonal line indicates a transition between main states. For example, according to the path given in Fig. 4.20, we can easily write down the probability of observing the path and the probability of generating the sequence conditional on the path as

$$P\,(\text{red path}) = i\,(1-e-d)\,d\,(1-i-\delta)\,(1-i-d)\,(1-e-d)$$
$$\text{and} \quad P\,(\text{ATCTTG}|\text{red path}) = p_{0a}\,p_{1t}\,p_{3c}\,p_{4t}\,p_{0t}\,p_{0t}.$$

Conditional on the HMM parameters Θ, which includes both emission probability of each state and transition probabilities between states, the probability for generating sequence s is given by

$$P\,(s|\Theta) = \sum_{\textbf{path}} P\,(s|\Theta, \text{path})\,P\,(\text{path}|\Theta).$$

The above probability can be calculated efficiently using a dynamic programming algorithm. Assuming that the transition probability from state k to state l is $\tau_{k,l}$, we initialize the algorithm by defining three functions and assigning their starting values: $F_m\,(0,0) = 0$, $F_i\,(0,0) = 0$, and $F_d\,(0,0) = 0$. Then, the latter elements $F_m\,(j,k)$, $F_i\,(j,k)$, and $F_d\,(j,k)$ are computed according to the following recursion:

$$\begin{aligned} F_m(j,k) &= p_{k,s_j}\left[F_m(j-1,k-1)\tau_{m_{k-1},m_k} + F_i(j-1,k-1)\tau_{i_{k-1},m_k}\right. \\ &\quad \left. + F_d(j-1,k-1)\tau_{d_{k-1},m_k}\right] \\ F_i(j,k) &= p_{0,s_j}\left[F_m(j-1,k)\tau_{m_k,i_k} + F_i(j-1,k)\tau_{i_k,i_k} + F_d(j-1,k)\tau_{d_k,i_k}\right], \\ F_d(j,k) &= F_m(j,k-1)\tau_{m_{k-1},d_k} + F_i(j,k-1)\tau_{i_{k-1},d_k} + F_d(j,k-1)\tau_{d_{k-1},d_k}. \end{aligned}$$

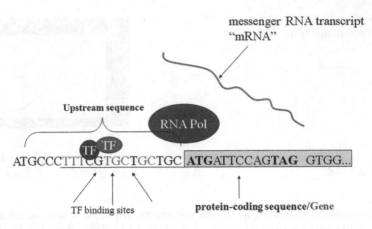

Fig. 4.21 Illustration of transcription process

The final probability is given by $F(l, n)$, where l is the length of sequence s and n is the number of main states.

In order to detect more remote homologs and to align subtly related proteins, one often has to further constrain the HMM model in a biologically meaningful way. The propagation model [30] achieves this type of parameter reduction by focusing on the alignment composed of multiple block motifs. A block motif is a special HMM that allows no insertions or deletions between match states. The propagation model consists of a number of block motifs as shown in Fig. 4.21. Similar to HMM, the propagation model also assumes that there are L-conserved model positions for each sequence to be aligned, with the only difference being that each model position represents a block of width w_l. We can imagine that L motif elements propagate along a sequence. Insertions are reflected by gaps between adjacent motif elements. No deletions are allowed, but it is possible to relax this constraint by treating each block as a mini-HMM.

Let Θ be the collection of all parameters including the insertion and emission probabilities as well as the transition probabilities; let Δ be the fragmentation indicators (for which column is in model and which is not), and let the alignment variable $\mathbf{A} = (a_{k,l})_{K \times L}$ be a matrix with $a_{k,l}$ indicating the starting position of the lth motif element in sequence k. Here, L represents the total number of motifs for the alignment, W is the total number of alignment columns, and $\mathbf{R} = (R_1, \ldots, R_K)$ denotes the set of protein sequences to be aligned.

The alignment variable $\mathbf{A}$ as represented by the locations of the motif elements is not observed. But once it is known, we can write down the conditional distribution $P(\mathbf{R}|\mathbf{A}, \Theta, \Delta, L, W)$ easily, from which we can make Bayesian inference on Θ and Δ easily. On the other hand, once Θ and Δ are given, we can sample $\mathbf{A}$ by a draw from $P(\mathbf{A}|\mathbf{R}, \Theta, \Delta, L, W)$, which can be achieved using a forward–backward strategy similar to that in Sect. 4.2.2. The number of motifs L and the total number of motif columns W are treated as random variables and can be selected by using a genetic algorithm or an MCMC-based sampling approach.

The Bayesian posterior in the above model can be written as

$$cP\ (\mathbf{R}|\mathbf{A}, \Theta, \Delta, L, W)\ P\ (\mathbf{A}, \Theta, \Delta|L, W),$$

where c is the normalizing constant and $P\ (\mathbf{A}, \Theta, \Delta|L, W)$ is the prior. By assigning a product Dirichlet distribution for Θ and an "equally likely" prior for $\mathbf{A}$ and Δ, we can employ a "collapsing" technique [31] to further improve the computational efficiency.

Several HMM-based software packages for producing multiple sequence alignments are available. HMMER was developed by Eddy [32] and is available at http://HMMer.wustl.edu/. SAM was developed by the UCSC group and is available at http://www.soe.ucsc.edu/projects/compbio/HMM-apps/HMM-applications.html.

4.5 Sequence Pattern Discovery

In eukaryotes, transcription is initiated by the binding of RNA polymerase II to the core promoters of genes, which reads the sequence of one strand of the DNA and synthesizes messenger RNA (mRNA). The efficiency of transcription is regulated by proteins called transcription factors (TFs) binding to their recognition sites located mostly in upstreams of the genes (promoter regions), but also not infrequently in downstreams or intronic regions (Fig. 4.21). Transcription factor-binding sites (TFBSs) are short sequence segments ($\sim$10 bp) located near genes' transcription start sites (TSSs). TFBSs usually show a conserved pattern, which is often called a TF-binding motif (TFBM). Laboratory assays such as electrophoretic mobility shift assays and DNase footprinting have been developed to locate TFBSs on gene-by-gene and site-by-site bases, but these methods are laborious, time-consuming, and unsuitable for large-scale studies. Computational methods thus have become necessary for finding "patterns" in biological sequence.

4.5.1 Basic Models and Approaches

The goal of motif finding is to look for common sequence segments enriched in a set of co-regulated genes (compared to the genome background). The simplest method for finding a motif is to check for the overrepresentation of every oligonucleotide of a given length (i.e., every k-mers). However, binding sites of a TF are usually "very badly spelled" and can tolerate many "typos." Hence, degenerated IUPAC symbols for ambiguous bases are frequently used in the consensus analysis.

Unlike consensus analysis, which only reflects most frequently occurring base types at each motif position without an explicit account of frequencies, statistical models based on a probabilistic representation of the preference of nucleotides at each position are generally more informative. In vivo, TF binds to any DNA

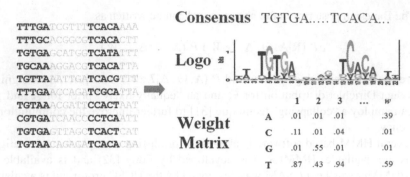

Fig. 4.22 Representation of a motif

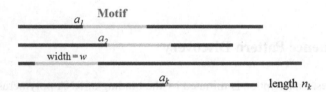

Fig. 4.23 A graphical illustration of block-motif model

sequence with a probability dependent on energy. Thus, it is desirable to model TF/DNA interaction based on statistical mechanics, which provides a theoretical support for the use of a probabilistic representation. The most widely used statistical model of motif is position-specific weight matrix (PWM), or a product multinomial distribution (see [34]). The columns of a PWM describe the occurrence frequencies of the four nucleotides in the corresponding motif position. Figure 4.22 summarizes several models to represent a motif, in which the sequence logo plotted provides a convenient graphical representation of the PWM: the height of each column is proportional to its information content and the size of each letter is proportional to its probability at that position.

In general, there are two types of approaches for motif finding based on different representations of motifs: those proceed by enumerating regular expressions, such as MobyDick, and those proceed by iteratively updating PWMs, such as Consensus, MEME, and Gibbs motif sampler. By checking the overrepresentation of every word pair in a dictionary of motif words, MobyDick algorithm reassigns word probability and considers every new word pair to build even longer words. Consensus [33] uses a greedy heuristic search strategy to identify motif modeled by PWM. It stars from the first sequence in the data set and adds one sequence at a time to look for the best motifs by adding the sequence with the current set of "best motifs."

The problem of motif finding can be formulated as a missing data problem under the block-motif model [34] in Fig. 4.23. The model says that at unknown ("missing") locations $\mathbf{A} = (a_1, \ldots, a_k)$, there are repeated occurrences of a motif. So the sequence segments at these locations should look similar to each other. In other parts of the sequence, the residues (or base pairs) are modeled as iid observations from a multinomial distribution (these background residues can also be modeled as a kth-order Markov chain). The background frequency $\theta_0 = (\theta_{0,a}, \ldots, \theta_{0,t})$ can be assumed known or estimated from the data in advance because the motif site positions are only a very tiny fraction of all the sequence positions. For a motif of width w, we let $\Theta = (\theta_1, \ldots, \theta_w)$, where each describes the base frequency at position j of the motif, and the matrix Θ is often referred to as the PWM for the motif. For a particular sequence s of length n, given the motif location a, the likelihood can be written as

$$
p\left(s|\Theta, \theta_0, A\right) = p\left(s_{[1:(a-1)]}|\theta_0\right) \times p\left(s_{[a:(a+w-1)]}|\Theta\right) \times p\left(s_{[(a+w):n]}|\theta_0\right)
$$

$$
= p\left(s_{[1:n]}|\theta_0\right) \times \frac{p\left(s_{[a:(a+w-1)]}|\Theta\right)}{p\left(s_{[a:(a+w-1)]}|\theta_0\right)} \tag{4.5}
$$

Suppose we are given a set of sequences $\mathbf{S} = (s_1, \ldots, s_k)$, which share a common repeated motif. By integrating out Θ after incorporating the prior, we can get a function for alignment vector $A = (a_1, \ldots, a_k)$:

$$
p\left(\mathbf{S}|\mathbf{A}\right) = \int \left\{ \prod_{j=1}^{k} p\left(s_i | \mathbf{A}, \Theta, \theta_0\right) \right\} P\left(\Theta, \theta_0\right) d\Theta.
$$

To maximize the posterior probability given above, Expectation Maximization (EM) algorithm provides an efficient iterative procedure in the presence of missing data (i.e., the alignment vector $\mathbf{A}$). The intuition behind the EM is to alternate between taking expectation of the complete-data log-likelihood function over the distribution of the missing variable (E-step), i.e., the computation of the Q-function and estimating the unknown parameters through the maximization of the eQ-function (M-step). According to (4.5), the complete-data log-likelihood for sequence s in the block-motif model can be written as

$$
\log p\left(s|A, \Theta, \theta_0\right) = \sum_{j=1}^{w} \log \left(\frac{\theta_{j,s_{a+j-1}}}{\theta_{0,s_{a+j-1}}}\right) + \log p\left(s|\theta_0\right).
$$

Let $p_{t,a}^{(0)}$ be the probability for the ath position in sequence t being a site based on current estimate $\Theta^{(0)}$ of the motif PWM; we can calculate Q-function in the E-step as

$$
Q\left(\Theta|\Theta^{(0)}\right) = \sum_{t=1}^{k} \sum_{a=1}^{n_k-w+1} p_{t,a}^{(0)} \sum_{j=1}^{w} \log \left(\frac{\theta_{j,s_{k,a+j-1}}}{\theta_{0,s_{k,a+j-1}}}\right).
$$

Parameter Θ can thus be updated by maximizing $Q\left(\Theta|\Theta^{(0)}\right)$ in the M-step, resulting in

$$\theta_{j,x} = \sum_{t=1}^{k} \sum_{a=1}^{n_k-w+1} p_{t,a}^{(0)} I\left(s_{k,a+j-1} = x\right), \tag{4.6}$$

where $j = 1,\ldots,w$ and $x = a,c,g,$ and t. The above procedure is summarized below.

Initialize a random matrix $\mathbf{P}$ of size $4 \times w$ and an empty matrix $\mathbf{L}$ of equal size.

Score all the segments in the set of sequences with $\mathbf{P}$.

Calculate weighted sum score in matrix $\mathbf{L}$ using (4.6).

Rewrite $\mathbf{P}$ with normalized $\mathbf{L}$.

Repeat steps 2–4 until convergence.

4.5.2   Gibbs Motif Sampler

The basics of Gibbs sampling algorithms have been introduced in Sect. 4.2.3. In this subsection, we will introduce the application of Gibbs sampling in finding recurring sequence motifs – the Gibbs motif sampler [35, 36].

For the block-motif model introduced in Sect. 4.5.1, Gibbs sampling can be done in two different ways: iterative sampling and predictive updating. The procedure of iterative sampling proceeds by iterating between sampling $\Theta^{(t)}$ from $p\left(\Theta|\mathbf{A}^{(t-1)}, \mathbf{S}\right)$ and drawing $\mathbf{A}^{(t)}$ from $p\left(\mathbf{A}|\Theta^{(t)}, \mathbf{S}\right)$. The predictive updating (PU) procedure is based on a "collapsed Gibbs sampler" that integrates out Θ and iteratively updates a_j. More precisely, one can pretend that binding sites in all but the jth sequence have been found and predict the binding site's location in the jth sequence by drawing $a_j^{(t)}$ from the predictive distribution:

$$P\left(a_j|\mathbf{A}_j^{(t-1)}, \mathbf{S}\right) \propto \prod_{i=1}^{w} \frac{q_{i,s_{j,a_j}+i-1}}{q_{0,s_{j,a_j}+i-1}},$$

where $q_{i,x} = (c_{ix} + b_x)/\left(\sum_x (c_{ix} + b_x)\right)$, c_{ix} is the count of nucleotide type x at position i, c_{0x} is the count of nucleotide type x in all non-site positions, and b_x is the "pseudo-count" for nucleotide type x.

The above model is not satisfactory in several aspects. Since the sequences for motif finding are put together mostly based on certain derived and imprecise information, some sequences often have multiple motif sites and some other sequences do not have any site at all. Thus, it is more reasonable to view the data set as a long sequence that houses an unknown number of TF-binding sites (possibly corresponding to multiple TFs). The goal of the TFBS prediction can then be formulated as a partition of the sequence into segments that correspond

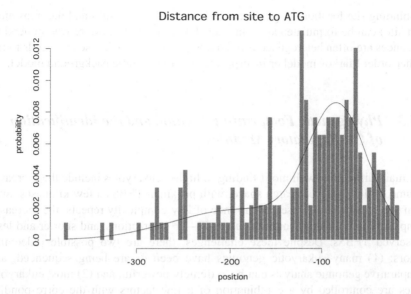

Fig. 4.24 Analysis of ATG to TF sites in ~150 *E. coli* genes. The *curve* is fitted using Gaussian Mixture Model

to different (unknown) motif models. Gibbs motif sampler is based on such a mixture model of motifs and background and implements a predictive updating scheme.

Suppose that all the sequences in the data set are concatenated as one sequence $\mathbf{R} = r_1 r_2 \ldots r_N$ with length N. We use indicator vector $\Delta = \delta_1 \delta_2 \ldots \delta_N$ to indicate the occurrence of motif sites in sequence $\mathbf{R}$:

$$\delta = \begin{cases} 1 \text{ if it is the start of a motif site} \\ 0 \text{ if not} \end{cases}.$$

By using the Bayes formulation, the predictive update is given by

$$\frac{\pi\left(\delta_k = 1 | \Delta_{[-k]}, \mathbf{R}\right)}{\pi\left(\delta_k = 0 | \Delta_{[-k]}, \mathbf{R}\right)} = \frac{\hat{\varepsilon}}{1 - \hat{\varepsilon}} \prod_{t=1}^{w} \left(\frac{\hat{p}_{i, r_{k+i-1}}}{\hat{p}_{0, r_{k+i-1}}}\right),$$

where ε is the prior probability for $\delta = 1$ and is given in advance, $\hat{\varepsilon}$ $\hat{p}_{0, r_{k+i-1}}$ and $\hat{p}_{i, r_{k+i-1}}$ are empirical estimates based on all the positions except position k.

Further modeling efforts can be made to incorporate biological considerations. For example, if we want to find palindromic patterns, we can set $\theta_{1a} = \theta_{wt}$, $\theta_{1t} = \theta_{wa}$, $\theta_{1c} = \theta_{wg}$, $\theta_{1g} = \theta_{wc}, \ldots$, etc., in modeling the base frequencies in the motif. Furthermore, as shown in Fig. 4.24, not all locations upstream of a gene are equally likely to be bound by a transcription factor. The distribution of the known

TF-binding site locations (e.g., distances between binding sites and the translation start site) can be formulated as an informed alignment prior. Moreover, noncoding sequences are often heterogeneous in compositions, in which case one needs to use higher-order Markov model or incorporate position-specific background model.

4.5.3   Phylogenetic Footprinting Method and the Identification of Cis-Regulatory Modules

The main difficulties with motif finding in higher eukaryotes include the increased volume of the sequence search space, with proximal TFBSs a few kilobases away from the TSSs; the increased occurrence of low-complexity repeats; the increased complexity in regulatory controls due to TF–TF interactions; and shorter and less-conserved TFBSs. Despite these challenges, there are two possible redeeming factors: (1) many eukaryotic genomes have been or are being sequenced, and comparative genomic analysis can be extremely powerful, and (2) most eukaryotic genes are controlled by a combination of a few factors with the corresponding binding sites forming homotypic or heterotypic clusters known as "*cis*-regulatory modules" (CRMs).

Transcription factor-binding sites across species are more conserved than random background due to functional constraints. With the advent of whole-genome sequencing, computational phylogenetic footprinting methods, involving cross-species comparison of DNA sequences, have emerged. Traditional "horizontal" approach introduced in Sects. 4.5.1 and 4.5.2 requires a set of co-regulated genes to identify common motifs. In contrast, phylogenetic footprinting is a "vertical" approach, which uses orthologous genes in related species to identify common motifs for this gene across multiple species.

McCue et al. [37] introduced a phylogenetic footprinting method with application in proteobacterial genomes. The method begins with *E. coli*-annotated gene and applies tBlastn with stringent criteria to identify orthologous genes in eight other gamma proteobacterial species (Fig. 4.25). Upstream intergenic regions from nine species are extracted, and a Gibbs motif sampler with the following important extensions was utilized. First, motif model that accounts for palindromic patterns in TF-binding sites was employed. Because DNA sequences tend to have varying composition, a position-specific background model, estimated with a Bayesian segmentation algorithm, was used to contrast with the motif PWM. Furthermore, the empirical distribution of spacing between TF-binding sites and the translation start site, observed from the *E. coli* genome sequence, was incorporated to improve the algorithm's focus on more probable locations of binding sites. Lastly, the algorithm was configured to detect 0, 1, or 2 sites in each upstream region in a data set.

The algorithm was applied to a study set of 184 *E. coli* genes whose promoters contain documented binding sites for 53 different TFs. Among the 184 most

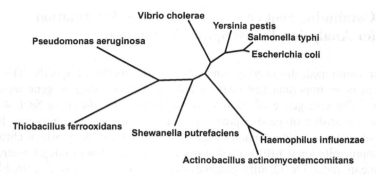

Fig. 4.25 Phylogenetic tree of gamma proteobacterial species used in phylogenetic footprinting study

probable motif predictions, 146 corresponds to known binding sites. The remaining data sets contain several predictions with larger maximum a posteriori probability (MAP) values than the true sites, suggesting the possibility of undocumented regulatory sites in these data. The documented TF-binding sites for these data were frequently detected as the second or third most probable motif. Among the 32 undocumented sites identified in the study were several strongly predicted sites, including one upstream of the *fabA* gene. A scan of the *E. coli* genome with the novel motif revealed two additional occurrences of this site in intergenic regions: one upstream of *fabB* and one upstream of *yqfA*. The protein bound to each of the predicted sites was then identified through affinity purification and mass spectrometry analysis as YijC, an uncharacterized member of the TetR family of transcription factors and likely repressor of *fabA* and *fabB*.

Transcription regulation is controlled by coordinated binding of one or more transcription factors in the promoter regions of genes. In many species, especially higher eukaryotes, transcription factor-binding sites tend to occur as homotypic or heterotypic clusters, also known as *cis*-regulatory modules. The number of sites and distances between the sites, however, vary greatly in a module. Several statistical models have been proposed to utilize the clustering effect in de novo motif finding (Thompson et al. [48], Zhou et al. [49]; and Gupta and Liu [50]). In particular, Gupta and Liu [38] describe the underlying cluster structure of a CRM as a hidden Markov model and design a Markov chain Monte Carlo strategy for predicting novel CRMs in upstream sequences of co-regulated genes. The method starts with an initial collection of putative TFBMs obtained from both databases such as TRANSFAC and JASPAR and de novo motif searches by using existing algorithms. It then iteratively selects motif types that are likely members of the CRM and updates the corresponding motif sites and parameters. The method employs the evolutionary Monte Carlo (EMC) method for screening the motif candidates and a dynamic programming-based recursion for locating the motif sites.

4.6 Combining Sequence and Expression Information for Analyzing Transcription Regulation

As aforementioned, discovering binding sites and motifs of specific TFs of an organism is an important first step toward the understanding of gene regulation circuitry. The emergence of microarray techniques introduced in Sect. 4.4 can give us a snapshot of gene expression in a cell on the genomic scale. Recent years have seen a rapid adoption of the ChIP-array technology, where chromatin immunoprecipitation (ChIP) is carried out in conjunction with mRNA microarray experiments (array) to identify genome-wide interaction sites of a DNA-binding protein. However, this method only yields a resolution of hundreds to thousands of base pairs, whereas the actual binding sites are only 10–20 bp long. Computational methods are crucially important for pinning down the exact binding-site locations. Furthermore, the incorporation of information from gene expression data has been demonstrated to improve upon computational TFBM finding methods. It also enables us to evaluate contributions of specific *cis*-regulatory elements and other genomic factors to the regulation of gene expression. In this section we will discuss how to use statistical methods for genome-wide analyses of gene expression and transcription regulation.

4.6.1 Motif Discovery in ChIP-Array Experiment

Chromatin immunoprecipitation followed by mRNA microarray analysis (ChIP-array) has become a popular procedure for studying genome-wide protein–DNA interactions and transcription regulation. However, it can only map the probable protein–DNA interaction loci within 1–2 kilobases resolution. Liu et al. [39] introduced a computational method, Motif Discovery scan (MDscan), that examines the ChIP-array selected sequences and searches for DNA sequence motifs representing the protein–DNA interaction sites. MDscan combines the advantages of two widely adopted motif search strategies, word enumeration, and position-specific weight matrix updating and incorporates the ChIP-array ranking information to accelerate searches and enhance their success rates.

Consider a set of n DNA sequences selected from ChIP-array experiments, ranked according to their ChIP-array enhancement scores, from the highest to the lowest. MDscan first scrutinizes the top t (e.g., 5–50) sequences in the ranking to form a set of promising candidates. Assuming the protein-binding motif to be of width w, MDscan enumerates each non-redundant w-mer (seed) that appears in both strands of the top t sequences and searches for all w-mers in the top t sequences with at least m base pairs matching the seed (called m-matches). m is determined so that the chance that two randomly generated w-mers are m-matches of each other is smaller than a certain threshold, such as 2 %. For each seed, MDscan finds all the m-matches in the top t sequences and uses them to form a motif weight matrix. If the expected number of bases per motif site in the top sequences can be estimated, the following approximate maximum a posteriori (MAP) scoring function can be used to evaluate a matrix:

$$\frac{x_m}{w} \times \left[\sum_{i=1}^{w} \sum_{j=a}^{t} p_{ij} \log p_{ij} - \frac{1}{x_m} \sum_{\text{all segments}} \log \left(p_0(s) \right) - \log \left(\text{expected bases} / \text{sites} \right) \right],$$

where x_m is the number of m-matches aligned in the motif, p_{ij} is the frequency of nucleotide j at position i of the motif matrix, and $p_0(s)$ is the probability of generating the m-match s from the background model. A Markov background model is used and estimated from all the intergenic regions of a genome. When the expected number of sites in the top sequences is unknown, the motif matrix can also be evaluated by

$$\frac{\log (x_m)}{w} \times \left[\sum_{i=1}^{w} \sum_{j=a}^{t} p_{ij} \log p_{ij} - \frac{1}{x_m} \sum_{\text{all segments}} \log \left(p_0(s) \right) \right].$$

After computing the scores for all the w-mer motifs established in this step, the highest 10–50 "seed" candidate motifs are retained for further improvement in the next step. In the motif improvement step, every retained candidate motif weight matrix is used to scan all the w-mers in the remaining sequences. A new w-mer is added into a candidate weight matrix if and only if the motif score of that matrix is increased. Each candidate motif is further refined by reexamining all the segments that are already included in the motif matrix during the updating step. A segment is removed from the matrix if doing so increases the motif score. The aligned segments for each motif usually stabilize within ten refinement iterations. MDscan reports the highest-scoring candidate motifs as the protein–DNA interaction motif.

4.6.2   Regression Analysis of Transcription Regulation

In Sect. 4.3, we talked about genome expression data and clustering analysis using these data. In Sect. 4.5 we asked how co-expression and co-regulation agree with each other and then explained some methods for discovering significant "words/motifs" in upstream sequences of co-regulated genes. In addition to discovering motifs based on sequence and expression data, we now want to know how well the upstream sequence motifs or other features can predict genes' expressions.

Bussemaker et al. [40] proposed a novel method for TFBM discovery via the association of gene expression values with abundance of certain oligomers. They first conducted word enumeration and then used regression to check whether the genes whose upstream sequences contain a set of words have significant changes in their expression.

Conlon et al. [41] presented an alternative approach, MotifRegressor, operating under the assumption that, in response to a given biological condition, the effect of a TFBM is approximately linear, the strongest among genes with the most dramatic increase or decrease in mRNA expression. The method combines the advantages of matrix-based motif finding and oligomer motif-expression analysis, resulting in high sensitivity and specificity. MDscan introduced in Sect. 4.6.1 is first used to generate a large set of candidate motifs that are enriched (maybe only slightly; it is not necessary to be stringent here) in the promoter regions (DNA sequences) of genes with the highest fold change in mRNA level relative to a control condition. How well the upstream sequence of a gene g matches a motif m, in terms of both degree of matching and number of sites, is determined by the following likelihood ratio function:

$$S_{mg} = \log_2 \left[\sum_{x \in X_{wg}} \Pr(x \text{ from } \theta_m) / \Pr(x \text{ from } \theta_0) \right], \tag{4.7}$$

where θ_m is the probability matrix of width, θ_0 represents the third-order Markov model estimated from all of the intergenic sequences (or all the sequences in the given data set), and X_{mg} is the set of all w-mers in the upstream sequence of gene g. For each motif reported by MDscan, MotifRegressor first fits the simple linear regression:

$$Y_g = \alpha + \beta_m S_{mg} + \varepsilon_g,$$

where Y_g is the $\log_2$-expression value of gene g, S_{mg} is defined in (4.7), and ε_g is the gene-specific error term. The baseline expression α and the regression coefficient β_m will be estimated from the data. A significantly positive or negative β_m indicates that motif m (and its corresponding binding TF) is very likely responsible for the observed gene expression changes.

The candidate motifs with significant p-values ($p \le 0.01$) for the simple linear regression coefficient β_m are retained and used by the stepwise regression procedure to fit a multiple regression model:

$$Y_g = \alpha + \sum_{m=1}^{M} \beta_m S_{mg} + \varepsilon_g. \tag{4.8}$$

Stepwise regression begins with model (4.8) with only the intercept term and adds at each step the motif that gives the largest reduction in residual error. After adding each new motif m, the model is checked to remove the ones whose effects have been sufficiently explained by m. The final model is reached when no motif can be added with a statistically significant coefficient.

Since the above procedure involves three steps, motif finding, simple linear regression, and stepwise regression, it is infeasible to compute statistical significance of the final regression result analytically. We resort to a permutation-type testing procedure. More precisely, simulation data sets (1,500) are first constructed from the original data set by only shuffling the gene names. Then, we proceed to "discovery" about 300 motifs in the "top" 100 genes using MDscan. Next, MotifRegressor is used to "reconfirm" the finding. Finally, 1,398 of the 40,324 (3.5 %) motifs have p-values less than 0.01, which is a slight inflation (expected to have only 1 % if there is no selection bias) due to the motif finding in top sequences. In contrast, 235 out of the 322 (73 %) motifs found in the read data have p-values less than 0.01.

There are some additional challenges that require for more sophisticated methods than linear regression, such as nonlinear relationship, large number of possible variables, and the generation of meaningful variables. Refined motif model accounting for within-motif dependence [42] and machine learning strategies like the boosting method [43] have been used to build more accurate binding model for transcription factors. Zhong et al. [44] proposed to use regularized sliced inverse regression (RSIR) to select relevant motifs. Compared with linear regression, RSIR is efficient in computation, very stable for data with high dimensionality and high collinearity, and improves motif detection sensitivities and specificities by avoiding inappropriate model specifications.

4.6.3   Regulatory Role of Histone Modification

Gene activities in eukaryotic cells are regulated by a concerted action of and interaction between transcription factors and chromatin structure. The basic repeating unit of chromatin is the nucleosome, an octamer containing two copies each of four core histone proteins. While nucleosome occupancy in promoter regions typically occludes transcription factor binding, thereby repressing global gene expression, the role of histone modification is more complex. Histone tails can be modified in various ways, including acetylation, methylation, phosphorylation, and ubiquitination. Even the regulatory role of histone acetylation, the best characterized modification to date, is still not fully understood. It is thus important to assess the global impact of histone acetylation on gene expression, especially the combinatory effect of histone acetylation sites and the confounding effect of sequence-dependent gene regulation and histone occupancy.

Yuan et al. [45] proposed a statistical approach to evaluate the regulatory effect of histone acetylation by combining available genome-wide data from histone acetylation, nucleosome occupancy, and transcriptional rate. The combined transcriptional control by TFBMs, nucleosome occupancy, and histone acetylation is modeled as follows:

$$y_i = \alpha + \sum_j \beta_j x_{ij} + \sum_k \eta_k z_{ik} + \delta w_i + \varepsilon_j,$$

where y_i is the transcription rate of gene i, the x_{ij} values are the three histone acetylation levels (corresponding to H_3K_9, H_3K_{14}, and H_4, respectively), the z_{ik}

Table 4.5 Model performance (adjusted R^2) with different covariates

	Adjusted R-square statistic		
	Acetyl alone	Acetyl + MD scan motifs	Acetyl + Align ACE motifs
NoH$_3$ or H$_4$	0	0.2094	0.2223
H$_3$ (K$_9$, K$_{14}$)	0.2443	0.3540	0.2546
H$_4$	0.1342	0.2979	0.3128
H$_3$ and H$_4$	0.2460	0.3535	0.3543

values are the corresponding scores to the selected motifs, and w_i is the nucleosome occupancy level. RSIR can also be used for model-free motif (feature) selection. It assumes that gene i's transcription rate y_i and its sequence motif scores $\mathbf{x}_i = (x_{i1}, \ldots, x_{iM})^{\mathrm{T}}$ are related as

$$y_i = f\left(\boldsymbol{\beta}_1^{\mathrm{T}} \mathbf{x}_1, \ldots, \boldsymbol{\beta}_k^{\mathrm{T}} \mathbf{x}_k, \varepsilon_i\right), \qquad (4.9)$$

where $f(\bullet)$ is an unknown (and possibly nonlinear) function, $\boldsymbol{\beta}_l = (\beta_{l1}, \ldots, \beta_{lM})^{\mathrm{T}}$ are vectors of linear coefficients, and ε_i represents the noise. The number k is called the dimension of the model. A linear regression model is a special one-dimensional case of (4.9). RSIR estimates both k and the $\boldsymbol{\beta}_l$ values without estimating $f(\bullet)$. Since many entries of the β_{lj} values are close to 0, which implies that the corresponding motif scores contribute very little, only those motifs whose coefficient β_{lj} is significantly nonzero are retained.

Table 4.5 shows the R^2 (referring to the adjusted R-square statistic, which measures the fraction of explained variance after an adjustment of the number of parameters fitted) of the various models. One can see that a simple regression of transcription rates against histone acetylation without considering any other factors gave an R^2 of 0.2049, implying that about 20 % of the variation of the transcription rates is attributable to histone acetylation. In contrast, the comprehensive model with all the variables we considered bumped up the R^2 to 0.3535, indicating that the histone acetylation does have a significant effect on the transcription rate, although not as high as that in the naïve model.

To confirm that the above results indicate intrinsic statistical associations rather than artifacts of the statistical procedure, the model is validated by testing whether applying the above procedure to the random inputs would yield substantially worse performance than applying it to the real data. We generated 50 independent samples by random permutations of the original data. As shown in Fig. 4.26, the R^2 for these randomized data are much smaller than for the real data.

4.7   Protein Structure and Proteomics

Despite the success of DNA microarrays in gene expression profiling, it is the activity of encoded proteins that directly manifest gene function. It is well known that protein amino acid sequence determines the structure and the function of the

Fig. 4.26 Model validation
by randomization

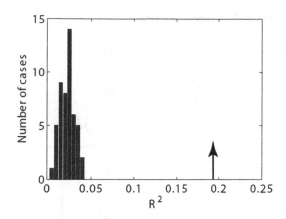

protein. Many protein sequences fold spontaneously to native structure having lowest energy conformation. In this section, we will briefly discuss the problem of protein structure prediction as well as some recent development of protein chip technology.

4.7.1   Protein Structure Prediction

Predicting a protein's tertiary structure from its primary amino acid sequences is a long-standing problem in biology. Two major difficulties have been challenging researchers: the design of appropriate energy functions and the exploration of the vast space of all possible structures.

In general, there are mainly three kinds of structure prediction methods: homology modeling, threading, and ab initio prediction. Homology modeling is used to predict the structure of sequences with high homology to sequences with known structures. When sequence homology is >70 %, high-resolution models are possible. More specifically, using BLAST or other sequence comparison algorithms, we first find for the target sequence a protein with known structure in Protein Data Bank (PDB) and with the highest possible sequence homology (>25–30 %) to the target. The known structure is then used as the starting point to create a three-dimensional structure model of the sequence. Threading method is applicable for sequence with no identity (≤30 %) to sequences of known structure. Given the sequence and a set of folds observed in PDB, one needs to see if any of the sequences could adopt one of the known folds. Ab initio prediction is used for sequence having no homology with sequence of known structure. The three-dimensional structure is predicted from "first principles" based on energetic or statistical principles. The ab initio method is difficult to implement because of the enormous number of protein conformations needed to be searched. Main strategies for exploring a complex configuration space include molecular dynamics simulations, Markov chain Monte Carlo methods, and other heuristics-based approaches.

Fig. 4.27 A sample spectrum

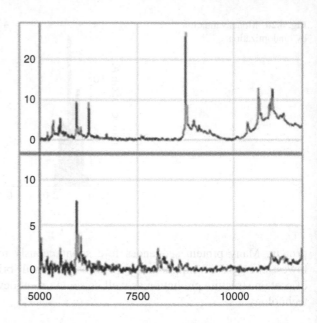

4.7.2 Protein Chip Data Analysis

Protein chip techniques, or as is often termed "matrix-assisted/surface-enhanced laser desorption/ionization time-of-flight mass spectrometry (MALDI/SELDI-TOF-MS)," is a rapid, high-throughput biomarker tool for biomedical scientists. To run an experiment on a MALDI-TOF instrument, the biological sample is first mixed with an energy-absorbing matrix (EAM). This mixture is crystallized onto a metal plate. (The commonly used surface-enhanced lasers desorption and ionization (SELDI) is a variant of MALDI that incorporates additional chemistry on the surface of the metal plate to bind specific classes of proteins.) The plate is inserted into a vacuum chamber, and the matrix crystals are struck with pulses from a nitrogen laser. The matrix molecules absorb energy from the laser, transfer it to the proteins causing them to desorb and ionize, and produce a plume of ions in the gas phase. This process takes place in the presence of an electric field, which accelerates the ions into a flight tube where they drift until they strike a detector that records the time of flight.

A typical data set arising in a clinical application of mass spectrometry contains tens or hundreds of spectra; each spectrum contains many thousands of intensity measurements representing an unknown number of protein peaks (Fig. 4.27). Any attempt to make sense of this volume of data requires extensive processing in order to identify the locations of peaks and to quantify their sizes accurately. Typical processing steps are as follows: filtering or denoizing removes random noise, typically electronic or chemical in origin. Peak detection identifies locations on the time or m/z scale that correspond to specific proteins or peptides striking the detector. Peak alignment across samples is required because neither calibration nor peak detection is perfect. Finally, samples are clustered based on tens to hundreds of peaks (features).

References

1. Metropolis N, Rosenbluth AW, Rosenbluth MN et al (1953) Equation of state calculations by fast computing machines. J Chem Phys 21(6):1087–1092
2. Hasting WK (1970) Monte Carlo sampling methods using Markov chains and their applications. Biometrika 57(1):97–109
3. Geman S, Geman D (1984) Stochastic relaxation, Gibbs distributions and the Bayesian restoration of images. IEEE Trans Pattern Anal Mach Intell 6:721–741
4. Liu JS (2001) Monte Carlo strategies in scientific computing. Springer, New York
5. Dudoit S, Yang YH, Callow MJ et al (2002) Statistical methods for identifying genes with differential expression in replicated cDNA microarray experiments. Stat Sin 12:111–139
6. Yang YH, Dudoit S, Luu P et al (2002) Normalization for cDNA microarray data: a robust composite method addressing single and multiple slide systematic variation. Nucleic Acids Res 30(4):e15
7. Tseng GC, Oh M-K, Rohlin L et al (2001) Issues in cDNA microarray analysis: quality filtering, channel normalization, models of variations and assessment of gene effects. Nucleic Acids Res 29(12):2549–2557
8. Li C, Wong WH (2001) Model-based analysis of oligonucleotide arrays: expression index computation and outlier detection. Proc Natl Acad Sci USA 98(1):31–36
9. Benjamini Y, Hochberg Y (1995) Controlling the false discovery rate: a practical and powerful approach to multiple testing. J R Stat Soc Ser B (Methodol) 57(1):289–300
10. Efron B, Tibshirani R, Storey JD et al (2001) Empirical Bayes analysis of a microarray experiment. J Am Stat Assoc 96:1151–1160
11. Calinski T, Harabasz J (1998) A dendrite method for cluster analysis. Commun Stat 3:1–27
12. Hartigan J (1975) Clustering algorithms. Wiley, New York
13. Tibshirani R, Walther G, Hastie T (2001) Estimating the number of clusters in a Data Set via the Gap statistic. J R Stat Soc Ser B (Stat Methodol) 63(2):411–423
14. Tseng GC, Wong WH (2005) Tight clustering: a resampling-based approach for identifying stable and tight patterns in data. Biometrics 61(1):10–16
15. Kohonen T (1989) Self-organization and associative memory, 3rd edn. Springer, Berlin
16. Bhattacharjee A, Richards WG, Staunton J et al (2001) Classification of human lung carcinomas by mRNA expression profiling reveals distinct adenocarcinoma subclasses. Proc Natl Acad Sci USA 98(24):13790–13795
17. Lazzeroni L, Owen A (2002) Plaid models for gene expression data. Stat Sin 12:61–86
18. Cheng Y, Church G (2000) Biclustering of expression data. In: Proceedings of the 8th international conference on intelligent system for molecular biology (ISMB2000), San Diego, 19–23 Aug 2000, pp 93–103
19. Bergmann S, Ihmels J, Barkai N (2003) Iterative signature algorithm for the analysis of large-scale gene expression data. Phys Rev E 67(3):031902
20. Hastie T, Tibshirani R, Eisen M et al (2000) 'Gene shaving' as a method for identifying distinct sets of genes with similar expression patterns. Genome Biol 1(2):RESEARCH0003
21. Fix E, Hodges JL (1951) Discriminatory analysis: non-parametric discrimination: consistency properties. USAF School of Aviation Medicine, Randolph Field
22. Brunak S, Engelbrecht J, Knudsen S (1991) Prediction of human mRNA donor and acceptor sites from the DNA sequence. J Mol Biol 220(1):49–65

23. Hebsgaard SM, Korning PG, Tolstrup N et al (1996) Splice site prediction in *Arabidopsis thaliana* pre-mRNA by combining local and global sequence information. Nucleic Acids Res 24(17):3439–3452
24. Khan J, Wei JS, Ringner M et al (2001) Classification and diagnostic prediction of cancers using gene expression profiling and artificial neural networks. Nat Med 7(6):673–679
25. Dayhoff MO (1969) Atlas of protein sequence and structure. National Biomedical Research Foundation, Washington, DC
26. Henikoff S, Henikoff JG (1992) Amino acid substitution matrices from protein blocks. Proc Natl Acad Sci USA 89(22):10915–10919
27. Bairoch A (1991) PROSITE: a dictionary of sites and patterns in proteins. Nucleic Acids Res 19:2241–2245
28. Altschul SF, Gish W, Miller W et al (1990) Basic local alignment search tool. J Mol Biol 215(3):403–410
29. Thompson JD, Higgins DG, Gibson TJ (1994) CLUSTAL W: improving the sensitivity of progressive multiple sequence alignment through sequence weighting, position-specific gap penalties and weight matrix choice. Nucleic Acids Res 22(22):4673–4680
30. Liu JS, Neuwald AF, Lawrence CE (1999) Markovian structures in biological sequence alignments. J Am Stat Assoc 94:1–15
31. Liu JS (1998) The collapsed Gibbs sampler with applications to a gene regulation problem. J Am Stat Assoc 89:958–966
32. Eddy SR (1998) Profile hidden Markov models. Bioinformatics 14(9):755–763
33. Hertz GZ, Hartzell GW III, Stormo GD (1990) Identification of consensus patterns in unaligned DNA sequences known to be functionally related. Bioinformatics 6(2):81–92
34. Liu JS, Neuwald AF, Lawrence CE (1995) Bayesian models for multiple local sequence alignment and Gibbs sampling strategies. J Am Stat Assoc 90:1156–1170
35. Lawrence CE, Altschul SF, Boguski MS et al (1993) Detecting subtle sequence signals: a Gibbs sampling strategy for multiple alignment. Science 262(5131):208–214
36. Liu JS, Lawrence CE (1999) Bayesian inference on biopolymer models. Bioinformatics 15(1):38–52
37. McCue LA, Thompson W, Carmack CS et al (2001) Phylogenetic footprinting of transcription factor binding sites in proteobacterial genomes. Nucleic Acids Res 29(3):774–782
38. Gupta M, Liu JS (2005) De novo cis-regulatory module elicitation for eukaryotic genomes. Proc Natl Acad Sci USA 102(20):7079–7084
39. Liu XS, Brutlag DL, Liu JS (2002) An algorithm for finding protein-DNA binding sites with applications to chromatin-immunoprecipitation microarray experiments. Nat Biotechnol 20(8):835–839
40. Bussemaker HJ, Li H, Siggia ED (2001) Regulatory element detection using correlation with expression. Nat Genet 27(2):167–174
41. Conlon EM, Liu XS, Lieb JD et al (2003) Integrating regulatory motif discovery and genome-wide expression analysis. Proc Natl Acad Sci USA 100(6):3339–3344
42. Zhou Q, Liu JS (2004) Modeling within-motif dependence for transcription factor binding site predictions. Bioinformatics 20(6):909–916
43. Hong P, Liu XS, Zhou Q et al (2005) A boosting approach for motif modeling using ChIP-chip data. Bioinformatics 21(11):2636–2643
44. Zhong W, Zeng P, Ma P et al (2005) RSIR: regularized sliced inverse regression for motif discovery. Bioinformatics 21(22):4169–4175
45. Yuan G-C, Ma P, Zhong W et al (2006) Statistical assessment of the global regulatory role of histone acetylation in *Saccharomyces cerevisiae*. Genome Biol 7(8):70
46. Needleman SB, Wunsch CD (1970) A general method applicable to the search for similarities in the amino acid sequence of two proteins. J Mol Biol 48(3):443–453
47. Smith TF, Waterman MS (1981) Identification of common molecular subsequences. J Mol Biol 147(1):195–197

48. Thompson W et al (2004) Decoding human regulatory circuits. Genome Res 14(10a): 1967–1974
49. Zhou T et al (2004) Genome-wide identification of NBS genes in japonica rice reveals significant expansion of divergent non-TIR NBS-LRR genes. Mol Genet Genom 271(4):402–415
50. Gupta M, Liu JS (2005) De novo cis-regulatory module elicitation for eukaryotic genomes. Proc Natl Acad Sci U S A 102(20):7079–7084

Chapter 5
Algorithms in Computational Biology

Tao Jiang and Jianxing Feng

5.1 Introduction

What is an algorithm? An algorithm is a sequence of unambiguous instructions for solving a problem, i.e., for obtaining a required output for any legitimate input in a finite amount of time. Figure 5.1 gives illustrative description of the relation between problem, algorithm and, the input and output of an algorithm.

Let's check these four components of a famous algorithm, the Euclid's Algorithm. This algorithm solves the problem of finding $gcd(m, n)$, the greatest common divisor of two nonnegative, not both zero integers m and n. For example, gcd(60,24)=12, gcd(60,0)=60, gcd(0,0)=?.

The Euclid's algorithm is based on repeated applications of equality $gcd(m,n) = gcd(n, m \bmod n)$ until the second number becomes 0, which makes the problem trivial. For example, gcd(60,24)=gcd(24,12)=gcd(12,0)=12.

We give two descriptions of Euclid's algorithm.

Algorithm 1 Euclid's algorithm (description 1)

1: If n=0, return m and stop; otherwise go to step 2.
2: Divide m by n and assign the value of the remainder to r.
3: Assign the value of n to m and the value of r to n. Go to step 1.

As we can see, an algorithm could have different descriptions. These descriptions could be stated in different languages but bear the same essence. However,

---

T. Jiang (✉)
Department of Computer Science and Engineering, University of California, Riverside, CA 92521, USA
e-mail: jiang@cs.ucr.edu

J. Feng
Department of Computer Science and Technology, Tsinghua University, Beijing 100084, China

R. Jiang et al. (eds.), *Basics of Bioinformatics: Lecture Notes of the Graduate Summer School on Bioinformatics of China*, DOI 10.1007/978-3-642-38951-1_5, © Tsinghua University Press, Beijing and Springer-Verlag Berlin Heidelberg 2013

Fig. 5.1 Illustration of an
algorithm

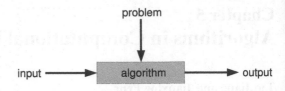

Algorithm 2 Euclid's algorithm (description 2)
___
1: **while** $n \neq 0$ **do**
2:     $r \leftarrow m \bmod n$
3:     $m \leftarrow n$
4:     $n \leftarrow r$
5: **end while**
___

in either case, the algorithm should be unambiguous. That is, we know exactly what's the effect of every step in the algorithm.

Here is a question: how efficient is the algorithm? Before we answer this question, let's look at another way of calculating gcd(m,n).

Algorithm 3 Naive algorithm for gcd(m,n)
___
1: Assign the value of min m,n to t.
2: Divide m by t. If the remainder is 0, go to step 3; otherwise, go to step 4.
3: Divide n by t. If the remainder is 0, return t and stop; otherwise, go to step 4.
4: Decrease t by 1 and go to step 2.
___

Suppose $m > n$, it is not difficult to see that we have to decrease t m times if gcd(m,n) = 1. In this case, we have to do division $2m$ times in steps 2 and 3. How about the Euclid's algorithm? A little math could show that in every two rounds, the number m would decrease at least half of its size, which means that the algorithm would terminate in $2\log_2 m$ rounds approximately. So we have to do $2\log_2 m$ times of divisions at most. We only count the number of divisions because it is the most time-consuming part in the algorithm compared to addition or subtraction.

In most cases, the constant involved in the running time of the algorithm is not important. The running time also has formal parasynonyms and, time complexity. In above examples, the time complexity of the Euclid and naive algorithms are $O(m)$ and $O(\log m)$, respectively. The formal definition of O-notation is as follows: $f(n) = O(g(n))$ if and only if there exist two positive constants c and n_0 such that $0 \leq f(n) \leq cg(n)$ for all $n \geq n_0$.

For example, $5n + 108 = O(n)$ and $2n = O(n\log n)$. O-notation is an asymptotic upper bound. It is more than just omitting a constant in the formula. More specifically, we write $2n = O(n\log n)$, because in some cases, we can only know that the time complexity of an algorithm is less than $cn\log n$ for some constant c, but we don't know the exact time complexity which may be $2n$. With the O-notation, we can safely say that the time complexity of the algorithm is $O(n\log n)$.

Fig. 5.2 Time complexity
of an algorithm

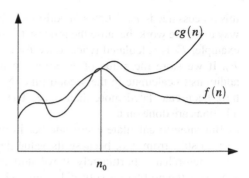

The time complexity measures the efficiency of algorithms. Euclid's algorithm is more efficient than the naive algorithm because it has a small time complexity.

To summarize, an algorithm has the following important properties:

1. Can be represented in various forms
2. Unambiguity/clearness
3. Effectiveness
4. Finiteness/termination
5. Correctness

Finiteness and correctness of an algorithm is self-clear. No one wants an algorithm to give a wrong result or an algorithm that runs forever without giving a final result. Figure 5.2 illustrates different complexities of an algorithm.

5.2 Dynamic Programming and Sequence Alignment

5.2.1 The Paradigm of Dynamic Programming

5.2.1.1 Overlapping Subproblems

Before an abstract description is provided, let's go through an illustrative example. The example is to calculate the famous Fibonacci number. Fibonacci numbers are defined by the following recurrences:

$$F_n = \begin{cases} 0 & \text{if } n = 0 \\ 1 & \text{if } n = 1 \\ F_{n-1} + F_{n-2} & \text{if } n > 1 \end{cases} \tag{5.1}$$

The problem is how to compute any specified number in this sequence, for example. To calculate F_{10}, we have to know F_9 and F_8 by the third recurrence. Recursively, to calculate F_9, we have to know F_8 and F_7, as showed in picture. To calculate F_n, this process needs F_{n-1} times of addition, which is $O(2^n)$. Clearly, it is not a clever way to do the work, because the process has done many redundant calculations. For example, F_8 is calculated twice: once for calculating F_{10} and once for calculating F_9. If we store the value of F_8 after the first calculation, we can avoid all the additions to calculate in the second time. Not only F_8, from F_2 to F_8, all of them have been calculated more than once. The less the index is, the more redundant additions are done on it.

But, how to calculate every number in the sequence only once? It is not very easy to start from F_{10}, because its value depends on two numbers that we don't know beforehand. Fortunately, if we start from calculating small numbers in the sequence, the problem is solved, because whenever we want to calculate a number, all the numbers before it have been calculated. More specifically, based on F_0 and F_1, F_2 is calculated, and then F_3 is calculated based on F_1 and F_2, and so on.

The second method to calculate a Fibonacci number is dynamic programming. One gradient of DP is clear from the above example. That is overlapping subproblems. In the word of above example, the overlapping subproblem of F_{10} and F_9 is F_8. A dynamic programming algorithm tries to calculate every subproblem once.

5.2.1.2 Optimal Substructure

The second gradient of DP is optimal substructure. We will explain it by another problem, maximum-sum interval. Given a sequence of real numbers $a_1, a_2, \ldots a_n$, the problem is to find a consecutive subsequence with the maximum sum. For each position, we can compute the maximum-sum interval starting at that position in $O(n)$ time. Therefore, a naive algorithm runs in $O(n^2)$ time. Using DP, we can get a faster algorithm as follows.

Define $S(i)$ to be the maximum sum of the intervals ending at position i. We have:

$$S(i) = a_i + \max\{S(i-1), 0\} \tag{5.2}$$

The final result is given by $\max_i\{S(i)\}$. The time complexity is $O(n)$. This recurrence says that if $S(i-1) < 0$, concatenating a_i with its previous interval gives less sum than a_i itself. If $S(i-1) \geq 0$, the optimal solution in the problem of $S(i)$ contains the optimal solution of the subproblem of $S(i-1)$, say, the optimal substructure. Here, $S(i-1)$ is considered as a subproblem of $S(i)$. Because of optimal substructure, the DP could calculate the optimal solutions of subproblems first and then merge them to get the optimal solution of larger problems.

As we can see, the optimal substructure is closely related to the recurrence set up for the DP. Once we have found one of them, we found the other. In the problem of calculating the Fibonacci number, this recurrence is given alone with the problem; however, in the second problem, we have to find such a recurrence, which is the general situation of designing a DP algorithm.

Table 5.1 Example of maximum-sum interval

	9	−3	1	7	−15	2	3	−4	2	−7	6	−2	8	4	−9
$S(i)$	9	6	7	14	−1	2	5	1	3	−4	6	4	12	16	7
	↑	←	←	←	←	↑	←	←	←	←	↑	←	←	←	←

5.2.1.3 Designing a Dynamic Programming Algorithm

The development of a dynamic programming algorithm has three basic components:

1. A recurrence relation (for defining the value/cost of an optimal solution). An easy but important part of the recurrence is the initial assignment. The recurrence divides a problem into subproblems, but this process cannot go on forever. It will reach some trivial subproblems, i.e., F_0 and F_1 in the first problem and $S(1)$ in the second problem. Generally, the solution of these trivial problems are provided alone in the input of the problem or can be solved directly.
2. A tabular computation (for computing the value of an optimal solution). The calculation of the recurrence could be considered as a process of filling a table. Starting from giving initial assignment for trivial problems, this process calculates a larger problem based on the solution of its subproblems which are already solved.
3. A backtracing procedure (for delivering an optimal solution). In most situations, we not only need to calculate the optimal value but also need to get the optimal solution which leads to the optimal value. This could be easily handled by remembering which optimal solution of the subproblems lead to the optimal solution of the larger problem when the recurrence is calculated.

We use an example of maximum-sum interval to illustrate these three components. Suppose the input of the problem is the sequence: 9 −3 1 7 −15 2 3 −4 2 −7 6 −2 8 4 −9. The recurrence has been established as the last subsection. $S(1)$ is assigned directly and $S(2)$ is calculated based on $S(1)$ and so on. ↑ means that the optimal solution of the corresponding subproblem starts from scratch and ← means that the optimal solution of the corresponding subproblem concatenates to the optimal solution of the former subproblem. So the optimal solution of the above example is the subsequence of 6 −2 8 4, which gives the optimal value 16 (Table 5.1).

5.2.2 Sequence Alignment

In bioinformatics, sequence alignment is a way of estimating the similarity between two or more DNA, RNA, or protein sequences. Take DNA sequences as an example. A DNA sequence is a string consisting of four characters: A, C, G and T. As a species evolves, the DNA sequence may change by deleting or inserting one or more

Table 5.2 Example of sequence alignment of two toy DNA sequences

C	–	–	–	T	T	A	A	C	T		
C	G	G	A	T	C	A	–	–	T		
+8	−3	−3	−3	+8	−5	+8	−3	−3	+8	=	+12

of the four characters, which are simulated by "gaps" in the alignment; the DNA sequence may also mute at some positions, which are simulated by "mismatches" in the alignment. Sequence alignment aims to set up a mapping between two or more sequences such that the "gaps" and "mismatches" are minimized locally or globally. Table 5.2 gives an example of aligning two toy DNA sequences: CTTAACT and CGGATCAT.

In this example, we give different scores when two different characters are aligned. Matches are given positive scores while mismatches and gaps are given negative scores such that more similar sequences could be aligned with a higher score.

- Match: +8 ($w(x, y) = 8$, if $x = y$)
- Mismatch: −5 ($w(x, y) = -5$, if $x \neq y$)
- Each gap symbol: −3 ($w(-, x) = w(x, -) = -3$)

5.2.2.1 Scoring Matrices: PAM and BLOSUM

How to score the alignment decides the quality of the alignment that maximizes the score. Because DNA sequences are less conserved than protein sequences, it is less effective to comparing coding regions at nucleotide level. Therefore, the emphasis is on scoring the amino acid substitution. Two types of substitution (scoring) matrices are widely used: point accepted mutation (PAM) introduced by Dayhoff and Blocks substitution matrix (BLOSUM).

PAM is a series of 20-by-20 matrices. Each row or column corresponds to an amino acid. The value in a cell (X, Y) is the probability of X changes to (substituted by) Y. The substitution is also known as point mutation.

5.2.2.2 Global Alignment Versus Local Alignment

Once the scoring matrix is fixed, global alignment asks for the alignment such that the score is maximized globally. Let $A = a_1 a_2 \ldots a_m$ and $B = b_1 b_2 \ldots b_n$ and $S_{i,j}$ be the score of an optimal alignment between $a_1 a_2 \ldots a_i$ and $b_1 b_2 \ldots b_j$. $S_{i,j}$ can be computed as follows:

$$S_{i,j} = \max \begin{cases} S_{i-1,j} & + w(a_i, -) \ i > 0 \\ S_{i,j-1} & + w(-, b_j) \ j > 0 \\ S_{i-1,j-1} + w(a_i, b_j) \ i > 0, j > 0 \end{cases} \quad (5.3)$$

$S_{0,j}$ and $S_{i,0}$ are not calculated by the recurrence but assign initially. For example, $S_{0,j}$ could be $\sum_{1 \leq k \leq j} w(-, b_k)$ and $S_{i,0}$ could be $\sum_{1 \leq k \leq i} w(a_k, -)$. $S_{m,n}$ gives the score value of the optimal alignment.

Local alignment is very similar to global alignment on account of the recurrence. Let A and B be defined as before, and $S_{i,j}$ is the score of an optimal local alignment ending at a_i and b_j. With proper initializations, $S_{i,j}$ can be computed by

$$S_{i,j} = \max \begin{cases} 0 \\ S_{i-1,j} & + w(a_i, -) \ i > 0 \\ S_{i,j-1} & + w(-, b_j) \ j > 0 \\ S_{i-1,j-1} & + w(a_i, b_j) \ i > 0, j > 0 \end{cases} \tag{5.4}$$

The score of the optimal solution is no longer $S_{m,n}$ but $\max_{i,j} S_{i,j}$. The essence of local alignment is the same as that of the problem of maximum-sum interval.

5.2.2.3   Practical Tools

The dynamic programming algorithm could find the best k local/global alignments, if the best k local/global alignments are recorded in each step. The famous Smith–Waterman algorithm is a dynamic programming algorithm. Unfortunately, dynamic programming-based algorithm performs poorly on large dataset, for example, database searching. On such large dataset, heuristic algorithms provide good options, which run significantly faster than the dynamic programming-based algorithm under the cost of the suboptimality of the solution. Two such widely used algorithms are FASTA and BLAST.

5.3   Greedy Algorithms for Genome Rearrangement

In the 1980s, Jeffrey Palmer studied evolution of plant organelles by comparing mitochondrial genomes of the cabbage and turnip. They found 99 % similarity between genes from the two plants. However, these surprisingly identical gene sequences differed in gene order. This study helped pave the way to analyzing genome rearrangements in molecular evolution.

5.3.1   Genome Rearrangements

Abstractly, a genome with n genes could be represented by a permutation of n numbers with each number representing a gene. Two genomes from different species with the same set of genes in different orders could be represented by

Table 5.3 Example
of reversal

0	3	−5	−4	2	−1	6
0	1	−2	4	5	−3	6
0	1	−2	3	−5	−4	6
0	1	2	3	−5	−4	6
0	1	2	3	4	5	6

two permutations. The problem of genome rearrangement is to find the minimum number of rearrangement operations needed to transform from one permutation to the other.

Some rearrangements would change the direction of genes in genomes. To further capture the reality in genome rearrangement, each number in the permutation could be leaded by a sign, ±, to reflect the direction of a gene in the genome. Table 5.3 shows a toy example. In this example, two genomes are represented by permutation $3, -5, -4, 2, -1$ and $1, 2, 3, 4, 5$ and the rearrangement operation is reversal, which cuts out a consecutive segment of the permutation, reverses the direction of this segment by reversing the order and the signs of numbers in the segment, and then pastes this segment back.

This example also shows two conventions:

1. The leading 0 and ending 6 in each row: They are dummy genes. The only function of them is to simplify the discussion such that the first and last number in the permutation could be handled easily.
2. The identity permutation in the last row: A simple substitution could simplify the problem of transforming permutation π_1 into π_2 by another problem of transforming π into identity permutation, where π comes from substituting every number and sign in π_1 by another number and sign, guided by π_2.

This simplified model asks to minimize the number of rearrangements. Nature, however, may not follow this parsimony strategy exactly. Even though, the parsimony result provides enough information of the relation between two concerned genomes. This model also doesn't take duplicated genes into consideration. If duplicated numbers are allowed, most of the corresponding problems become much harder.

Different rearrangements correspond to different problems. Some of them have polynomial algorithms but some of them do not. In the following, we sketch the result of three well-studied problems. Except the following three operations, there are several other operations: deletion, insertion, fusion, fission, transreversal, block interchange, etc. Various combinations of these operations were considered too.

5.3.1.1 Reversal

Given a signed permutation $\pi = \pi_1, \pi_2, \ldots \pi_n$, a reversal $\rho(i, j)$ $(i < j)$ transforms it to $\pi' = \pi'_1, \pi'_2, \ldots \pi'_n$, where $\pi'_k = \pi_k$ for $1 \le k < i$ or $k > j$; $\pi'_k = -\pi_{j+i-k}$ for $i \le k \le j$. For the unsigned reversal, we just omit all the signs involved in the definition. The problem of sorting by signed (unsigned) reversal asks a sequence of reversals which transform a given permutation to the identity permutation $1, 2, 3, \ldots, n$, such that the number of reversals is minimized.

A lot of work were devoted to this problem. To name a few. For the signed version of this problem, Hannenhalli and Pevzner gave the first polynomial algorithm with time complexity $O(n^4)$[1]. A series of later work improved the time complexity [2, 3] and got an algorithm running in $O(n^{3/2} \sqrt{\log n})$. If we don't require to get the reversals but care about the number of reversals only, an $O(n)$ algorithm handled this [4].

For the unsigned version, Caprara proved that it is NP-hard. Berman provided an 1.375-approximation algorithm, which improved earlier work, for example, Bafna and Pevzner [5] and Christie [6].

5.3.1.2 Translocation

Let $\pi = \pi_1, \pi_2, \ldots, \pi_n$ be a permutation of n integers; a transposition $\rho(i, j, k)$ acting on π inserts the interval $[\pi_i, \pi_{i+1}, \ldots, \pi_{j-1}]$ between π_{k-1} and π_k, $1 \leq i < j < k \leq n + 1$ and thus transforms π into $\pi' = \pi_1, \ldots, \pi_{i-1}, \pi_j, \ldots, \pi_{k-1}, \pi_i, \ldots \pi_{j-1}, \pi_k, \ldots, \pi_n$.

Bafna and Pevzner gave the first 1.5-approximation algorithm for this problem [7]. This approximation ratio was not improved for a long time and many work devoted to simply the underlying structure of this problem and provided faster approximation algorithm [8–10], until a 1.375-approximation proposed by Elias and Hartman [11]. The complexity of this problem is still open.

5.3.2 Breakpoint Graph, Greedy Algorithm and Approximation Algorithm

To study all the related problems of sorting permutations mentioned above, breakpoint graph is the most powerful, dominating, and widely used tool. Here, we introduce the basic definition of it. By applying breakpoint graph on sorting by transposition, we give an introductory explanation of greedy algorithm and approximation algorithm.

Let $\pi = \pi_1, \pi_2, \ldots, \pi_n$ be the permutation under consideration, where $\pi \in \{1..n\}$. Construct a graph $G(\pi)$ with node sets $\{0, 1, \ldots, 2n + 1\}$, black edge set $\{(2\pi_i, 2\pi_{i+1} - 1)\}$, and gray edge set $\{(2i, 2i + 1)\}$, where $0 \leq i \leq n$ and $\pi_{n+1} = n + 1$. Figure 5.3 is the breakpoint graph for an example permutation $4, 3, 2, 5, 1$.

Following the definition, in $G(\pi)$, there are $n + 1$ black edges and $n + 1$ gray edges. Every node in $G(\pi)$ is adjacent to one black and one gray edge; thus this graph could be uniquely decomposed into cycles with alternate color edges. The breakpoint graph for identity permutation contains $n + 1$ cycles, which is the

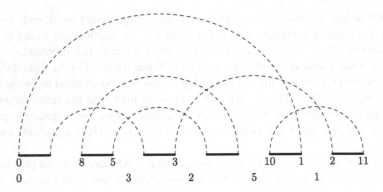

Fig. 5.3 The breakpoint graph for permutation 4, 3, 2, 5, 1. The first row of numbers correspond to nodes in the graph. The second row of numbers are the elements in the permutation with the leading 0 and ending 6, which are the dummy numbers. Each number in the permutation corresponds to two numbers in the first row

maximum number of cycles we can expect in a breakpoint graph corresponding to permutation with n elements. Denote the number of cycles in $G(\pi)$ with $c(\pi)$. If a transposition ρ transforms π into π', a case by case analysis [7] shows that $c(\pi') - c(\pi) \in \{-2, 0, 2\}$, which means that each time we can increase at most 2 cycles in the breakpoint graph by a transposition. Therefore, we get a lower bound on the minimum number $d(\pi)$ of transpositions needed to transform π into identity permutation. That is $d(\pi) \geq \frac{n+1-c(\pi)}{2}$. This lower bound serves as a critical gradient for an approximation algorithm.

However, on some "bad" permutation, none of the possible transposition can increase the number of cycles. Unfortunately, we cannot foresee it and cannot enumerate all the exponential-size possibilities either. A feasible strategy is to greedily select a transposition which increase the number of cycles if there is such a transposition and don't care about the possible bad consequences leaded by this choice. Generally, a greedy algorithm may not lead to the optimized solution. On this problem, however, one can prove that whenever a "bad" permutation is encountered, we can carefully select a transposition which keeps the number of cycles and guarantees that after the transposition, the permutation is not a "bad" one. In other words, this greedy strategy will increase 1 cycle by each transposition on average. Combined with the lower bound, this greedy algorithm is a 2-approximation algorithm.

Abstractly, on a minimization problem, let the optimal solution for an instance I of this problem be denoted by $S(I)$. Denote the solution of an approximation algorithm A with $A(I)$. In some cases, we don't know $S(I)$ but can calculate a lower bound $L(I)$ such that $L(I) \leq S(I)$. Then, the ratio $\frac{A(I)}{L(I)}$ is called the approximation ratio of algorithm A. On a maximization problem, we calculate the upper bound $U(I)$ instead of $L(I)$ and the ratio is reversed, i.e., $\frac{U(I)}{A(I)}$ such that it is always not less than 1.

Note that, for different rearrangement problems, the definitions of breakpoint graph differ slightly. But the essence is the same, i.e., transform the problem of rearrangement to the problem of calculating the changes on the breakpoint graph. To get a better approximation ratio on sorting by transposition, we need to further analyze the effect of a transposition operating on the breakpoint graph. For example, one can prove that whenever a "bad" permutation is encountered, a transposition could be carefully selected such that two transpositions which increase the number of cycles could be applied consecutively, which leads to a 1.5-approximation algorithm.

References

 1. Hannenhalli S, Pevzner PA (1999) Transforming cabbage into turnip: polynomial algorithm for sorting signed permutations by reversals. J ACM 46(1):1–27
 2. Kaplan H, Shamir R, Tarjan RE (1999) Faster and simpler algorithm for sorting signed permutations by reversals. SIAM J Comput 29:880–892
 3. Bergeron A (2005) A very elementary presentation of the Hannenhalli-Pevzner theory. Discret Appl Math 146(2):134–145
 4. Bader DA, Moret BME, Yan M (2001) A linear-time algorithm for computing inversion distance between signed permutations with an experimental study. J Comput Biol 8(5):483–491
 5. Bafna V, Pevzner PA (1996) Genome rearrangements and sorting by reversals. SIAM J Comput 25:272–289
 6. Christie DA (1998) A 3/2 approximation algorithm for sorting by reversals. In: Proceedings of the 9th annual ACM-SIAM symposium on discrete algorithms, Society for Industrial and Applied Mathematics Philadelphia, PA
 7. Bafna V, Pevzner PA (1998) Sorting by transpositions. SIAM J Discret Math 11(2):224–240
 8. Hartman T, Shamir R (2006) A simpler and faster 1.5-approximation algorithm for sorting by transpositions. Inf Comput 204(2):275–290
 9. Walter MET, Curado LRAF, Oliveira AG (2003) Working on the problem of sorting by transpositions on genome rearrangements. 14th annual symposium on combinatorial pattern matching (Morelia, Michoacn, Mexico). Lecture notes in computer science. New York, Springer, pp 372–383
10. Christie DA (1999) Genome rearrangement problems. PhD thesis, University of Glasgow
11. Elias I, Hartman T (2006) A 1.375-approximation algorithm for sorting by transpositions. IEEE/ACM Trans Comput Biol Bioinformatics 3(4):369–379

Chapter 6
Multivariate Statistical Methods in Bioinformatics Research

Lingsong Zhang and Xihong Lin

6.1 Introduction

In bioinformatics research, data sets usually contain tens of thousands of variables, such as different genes and proteins. Statistical methods for analyzing such multivariate data sets are important. In this chapter, multivariate statistical methods will be reviewed, and some challenges in bioinformatic research will be also discussed.

This chapter will be arranged as follows. Section 6.2 will review the multivariate normal distribution. Section 6.3 discusses classical multivariate hypothesis testing problems: one-sample and two-sample tests for multivariate data. Section 6.4 reviews principal component analysis (PCA), and Sect. 6.5 discusses factor analysis. Section 6.6 discusses linear discriminant analysis, and more classification methods are summarized in Sect. 6.7. Section 6.8 briefly reviews variable selection methods.

6.2 Multivariate Normal Distribution

The multivariate normal distribution is a multidimensional extension of the one-dimensional normal distribution, which is well known in basic statistical textbooks.

6.2.1 Definition and Notation

Assume $\mathbf{Y}_1, \ldots, \mathbf{Y}_n$ are n independent samples of a p-dimensional random vector with mean $\boldsymbol{\mu}$ and covariance matrix $\boldsymbol{\Sigma}$, where $\boldsymbol{\mu}$ is a $p \times 1$ vector and $\boldsymbol{\Sigma}$ is a $p \times p$

L. Zhang (✉) • X. Lin
Department of Biostatistics, Harvard University, Cambridge, MA 02115, USA
e-mail: xlin@hsph.harvard.edu

R. Jiang et al. (eds.), *Basics of Bioinformatics: Lecture Notes of the Graduate Summer School on Bioinformatics of China*, DOI 10.1007/978-3-642-38951-1_6,
© Tsinghua University Press, Beijing and Springer-Verlag Berlin Heidelberg 2013

positive definite symmetric matrix. The density function for the multivariate normal random variable **Y** is given by

$$f(\mathbf{Y}; \boldsymbol{\mu}, \boldsymbol{\Sigma}) = (2\pi)^{-p/2} |\boldsymbol{\Sigma}|^{-1/2} \exp\left\{ -\frac{1}{2}(\mathbf{Y} - \boldsymbol{\mu})^T \boldsymbol{\Sigma}^{-1}(\mathbf{Y} - \boldsymbol{\mu}) \right\}.$$

6.2.2 Properties of the Multivariate Normal Distribution

There are many attractive properties of the multivariate normal distribution. In this subsection, we will review some important properties. Other properties can be found in classical multivariate textbooks such as Anderson [1].

6.2.2.1 Normalize Multivariate Normal Random Vector

It is well known that a univariate normal random variable $Y \sim N(\mu, \sigma^2)$ can be normalized to have a standard normal distribution by $Z = (Y - \mu)/\sigma \sim N(0, 1)$. Similarly, a multivariate normal random vector $\mathbf{Y} \sim N(\boldsymbol{\mu}, \boldsymbol{\Sigma})$ can also be normalized to have a standard multivariate normal distribution. The normalization procedure is

$$\mathbf{Z} = \boldsymbol{\Sigma}^{-1/2}(\mathbf{Y} - \boldsymbol{\mu}) \sim N_p(0, \mathbf{I}).$$

6.2.2.2 Distribution of a Linear Transformation of a Multivariate Normal Random Vector

If **Y** is a random vector following a multivariate normal distribution, i.e., $\mathbf{Y} \sim N_p(\boldsymbol{\mu}, \boldsymbol{\Sigma})$, any linear transformation of **Y** is also multivariate normally distributed. Specifically, let **C** be a $c \times p$ matrix, then **CY** is a c-dimensional random vector from a multivariate normal distribution as

$$\mathbf{CY} \sim N_c(\mathbf{C}\boldsymbol{\mu}, \mathbf{C}\boldsymbol{\Sigma}\mathbf{C}^T).$$

6.2.2.3 The Moment Generating Function of the Multivariate Normal Distribution

The moment generating function usually brings us more understanding of the statistical properties of any given statistical distribution. For example, it can be used to calculate the moments. The moment generating function of a multivariate normal distribution can be easily derived as follows:

$$M(\mathbf{t}) = E(e^{\mathbf{Y}^T \mathbf{t}}) = \exp\{\mathbf{t}^T \boldsymbol{\mu} + \mathbf{t}^T \boldsymbol{\Sigma} \mathbf{t}\}.$$

6.2.2.4 The Mahalanobis Distance

The density function of $\mathbf{Y}$ is proportional to a function of a distance. It is easy to notice that

$$f(\mathbf{Y}) \propto \exp\left(\frac{-D^2}{2}\right), \tag{6.1}$$

where

$$D^2 = (\mathbf{Y} - \boldsymbol{\mu})^T \boldsymbol{\Sigma}^{-1}(\mathbf{Y} - \boldsymbol{\mu}). \tag{6.2}$$

Here, D^2 is called the *Mahalanobis distance* between $\mathbf{Y}$ and $\boldsymbol{\mu}$. Note that the Mahalanobis distance provides a very nice graphical (or geometric) interpretation in the high dimensional space. The contour of D^2 is a hyper-ellipsoid in the p-dimension space, which is centered at $\boldsymbol{\mu}$ with orientation of axes given by the eigenvectors of the population covariance matrix $\boldsymbol{\Sigma}$ (or the concentration matrix $\boldsymbol{\Sigma}^{-1}$).

6.2.3 Bivariate Normal Distribution

In this subsection, we will use the bivariate normal distribution as a special instance of a general multivariate normal distribution to illustrate its geometric (or graphical) properties. The bivariate normal distribution, as defined in Eqs. (6.1) and (6.2), has the following simple form:

$$D^2 = (1 - \rho^2)^{-1}\left\{\frac{(Y_1 - \mu_1)^2}{\sigma_{11}} + \frac{(Y_2 - \mu_2)^2}{\sigma_{22}} - 2\rho\frac{(Y_1 - \mu_1)(Y_2 - \mu_2)}{\sqrt{\sigma_{11}\sigma_{22}}}\right\}.$$

Note that σ_{ij} is the (i, j)th element of $\boldsymbol{\Sigma}$. Sometimes, we denote $\sigma_{ii} = \sigma_i^2$.

It is clear that D^2 modifies the Euclidean distance between $\mathbf{Y}$ and $\boldsymbol{\mu}$ by accounting for the variances of $\mathbf{Y}_1$ and $\mathbf{Y}_2$ and their correlation ρ. When the variance is large, D^2 down-weights the deviations from the center $\boldsymbol{\mu}$. If ρ is large, i.e., a stronger correlation, knowing Y_1 is "close" to μ_1 also tells us that Y_2 is close to μ_2.

Figure 6.1 provides the surface plots and the contour plots for the bivariate normal distribution of (X, Y) with different correlation coefficients. Panel (a) shows the standard bivariate normal distribution. The contours are spheral, which correspond to the same Mahalanobis distance to the center point $\boldsymbol{\mu}$. Panels (b) and (c) correspond to a strong negative correlation and a strong positive correlation, respectively. The contour plot indicates that when X is far away from its center, Y will be far away from its center as well. However, when $\rho < 0$, when X is larger than its center, Y will be smaller than its center. When $\rho > 0$, X and Y will be simultaneously larger or smaller than their corresponding centers.

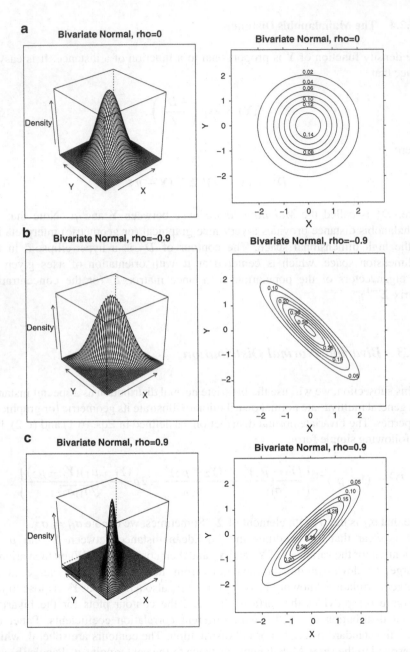

Fig. 6.1 The surface (*left panels*) and the contour (*right panels*) plots for bivariate normal distributions with different correlations: (**a**) $\rho = 0$, i.e., X and Y are independent. (**b**) $\rho = -0.9$, i.e., X and Y are strongly negatively correlated. (**c**) $\rho = 0.9$, i.e., X and Y are strongly positively correlated

6.2.4 Wishart Distribution

In the one-dimensional case, a standard normal distribution has a related χ^2 distribution. Assume that $Z_1, \ldots, Z_q$ are q random variables, which are identically independent distributed (i.i.d.) from the standard normal distribution $N(0, 1)$. Define a random variable W as

$$W = \sum_{j=1}^{q} Z_j^2,$$

which is distributed as a χ^2 distribution with q degrees of freedom. This is usually denoted as $W \sim \chi_q^2$.

In the multivariate context, the Wishart distribution parallels the univariate chi-square distribution. Assume there are q random vectors of dimension p ($q > p$), $\mathbf{Z}_1, \ldots, \mathbf{Z}_q$, which are i.i.d. from the centered multivariate normal distribution $N_p(\mathbf{0}, \boldsymbol{\Sigma})$. Define a random matrix $\mathbf{W}$ as

$$\mathbf{W} = \sum_{j=1}^{q} \mathbf{Z}_j \mathbf{Z}_j^T.$$

Then $\mathbf{W}$ follows a *Wishart distribution*, with parameters $(p, \boldsymbol{\Sigma}, q)$, where q is the degree of freedom. It is usually denoted as $\mathbf{W} \sim W_p(\boldsymbol{\Sigma}, q)$. The density function of a Wishart distribution is as follows:

$$f(\mathbf{W}) \propto |\boldsymbol{\Sigma}|^{-q/2} |\mathbf{W}|^{(q-p-1)/2} \exp\left\{-\frac{1}{2} tr(\mathbf{W}\boldsymbol{\Sigma}^{-1})\right\}.$$

The Wishart distribution has several attractive properties, see, e.g., Anderson[1]. We here list two commonly used properties:

1. $E(\mathbf{W}) = q\boldsymbol{\Sigma}$.
2. Assume that $\mathbf{W}_1$ and $\mathbf{W}_2$ are independent, $\mathbf{W}_1 \sim W_p(\boldsymbol{\Sigma}, q_1)$ and $\mathbf{W}_2 \sim W_p(\boldsymbol{\Sigma}, q_2)$, their sum $\mathbf{W}_1 + \mathbf{W}_2 \sim W_p(\boldsymbol{\Sigma}, q_1 + q_2)$.

Property 1 suggests that the sample covariance matrix gives an unbiased estimator of the population covariance matrix $\boldsymbol{\Sigma}$ (see the next subsection). Property 2 shows the additivity of the Wishart distribution if two random matrices are independent Wishart and share the same parameters.

6.2.5 Sample Mean and Covariance

In this section, the properties of the sample mean vector and the sample covariance matrix of a multivariate normal distribution will be discussed. Assume that there are n independent samples, each following the multivariate normal distribution $N_p(\boldsymbol{\mu}, \boldsymbol{\Sigma})$, i.e., $\mathbf{Y}_1, \ldots, \mathbf{Y}_n \overset{\text{i.i.d}}{\sim} N(\boldsymbol{\mu}, \boldsymbol{\Sigma})$, the sample mean vector and the sample covariance matrix are defined as

$$\overline{\mathbf{Y}} = n^{-1} \sum_{i=1}^{n} \mathbf{Y}_i,$$

$$\overline{\mathbf{S}} = \frac{1}{n-1} \sum_{i=1}^{n} (\mathbf{Y}_i - \overline{\mathbf{Y}})(\mathbf{Y}_i - \overline{\mathbf{Y}})^T.$$

Note that $\overline{\mathbf{Y}}$ is a $p \times 1$ vector and $\mathbf{S}$ is a $p \times p$ matrix.

The sample mean vector and the sample covariance matrix have several nice properties. Here we list a few:

1. $\overline{\mathbf{Y}}$ is an unbiased estimator of $\boldsymbol{\mu}$ and $\overline{\mathbf{Y}} \sim N_p(\boldsymbol{\mu}, \boldsymbol{\Sigma})$.
2. $\mathbf{S}$ is an unbiased estimator of $\boldsymbol{\Sigma}$, and $(n-1)\mathbf{S} \sim W_p(\boldsymbol{\Sigma}, n-1)$.
3. $\overline{\mathbf{Y}}$ and $\mathbf{S}$ are independent.

6.3 One-Sample and Two-Sample Multivariate Hypothesis Tests

In the one-sample problem, one is interested in testing whether the population mean $\boldsymbol{\mu}$ equals to a prespecified quantity. In this section, we will first review the one-dimensional t test and the F test, and then introduce the corresponding Hotelling's T^2 test.

6.3.1 One-Sample t Test for a Univariate Outcome

We first consider scalar outcomes $Y_1, \ldots, Y_n$ following i.i.d $N(\mu, \sigma^2)$. A typical one-sample hypothesis testing problem is

$$H_0 : \mu = \mu_0, \text{ vs. } H_1 : \mu \neq \mu_0. \tag{6.3}$$

Assume that σ is known. Under the null hypothesis, it is well known that $\overline{Y} - \mu_0 \sim N(0, \sigma^2/n)$. So we can use a Z statistic to test for Eq. (6.3):

$$Z = \frac{\overline{Y} - \mu_0}{\sigma/\sqrt{n}} \sim N(0, 1).$$

When σ is unknown, we need to estimate σ^2 by the sample variance

$$S = s^2 = (n-1)^{-1} \sum_{i=1}^{n} (Y_i - \overline{Y})^2.$$

The distribution of s^2 is that $(n-1)s^2/\sigma^2 \sim \chi^2_{n-1}$. In order to test for Eq. (6.3), we can construct the following statistic

$$T = \frac{\overline{Y} - \mu_0}{s/\sqrt{n}}.$$

When the null hypothesis in Eq. (6.3) is true, the statistic T follows a t distribution with $n-1$ degrees of freedom, i.e., $T \sim t_{n-1}$. When the alternative hypothesis in (6.3) is true, and let μ to be the true underlying mean, then T is from a noncentral t distribution with parameter δ and degree of freedom $n-1$, i.e., $T \sim t_{n-1}(\delta)$, where $\delta = \mu - \mu_0$ is the noncentrality parameter. This can be used to calculate the power of the t test. One can equivalently use the F statistics to test for Eq. (6.3) as follows:

$$F := T^2 = \frac{n(\overline{Y} - \mu_0)^2}{S}.$$

It can be shown that the F statistic is distributed as $F_{1,(n-1)}$ under the null hypothesis in Eq. (6.3).

6.3.2 Hotelling's T^2 Test for the Multivariate Outcome

In the multivaraite case, the Hotelling's T^2 test corresponds to the F test (or t test) in the univariate case. Assume that $\mathbf{Y}_1, \ldots, \mathbf{Y}_n$ are i.i.d. from $N_p(\boldsymbol{\mu}, \boldsymbol{\Sigma})$. In this subsection, the hypothesis we are interested in is

$$H_0 : \boldsymbol{\mu} = \boldsymbol{\mu}_0 \text{ vs. } H_1 : \boldsymbol{\mu} \neq \boldsymbol{\mu}_0. \tag{6.4}$$

As discussed in Sect. 6.2.5, under H_0, $\overline{\mathbf{Y}} - \boldsymbol{\mu}_0 \sim N_p(0, \boldsymbol{\Sigma}/n)$ when $\boldsymbol{\Sigma}$ is known. If $\boldsymbol{\Sigma}$ is unknown, the sample covariance matrix $\mathbf{S}$ has the following distribution: $(n-1)\mathbf{S} \sim W_p(\boldsymbol{\Sigma}, n-1)$. Furthermore, $\overline{\mathbf{Y}}$ and $\mathbf{S}$ are independent.

To test for the hypothesis in Eq. (6.4), the following statistic was proposed by Hotelling and is very similar to the F statistic in the one-dimensional case. The Hotelling's T^2 statistic is defined as

$$T^2 = n(\overline{\mathbf{Y}} - \boldsymbol{\mu}_0)^T \mathbf{S}^{-1}(\overline{\mathbf{Y}} - \boldsymbol{\mu}_0). \tag{6.5}$$

When the null hypothesis in Eq. (6.4) is true, the Hotelling's T^2 statistic has a generalized T^2 distribution with degree of freedom $(p, n-1)$, i.e., $T^2 \sim T^2_{p,(n-1)}$. When the alternative hypothesis in (6.4) is true, the Hotelling's T^2 statistic follows a noncentral generalized T^2 distribution, i.e., $T^2 \sim T^2_{p,(n-1)}(\delta)$, where $\delta = (\boldsymbol{\mu} - \boldsymbol{\mu}_0)^T \boldsymbol{\Sigma}^{-1}(\boldsymbol{\mu} - \boldsymbol{\mu}_0)$ is the noncentrality parameter. Here, $\boldsymbol{\mu}$ is the true underlying mean vector in the population. It can be shown that when n is large, T^2 is approximately χ^2_p.

Note that the Hotelling's T^2 statistic also has a strong connection to the usual F test. Specifically, by defining

$$F = \frac{n-p}{(n-1)p}T^2,$$

it can be shown that when H_0 is true, $F \sim F_{p,n-p}$; when H_1 is true, $F \sim F_{p,n-p}(\delta)$, where δ is the noncentrality parameter defined above. It is straightforward to see that when $p = 1$, the F statistic is the same as the T^2 statistic. To test for the hypothesis in Eq. (6.4) by using the T^2 statistic, we can define the following rejection region (i.e., reject H_0 when the observed T^2 falls into R):

$$R : T^2 > \frac{(n-1)p}{n-p}F_{p,n-p,1-\alpha}$$

at the significance level α.

6.3.3 Properties of Hotelling's T^2 Test

The Hotelling's T^2 test is connected with several popular tests, such as the likelihood ratio test. We discuss several of its major properties in this section.

6.3.3.1 Connections with the Likelihood Ratio Test

The T^2 test is equivalent to the likelihood ratio test for the hypothesis in Eq. (6.4). Define

$$\Lambda = \frac{\max\limits_{\Sigma} \ell(\boldsymbol{\mu}_0, \boldsymbol{\Sigma})}{\max\limits_{\boldsymbol{\mu},\boldsymbol{\Sigma}} \ell(\boldsymbol{\mu}, \boldsymbol{\Sigma})} = \frac{|\hat{\boldsymbol{\Sigma}}_{ML}|^{n/2}}{|\hat{\boldsymbol{\Sigma}}_0|^{n/2}},$$

where Λ is the Wilk's Lambda, and

$$\hat{\boldsymbol{\Sigma}}_{ML} = \frac{1}{n}\sum_{i=1}^{n}(\mathbf{Y}_i - \overline{\mathbf{Y}})(\mathbf{Y}_i - \overline{\mathbf{Y}})^T,$$

$$\hat{\boldsymbol{\Sigma}}_0 = \frac{1}{n}\sum_{i=1}^{n}(\mathbf{Y}_i - \boldsymbol{\mu}_0)(\mathbf{Y}_i - \boldsymbol{\mu}_0)^T.$$

The likelihood ratio statistic is defined as

$$LR = -2\log(\Lambda),$$

which follows χ_p^2 under H_0 for large n.

6.3.3.2   Connections with the Univariate T Test

The Hotelling's T^2 test is an omnibus test and has an attractive connection with the univariate t test. The T^2 test is the maximum of all possible univariate t tests constructed using any linear combination of the p components of $\mathbf{Y}$. Specifically, suppose $\mathbf{Y}_i \sim N_p(\boldsymbol{\mu}, \boldsymbol{\Sigma})$ and $X_i = \mathbf{a}^T \mathbf{Y}_i$ (a scalar). It is clear that $X_i \sim N(\mathbf{a}^T \boldsymbol{\mu}, \mathbf{a}^T \boldsymbol{\Sigma} \mathbf{a})$. The univariate t test for $H_0 : \mathbf{a}^T \boldsymbol{\mu} = \mathbf{a}^T \boldsymbol{\mu}_0$ is

$$T_{\mathbf{a}} = \frac{\sqrt{n}\mathbf{a}^T(\bar{\mathbf{Y}} - \boldsymbol{\mu}_0)}{(\mathbf{a}^T \mathbf{S} \mathbf{a})^{1/2}} = \frac{\sqrt{n}(\bar{X} - \mathbf{a}^T \boldsymbol{\mu}_0)}{(\mathbf{a}^T \mathbf{S} \mathbf{a})^{1/2}}.$$

By the Cauchy-Schwarz inequality, it is straightforward to show that

$$\sup_{\mathbf{a}} \frac{\mathbf{a}^T \mathbf{Z}}{\mathbf{a}^T \mathbf{D} \mathbf{a}} = \mathbf{Z}^T \mathbf{D}^{-1} \mathbf{Z},$$

for any fixed $\mathbf{a}$ and a positive definite matrix $\mathbf{D}$. Thus, we have

$$T^2 = n(\hat{\mathbf{Y}} - \boldsymbol{\mu}_0)^T \mathbf{S}^{-1}(\bar{\mathbf{Y}} - \boldsymbol{\mu}_0) = \sup_{\mathbf{a}}(T_{\mathbf{a}}^2).$$

This justifies the above statement. Essentially, it states that if the Hotelling's T^2 rejects $H_0 : \boldsymbol{\mu} = \boldsymbol{\mu}_0$ at the level α, for any $\mathbf{a}$ (a $p \times 1$ vector), the univariate t test of $H_0 : \mathbf{a}^T \boldsymbol{\mu} = \mathbf{a}^T \boldsymbol{\mu}_0$ will reject H_0 at the level α.

6.3.3.3   Invariance Under Full Rank Affine Transformations

The T^2 test is invariant under all full rank affine transformations of $\mathbf{Y}$. To see this, let $\mathbf{Y} \sim N_p(\boldsymbol{\mu}, \boldsymbol{\Sigma})$. Define a linearly transform variable $\mathbf{Y}^* = \mathbf{C}\mathbf{Y} + \mathbf{a}$, where $\mathbf{C}$ is a full rank $p \times p$ matrix and $\mathbf{a}$ is a $p \times 1$ vector. Thus,

$$\mathbf{Y}^* \sim N_p(\boldsymbol{\mu}^*, \boldsymbol{\Sigma}^*),$$

where $\boldsymbol{\mu}^* = \mathbf{C}\boldsymbol{\mu} + \mathbf{a}$ and $\boldsymbol{\Sigma}^* = \mathbf{C}\boldsymbol{\Sigma}\mathbf{C}^T$. Some simple algebra shows that the Hotelling's T^2 statistic to test for $H_0 : \boldsymbol{\mu} = \boldsymbol{\mu}_0$ based on the data $\mathbf{Y}$s is the same as the Hotelling's T^2 statistic to test for $H_0^* : \boldsymbol{\mu}^* = \boldsymbol{\mu}_0^*$ based on the data $\{\mathbf{Y}^*\}$s.

6.3.4   Paired Multivariate Hotelling's T^2 Test

In the univariate case, the t test can be used for testing for paired samples. This can be extended to the multivariate case, which is the paired Hotelling's T^2 test. Define $(\mathbf{Y}_{0i}, \mathbf{Y}_{1i})$ $(i = 1, \ldots, n)$ as paired multivariate observations, where $(\mathbf{Y}_{0i}, \mathbf{Y}_{1i})$ are from the same subject i and each is a $p \times 1$ vector. For example, the pretreatment and posttreatment measurements of the same subject. Such studies are common in biomedical research. Assume that $(\mathbf{Y}_{0i}, \mathbf{Y}_{1i})$ are MVN with means $\boldsymbol{\mu}_0$ and $\boldsymbol{\mu}_1$, respectively. The hypothesis we are interested in is

$$H_0 : \boldsymbol{\mu}_0 = \boldsymbol{\mu}_1 \text{ vs. } H_1 : \boldsymbol{\mu}_0 \neq \boldsymbol{\mu}_1.$$

This paired problem can be transformed into a one-sample test using the paired Hotelling's T^2 test as follows:

1. Calculate the difference $\mathbf{Y}_i = \mathbf{Y}_{1i} - \mathbf{Y}_{0i}$.
2. Change the above test to be

$$H_0 : \boldsymbol{\mu}_1 - \boldsymbol{\mu}_0 = \boldsymbol{\mu} = 0 \text{ vs. } H_1 : \boldsymbol{\mu}_1 - \boldsymbol{\mu}_0 = \boldsymbol{\mu} \neq 0,$$

   where $\boldsymbol{\mu} = E(\mathbf{Y})$.
3. Calculate the Hotelling's T^2 statistic based on the difference data $\mathbf{Y}$s as

$$T^2 = n \overline{\mathbf{Y}}^T \mathbf{S}_{\mathbf{Y}}^{-1} \overline{\mathbf{Y}}.$$

4. When H_0 is true, this $T^2 \sim T_{p,n-1}^2$. So we can reject H_0 if

$$R : T^2 > \frac{(n-1)p}{n-p} F_{p,n-p,1-\alpha}$$

   at a given significance level α.

6.3.5 Examples

In this subsection, we use the data from a welder study to illustrate the paired t test. The samples were collected in a welding school at Massachusetts, USA, to study how gene expressions were affected by exposure to heavy metal fumes. The data set contains nine welders exposed to metal fumes and seven unexposed subjects. Microarray gene expressions were measured at baseline (i.e., before welding) and 6 h later after work. The major objective in this study is to identify genes that are differentially expressed between the baseline and postexposure in welders. For illustration, we consider three genes ($p = 3$) in the immune responses pathway. The data are shown in Table 6.1.

To perform the paired Hotelling's T^2 test, let $(\mathbf{Y}_{\text{pre},i}, \mathbf{Y}_{\text{post},i})$ be the pre- and postexpressions of the three genes. The scientific problem we are interested in can be formulated as the following hypotheses:

$$H_0 : \boldsymbol{\mu}_{\text{post}} - \boldsymbol{\mu}_{\text{pre}} = 0 \text{ vs. } H_0 : \boldsymbol{\mu}_{\text{post}} - \boldsymbol{\mu}_{\text{pre}} \neq 0.$$

Table 6.1 The gene expressions for the 9 welders at baseline and 6 h after work

Subject	Time point	Expression value of		
		Gene 1	Gene 2	Gene 3
1	0	6.177	7.558	12.297
2	0	6.647	7.326	12.508
3	0	6.766	6.767	12.285
4	0	6.882	7.463	12.307
5	0	6.584	7.101	12.219
6	0	6.413	7.06	12.463
7	0	6.419	7.437	12.177
8	0	6.511	7.131	11.86
9	0	6.615	7.253	12.781
1	1	5.812	7.581	12.33
2	1	6.656	7.233	12.158
3	1	6.302	7.098	12.287
4	1	6.269	7.595	12.362
5	1	6.532	7.485	12.56
6	1	6.026	7.343	12.738
7	1	6.146	7.634	11.915
8	1	6.25	7.442	12.119
9	1	6.035	7.445	12.93

The results are given below. One can see that the mean expression levels are significantly different between pre- and postexposures to metal fume for gene 1 and gene 2 at the level of 0.05, and no significant difference is found for gene 3.

	Gene 1	Gene 2	Gene 3
Mean (difference)	−0.331	0.196	0.056
Std. Dev. (difference)	0.214	0.155	0.237
paired t statistic	−4.655	3.788	0.706
p-value	0.002	0.005	0.5

The R code for calculating the paired Hotelling's T^2 is:

```
> Y = Y1-Y0
> Ybar = apply(Y, 2, mean)
> S = cov(Y)
> T2 = n*Ybar%*%solve(S)%*%Ybar
> tf = (n-p)/((n-1)*p) *T2
> pval = 1-pf(tf, p, (n-p))
```

Note that the R function apply(Y, 2, mean) is to calculate the mean of all the rows in Y, solve(S) is used to calculate the inverse of the input matrix S, and pf(tf, p, (n-p)) is to calculate the probability that a F distribution with degree of freedoms of $(p, n-p)$ is less than tf. For more details about programming in R, see the R manual.

The output for the above *R* codes is

```
> Ybar
        G1          G2          G3
-0.33177778  0.19555556  0.05577778
> S
            G1            G2           G3
G1   0.045650694 -0.004450514 -0.01124194
G2  -0.004450514  0.023990528  0.02617176
G3  -0.011241944  0.026171764  0.05605619
> T2
        [,1]
[1,]  45.36926
> tf
        [,1]
[1,]  11.34231
> pval
          [,1]
[1,]  0.006940488
```

We can also use the MANOVA function in *R* to obtain the same results. The *R* code is

```
> model = manova(as.matrix(Y)~1)
```

The MANOVA code generates the following results:

```
> summary(model, intercept = T)
            Df  Pillai approx F num Df den Df  Pr(>F)
(Intercept)  1  0.8501  11.3423      3      6 0.00694 **
Residuals    8
---
Signif. codes:  0 *** 0.001 ** 0.01 * 0.05 . 0.1  1
```

6.3.6   Two-Sample Hotelling's T^2 Test

A common problem in biomedical research is to test whether two groups of subjects, e.g., disease and healthy subjects, have the same population mean. Take the above welder data as an example, we are interested in testing whether the means of the pre-post changes of gene expression levels are the same for the welders and the controls. Similar to the two-sample t test in the one-dimensional situation, the two-sample Hotelling's T^2 test can be used for the multivariate case. Assume there are two groups. Group 1 contains n_1 subjects, e.g., the welders. Assume $\mathbf{Y}_{1i}$ is a $p \times 1$ random vector from $N_p(\boldsymbol{\mu}_1, \boldsymbol{\Sigma})$ for $i = 1, \ldots, n_1$. Group 2 contains n_2 subjects, e.g., the controls. Assume $\mathbf{Y}_{2i}$ is also a $p \times 1$ random vector but from $N_p(\boldsymbol{\mu}_2, \boldsymbol{\Sigma})$

for $i = 1, \ldots, n_2$. Further assume that Y_{1i} and Y_{2i} are independent, and $\text{cov}(\mathbf{Y}_{1i}) = \text{cov}(\mathbf{Y}_{2i}) = \boldsymbol{\Sigma}$, i.e., the two populations share the same covariance matrix. We are interested in testing the following hypothesis:

$$H_0 : \boldsymbol{\mu}_1 = \boldsymbol{\mu}_2 \text{ vs. } H_1 : \boldsymbol{\mu}_1 \neq \boldsymbol{\mu}_2.$$

To proceed, define the pooled sample covariance matrix as

$$\mathbf{S}_{\text{pool}} = \frac{(n_1 - 1)\mathbf{S}_1 + (n_2 - 1)\mathbf{S}_2}{n_1 + n_2 - 2},$$

where $\mathbf{S}_k$ $(k = 1, 2)$ is the sample covariance matrix for group k,

$$\mathbf{S}_k = \frac{1}{n_k - 1} \sum_{i=1}^{n_k} (\mathbf{Y}_{ki} - \bar{\mathbf{Y}}_k)(\mathbf{Y}_{ki} - \bar{\mathbf{Y}}_k)^T.$$

The sampling distribution of the mean difference is

$$\bar{\mathbf{Y}}_1 - \bar{\mathbf{Y}}_2 \sim N_p \left(\boldsymbol{\mu}_1 - \boldsymbol{\mu}_2, \left[\frac{1}{n_1} + \frac{1}{n_2} \right] \boldsymbol{\Sigma} \right).$$

The two-sample Hotelling's T^2 statistic for testing H_0 is defined as

$$T^2 = \frac{n_1 n_2}{n_1 + n_2} (\bar{\mathbf{Y}}_1 - \bar{\mathbf{Y}}_2)^T \mathbf{S}^{-1} (\bar{\mathbf{Y}}_1 - \bar{\mathbf{Y}}_2).$$

When H_0 is true, it can be shown that T^2 has a T^2_{p,n_1+n_2-2} distribution with degrees of freedom $(p, n_1 + n_2 - 2)$. When H_1 is true, T^2 has a noncentral T^2 distribution, $T^2_{p,n_1+n_2-2}(\delta)$, where δ is the noncentrality parameter

$$\delta = \frac{n_1 n_2}{n_1 + n_2} (\boldsymbol{\mu}_1 - \boldsymbol{\mu}_2)^T \boldsymbol{\Sigma}^{-1} (\boldsymbol{\mu}_1 - \boldsymbol{\mu}_2).$$

To perform the Hotelling's T^2 test, we can perform an F test by

$$F = \frac{n_1 + n_2 - p - 1}{n_1 + n_2 - 2} T^2,$$

where F has a distribution F_{p,n_1+n_2-p-1} under H_0 and $F_{p,n_1+n_2-p-1}(\delta)$ under H_1.

6.3.6.1 The Welder Data Example

We use the welder data to illustrate the two-sample Hotelling's T^2 test. In this example, the question of interest is whether the changes in the mean expressions between post- and pre-shifts are differentially expressed between welders and

Table 6.2 The two-sample difference (post-pre) data

Subject	Group	Difference(post-pre) in expression of		
		Gene 1	Gene 2	Gene 3
1	0	−0.212	−0.177	0.558
2	0	0.181	0.057	0.082
3	0	−0.124	0.009	−0.133
4	0	0.336	0.277	−0.47
5	0	0.067	0.073	0.317
6	0	0.506	−0.171	0.07
7	0	0.234	0.028	0.084
8	1	−0.365	0.023	0.033
9	1	0.009	−0.093	−0.35
10	1	−0.464	0.332	0.001
11	1	−0.612	0.132	0.055
12	1	−0.052	0.384	0.341
13	1	−0.387	0.283	0.275
14	1	−0.273	0.197	−0.262
15	1	−0.26	0.311	0.259
16	1	−0.58	0.192	0.148

controls. Table 6.2 gives the data of the difference between post- and pre-work for the two groups for the three genes. We performed two-sample t tests for each individual gene, and the results are summarized in the following table.

	Gene 1	Gene 2	Gene 3
Mean (welders)	−0.331	0.196	0.056
Mean (controls)	0.141	0.014	0.073
SD (welders)	0.214	0.155	0.237
SD (controls)	0.252	0.156	0.325
t statistic	−3.97	2.323	−0.116
p value	0.002	0.037	0.909

This table shows that the mean post-pre changes are significantly different for genes 1 and 2 at the $\alpha = 0.05$ level, while no significant difference is found for gene 3. We next perform the two-sample Hotelling's T^2 test to study whether the overall gene expression profile in the immune response pathway is the same between the welders and the controls. This can proceed using the following R code:

```
> Ybar1 = apply(Y1, 2, mean)
> Ybar2 = apply(Y2, 2, mean)
> S1=cov(Y1)
> S2=cov(Y2)
> Sp=((n1-1)*S1+(n2-1)*S2)/(n1+n2-2)
> T2=t(Ybar1-Ybar2)%*%solve((1/n1+1/n2)*Sp)%
  *%(Ybar1-Ybar2)
> tf = (n1+n2-p-1)/((n1+n2-2)*p) *T2
> pval = 1-pf(tf, p, (n1+n2-p-1))
```

We get the following results:

```
> Ybar1
          G1              G2              G3
-0.33155556   0.19566667   0.05555556
> Ybar2
          G1              G2              G3
0.14114286   0.01371429   0.07257143
> S1
                G1              G2                G3
G1    0.045598278  -0.004460583  -0.01116903
G2   -0.004460583   0.024024500   0.02614846
G3   -0.011169028   0.026148458   0.05604653
> S2
                G1              G2              G3
G1    0.063634810   0.008527381  -0.04004460
G2    0.008527381   0.024237571  -0.03648714
G3   -0.040044595  -0.036487143   0.10537595
> Sp
                G1              G2              G3
G1    0.053328220   0.0011056871  -0.0235442710
G2    0.001105687   0.0241158163  -0.0006953707
G3   -0.023544271  -0.0006953707   0.0771877098
> T2
            [,1]
[1,]  25.46684
> tf
            [,1]
[1,]   7.276241
> pval
                [,1]
[1,]   0.00487202
```

This suggests that we will reject the null hypothesis and conclude that the overall
gene expression profile in the immune response pathway is significantly different
between the welders and the controls.

This two-sample Hotelling's T^2 test can also be performed using the MANOVA
code in R as

```
> model = manova(as.matrix(X)~group)
```

The results are as below and the same as above:

```
> summary(model)
            Df  Pillai  approx F  num Df  den Df     Pr(>F)
group        1  0.6453    7.2762       3      12   0.004872 **
Residuals   14
---
Signif. codes:  0 *** 0.001 ** 0.01 * 0.05 . 0.1  1
```

6.4   Principal Component Analysis

To analyze high dimensional multivariate data, it is common to reduce the dimension of the data. This allows one to draw inference based on the dimension-reduced data. Principal component analysis (PCA) is one common approach for dimension reduction and is one of the fundamental techniques in multivariate analysis. It summarizes the data by using a few linear combinations of the variables that can explain most variability of the data. A comprehensive review of the PCA method can be found in Jolliffe [2].

A typical microarray gene expression data have the following data structure

Subject	Gene	Group (clinical outcome)
1	$X_{11}, X_{12}, \ldots, X_{1p}$	Y_1
2	$X_{21}, X_{22}, \ldots, X_{2p}$	Y_2
$\vdots$	$\vdots$	$\vdots$
n	$X_{n1}, X_{n2}, \ldots, X_{np}$	Y_n

PCA can be used to calculate fewer summaries of the gene profiles $\mathbf{X}$ for each subject. Inference can then be drawn based on these summaries. The PCA method is an unsupervised method, so the class labels, $(Y_1, \ldots, Y_n)$, are not used in dimension reduction of $\mathbf{X}$ using PCA. Note that PC is usually an intermediate analysis step. After performing dimension reduction of $\mathbf{X}$ using PCA, one uses the resulting principal components (PCs) to perform supervised analysis by employing class labels. Examples of such methods include logistic regression of class labels on the PCs.

6.4.1   Definition of Principal Components

Dimension reduction is a general technique for high dimensional data analysis. Usually, our original data are in a space of dimension p, e.g., p genes. The goal of dimension reduction is to project the data into a subspace of dimension k ($k < p$) such that we retain most of the variability in the data (i.e., the information). One way to achieve this objective is to rotate the data so that the new axes represent the directions of the largest variability.

Before we introduce the PCA method, we introduce the following notation. Assume that $\mathbf{X} = (X_1, \ldots, X_p)^T$ is the original vector of p random variables, e.g., X_j is the expression of gene j. Denote by $\boldsymbol{\Sigma}$ the covariance matrix of $\mathbf{X}$, and $\lambda_1 \geq \lambda_2 \geq \cdots \lambda_p \geq 0$ are the ordered eigenvalues of $\boldsymbol{\Sigma}$. We want to find the rotated random variables $\mathbf{Z} = [\mathbf{Z}_1, \ldots, \mathbf{Z}_k]$ as linear combinations of the original data $\mathbf{X}$ as

$$Z_1 = \mathbf{a}_1^T \mathbf{X}$$

$$Z_2 = \mathbf{a}_2^T \mathbf{X}$$

$$\vdots \quad \vdots$$

$$Z_k = \mathbf{a}_k^T \mathbf{X}.$$

The principal component analysis is to find $\mathbf{a}_j$ such that $\text{var}(Z_j) = \mathbf{a}_j^T \boldsymbol{\Sigma} \mathbf{a}_j \geq \text{var}(Z_{j+1})$ is maximized, and $\text{cov}(Z_j, Z_{j'}) = \mathbf{a}_j^T \boldsymbol{\Sigma} \mathbf{a}_{j'} = 0$ for $j \neq j'$, i.e., Z_j and $Z_{j'}$, are orthogonal. In order to achieve the uniqueness of the maximizer of the above objective function, we will restrict $\mathbf{a}_j$s to have norm 1 for all j. We can rewrite the first principal component as $Z_1 = \mathbf{a}_1^T \mathbf{X}$ that maximizes $\text{var}(Z_1)$ subject to $\mathbf{a}^T \mathbf{a} = 1$. The jth principal component Z_j is defined as $\mathbf{a}_j^T \mathbf{X}$ that maximizes $\text{var}(Z_j)$ subject to $\mathbf{a}_j^T \mathbf{a}_j = 1$ and $\text{cov}(\mathbf{a}_j, \mathbf{a}_{j'}) = 0$ for $j \neq j'$.

6.4.2 Computing Principal Components

To calculate the principal components (PCs), we can use the eigen-analysis of the covariance matrix $\boldsymbol{\Sigma}$. Denote the eigenvalue-eigenvector pairs of $\boldsymbol{\Sigma}$ as $(\lambda_1, \mathbf{a}_1)$, $(\lambda_2, \mathbf{a}_2), \ldots, (\lambda_p, \mathbf{a}_p)$, with $\lambda_1 \geq \lambda_2 \geq \cdots \lambda_p \geq 0$. Then, the jth PC is given by

$$Z_j = \mathbf{a}_j^T \mathbf{X} = a_{j1} X_1 + a_{j2} X_2 + \cdots + a_{jp} X_p,$$

and

$$\text{var}(\mathbf{Z}_j) = \lambda_j \text{ and } \text{cov}(\mathbf{Z}_i, \mathbf{Z}_j) = 0.$$

The proof of the above results can be found in many multivariate analysis books (e.g., pp. 428 in Johnson and Wichern [3]).

There are several functions in R that can be used to calculate PCs, such as `eigen`, `svd`, `prcomp`, and `princomp`. Here, we will call $a_{j1}, \ldots, a_{jp}$ as the component *loadings* and $Z_{ij} = \mathbf{a}_j^T \mathbf{X}_i$ the component *score* for the jth component of the ith subject.

6.4.3 Variance Decomposition

The PCA method provides a special variance decomposition. It can be shown that

$$\sum_{j=1}^{p} \text{var}(\mathbf{X}_j) = \text{trace}(\boldsymbol{\Sigma}) = \sigma_{11}^2 + \sigma_{22}^2 + \cdots + \sigma_{pp}^2 = \sum_{j=1}^{p} \text{var}(\mathbf{Z}_j) = \lambda_1 + \lambda_2 + \cdots + \lambda_p.$$

This suggests that the total variance is equal to the sum of the eigenvalues. Thus, the proportion of variability due to (or explained by) the jth PC is

$$\frac{\lambda_j}{\lambda_1 + \lambda_2 + \cdots + \lambda_p}.$$

It is clear that if we use all p principal components, we have not performed any dimension reduction. So usually we will select the first k PCs, i.e., we will "throw away" some of the components with small eigenvalues. Obviously, the number of principal components that explain all the variability (which corresponds to the number of positive eigenvalues) is $d = \text{rank}(\boldsymbol{\Sigma})$. In order to reduce the dimensionality, we will select $k \leq d$, and make the remaining $d - k$ components explain a relatively small proportion of the total variability.

To interpret the PCA loadings, denote the jth loading by $\mathbf{a}_j = [a_{j1}, a_{j2}, \ldots, a_{jp}]^T$. Its rth coefficient a_{jr} is a measurement of the importance of the rth variable to the jth principal component independent of the other variables. In fact, using the definition of the PCA score Z defined above, one can easily show that

$$\text{cor}(\mathbf{Z}_j, \mathbf{X}_r) = \frac{a_{jr}\sqrt{\lambda_j}}{\sigma_{rr}}.$$

A larger a_{jr} means the rth variable X_j contributes more to the jth PC. Although both the correlation of the variable to the component and the variable's component loading can be used as measures of importance, the loading is generally used, though often the difference between the two is not great.

6.4.4 PCA with a Correlation Matrix

Typically, PCA can be viewed as the eigen-analysis of the covariance matrix, i.e., the covariates are not standardized. It is natural to standardize each predictor to avoid scaling difference between different variables. This leads to the correlation matrix based PCA method.

Suppose we have standardized each predictor to have mean 0 and variance 1 by setting

$$\mathbf{X}^{\text{std}} = \mathbf{V}^{-1/2}(\mathbf{X} - \boldsymbol{\mu}),$$

where $\mathbf{V}$ is $\text{diag}(\sigma_{11}^2, \sigma_{22}^2, \ldots, \sigma_{pp}^2)$. Define

$$\mathbf{R} = \text{cov}(\mathbf{X}^{\text{std}}) = \text{cor}(\mathbf{X}).$$

We can perform PCA onto $\mathbf{X}^{\text{std}}$. The jth PC of $\mathbf{X}^{\text{std}}$ is

$$\mathbf{Z}_j = \mathbf{a}_j^T \mathbf{X}^{\text{std}} = \mathbf{a}_j^T \mathbf{V}^{-1/2}(\mathbf{X} - \boldsymbol{\mu}).$$

One can easily show that

$$\sum_{j=1}^{p} \text{var}(\mathbf{Z}_j) = \sum_{j=1}^{p} \text{var}(\mathbf{X}_j^{\text{std}}) = \sum_{j=1}^{p} \lambda_j = p,$$

and

$$\text{cor}(\mathbf{Z}_j, \mathbf{X}_r^{\text{std}}) = a_{ir}\sqrt{\lambda_j},$$

where $(\lambda_j, \mathbf{a}_j)$ is the jth eigenvalue-eigenvector pair of $\mathbf{R}$ with $\lambda_1 \geq \lambda_2 \geq \cdots \geq \lambda_p \geq 0$.

In general, the principal components derived from the covariance matrix are not the same as those derived from the correlation matrix. Note that PCA with a covariance matrix is sensitive to different scales of outcomes.

6.4.4.1 An Example

Consider the following covariance matrix and correlation matrix.

$$\boldsymbol{\Sigma} = \begin{pmatrix} 1 & 4 \\ 4 & 25 \end{pmatrix}, \quad \mathbf{R} = \begin{pmatrix} 1 & 0.8 \\ 0.8 & 1 \end{pmatrix}.$$

The eigenvalues and eigenvectors of $\boldsymbol{\Sigma}$ are

$$\lambda_1 = 25.65, \quad \mathbf{a}_1^T = (0.16, 0.99)$$

$$\lambda_2 = 0.25, \quad \mathbf{a}_2^T = (0.987, -0.160).$$

The above decomposition suggests that the first principal component, which accounts for 99 % of the total variance, is fully dominated by X_2 because of its large variance.

The eigenvalues and eigenvectors of $\mathbf{R}$ are

$$\lambda_1 = 1.8, \quad \mathbf{a}_1^T = (0.707, 0.707)$$

$$\lambda_2 = 0.2, \quad \mathbf{a}_2^T = (0.707, -0.707).$$

This decomposition is different from the above decomposition based on the covariance matrix. Note that the first principal component for $\mathbf{R}$ corresponds to the average of the two standardized X variables and accounts for 90 % of the total variability.

6.4.5 Geometric Interpretation

In this section, we will show the geometric interpretation based on a 2-gene expression data set. Figure 6.2a shows the scatter plot between the two genes in the two-gene microarray data set. The blue line shows the first principal component

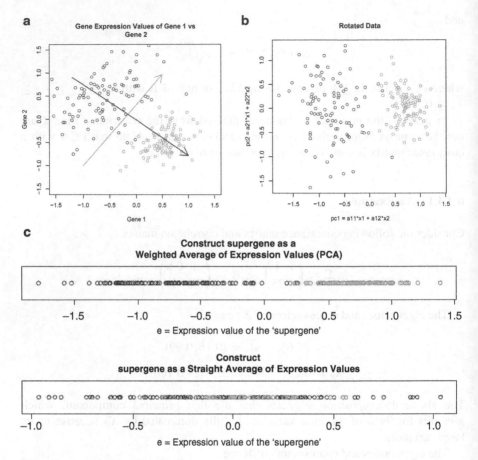

Fig. 6.2 The *left panel* in the first row shows the scatter plot of the two-gene microarray data set. The *blue arrow* shows the first PC direction, and the *orange arrow* is the second PC direction. The *right panel* in the first row shows the scatter plot between the projections of PC1 and PC2, which is essentially a rotation of the *left panel*. The *bottom panel* shows comparisons between the first principal component and the straight average of the two gene values

direction, and the orange line is the second principal component direction. After rotating the data set to the PC directions, the resulting scatter plot is shown in Fig. 6.2b. In this case, the first principal component captures the variability of interest: the contrast between the cases and the controls.

Figure 6.2c shows comparisons between the first principal component and just a straight average of the two gene values. For this particular data set, this figure suggests that the first PC provides a better contrast between the cases and the controls compared to the simple average.

However, there is no guarantee that the first component will capture the variability we are interested in. Figure 6.3 provides an example that the first PC direction does not provide a desirable contrast between cases and controls. The left panel in

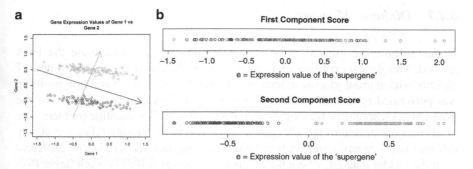

Fig. 6.3 This plot shows an example that the first PC does not provide a contrast between cases and controls. The *left panel* shows the scatter plot between the two genes. The *right panel* shows the two PCs, where the first PC does not show any contrast between cases and controls and the second PC does

Fig. 6.3 provides the scatter plot between the two genes. The blue arrow is the first PC, and the orange arrow corresponds to the second PC. In this example, we can see that the within-group variability is larger than the between-group variability. The right panel in Fig. 6.3 shows the dot plots for the first and second PCs. It shows that the first PC does not provide a good contrast between the cases and the controls, while the second component captures the differences between cases and controls well.

6.4.6 Choosing the Number of Principal Components

The above toy example only includes two genes. In practice, there is generally a large number of genes in a group. For example, in our diabetes data example, the number of genes p in the pathway is 105. One needs to select many PC components to use. There are several ways to select the number of PCs. Here we list a few. See more detailed discussions in Jolliffe [2].

1. Calculate the proportion of variability explained, and then select the smallest k that reaches a pre-defined threshold c.

$$k = \min_{k \in 1, \dots, d} \frac{\sum_{j=1}^{k} \lambda_j}{\sum_{j=1}^{d} \lambda_j} \geq c.$$

Usually, c is some cutoff between 0.7 and 0.95.
2. Retain the components whose eigenvalues are greater than the average eigenvalue $\sum_{j=1}^{d} \lambda_j / d$.
3. Scree plot: plot the eigenvalues and look for a "gap" (an elbow point).

6.4.7 Diabetes Microarray Data

In this section, we use the diabetes microarray data set to illustrate the PCA method. This data set consists of 17 patients with Type II diabetes (cases) and 17 patients with normal glucose tolerance (controls). The gene expression profiling was performed on muscle tissues. Suppose we are interested in a subset of the genes: genes responsible for oxidative phosphorylation. The scientific problem here is whether the profile of the 105 genes in the oxidative phosphorylation pathway different between cases and controls? Note that the data set here is very different from the welder data set, where the number of genes p (=105) is much larger than the sample size n (=34). We will use the PCA method to reduce the dimension of the data first, and then perform further analysis later. For now, we will ignore the class labels (case/control). In other words, we have 34 subjects, 105 genes, and the rank of the input data set is 33, i.e., $n = 34$, $p = 105$, and $d = 33$.

Read in data:

```
> data = read.table("PathwayData.txt", sep = "\t")
> group = data[,1]
> X = as.matrix(data[,2:ncol(data)])
```

Compute the eigenvalues and eigenvectors in R:

```
> S = cov(X)
> lambdas = eigen(S)$ values
> A=eigen(S)$vectors
```

OR

```
> pca = prcomp(X)
> lambdas = pca$sd**2
> A=pca$rotation
```

The eigenvalues we get:

$\hat{\lambda}_1 = 9.307$	$\hat{\lambda}_7 = 1.126$	$\hat{\lambda}_{13} = 0.453$	$\hat{\lambda}_{19} = 0.208$	$\hat{\lambda}_{25} = 0.119$	$\hat{\lambda}_{31} = 0.061$
$\hat{\lambda}_2 = 6.105$	$\hat{\lambda}_8 = 0.954$	$\hat{\lambda}_{14} = 0.436$	$\hat{\lambda}_{20} = 0.199$	$\hat{\lambda}_{26} = 0.104$	$\hat{\lambda}_{32} = 0.049$
$\hat{\lambda}_3 = 2.847$	$\hat{\lambda}_9 = 0.68$	$\hat{\lambda}_{15} = 0.341$	$\hat{\lambda}_{21} = 0.18$	$\hat{\lambda}_{27} = 0.095$	$\hat{\lambda}_{33} = 0.038$
$\hat{\lambda}_4 = 2.589$	$\hat{\lambda}_{10} = 0.6$	$\hat{\lambda}_{16} = 0.306$	$\hat{\lambda}_{22} = 0.165$	$\hat{\lambda}_{28} = 0.094$	$\hat{\lambda}_{34} = 0$
$\hat{\lambda}_5 = 1.667$	$\hat{\lambda}_{11} = 0.572$	$\hat{\lambda}_{17} = 0.265$	$\hat{\lambda}_{23} = 0.149$	$\hat{\lambda}_{29} = 0.079$	$\hat{\lambda}_{35} = 0$
$\hat{\lambda}_6 = 1.416$	$\hat{\lambda}_{12} = 0.483$	$\hat{\lambda}_{18} = 0.243$	$\hat{\lambda}_{24} = 0.134$	$\hat{\lambda}_{30} = 0.066$	$\hat{\lambda}_{36} = 0$

The proportion of variability explained by each component is achieved by

```
> proportion = cumsum(lambdas)/sum(lambdas)
```

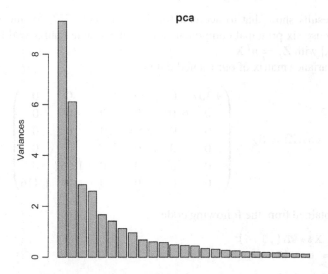

Fig. 6.4 The scree plot for the diabetes microarray data

The relative energy is summarized in the following table:

k	prop. var.	k	prop. var.	k	prop. var.
1	0.29	13	0.896	25	0.982
2	0.48	14	0.91	26	0.985
3	0.568	15	0.92	27	0.988
4	0.649	16	0.93	28	0.991
5	0.701	17	0.938	29	0.993
6	0.745	18	0.946	30	0.995
7	0.78	19	0.952	31	0.997
8	0.809	20	0.958	32	0.999
9	0.831	21	0.964	33	1
10	0.849	22	0.969	34	1
11	0.867	23	0.974	35	1
12	0.882	24	0.978	36	1

The scree plot can be plotted by

```
> barplot(lambdas[1:25])
```

OR

```
> screeplot(pca, npcs - 25)
```

Figure 6.4 is the scree plot for the diabetes microarray data set.

These results show that to account for approximately 75 % of the variability, we need to use six principal components. The rotated data matrix will be $\mathbf{Z}_{n\times6} = [\mathbf{Z}_1, \ldots, \mathbf{Z}_6]$ with $\mathbf{Z}_j = \hat{\mathbf{a}}^T \mathbf{X}$.

The covariance matrix of our rotated data is:

$$\text{cov}(\mathbf{Z}) = \mathbf{S}_{\mathbf{Z}} = \begin{pmatrix} 9.307 & 0 & 0 & 0 & 0 & 0 \\ 0 & 6.105 & 0 & 0 & 0 & 0 \\ 0 & 0 & 2.847 & 0 & 0 & 0 \\ 0 & 0 & 0 & 2.589 & 0 & 0 \\ 0 & 0 & 0 & 0 & 1.667 & 0 \\ 0 & 0 & 0 & 0 & 0 & 1.416 \end{pmatrix}$$

which is obtained from the following code

```
> Z = X%*%A[,1:6]
> Sz = cov(Z)
```

To test for whether the pathway is differentially expressed, we can test for a difference in the mean scores of the six PCs between diabetics and nondiabetics. The hypothesis testing problem becomes

$$H_0 : \mu_{\text{PC1,diabetic}} = \mu_{\text{PC1,normal}}; \quad \cdots \quad \mu_{\text{PC6,diabetic}} = \mu_{\text{PC6,normal}}$$

and

H_α is the general alternative that one of the components is not equal.

This can be done by using the Hotelling's T^2 test. We computed the Hotelling's T^2 statistic $T^2 = 13.168$, which corresponds to a p-value of $p = 0.126$, so we fail to reject the null at the significance level $\alpha = 0.05$ and conclude the pathway is significantly differentially expressed. When a large number of PCs is used, the global Hotelling's t-test is often lack of power because of the noise dimensions.

The R code of the Hotelling's T^2 test is

```
> Z1 = Z[group==1,]
> Z2 = Z[group==0,]
> Zbar1 = apply(Z1, 2, mean)
> Zbar2 = apply(Z2, 2, mean)
> S1 = cov(Z1)
> S2 = cov(Z2)
> Sp = S1*(n1-1)/(n1+n2-2)+S2*(n2-1)/(n1+n2-2)
> T2 = t(Zbar1-Zbar2)%*%solve((1/n1+1/n2)*Sp)
  %*%(Zbar1-Zbar2)
> tf = (n1+n2-p-1)/((n1+n2-2)*p) *T2
> pval = 1- pf(tf, p, (n1+n2-p-1))
```

If we use the scree plot to choose the number of components, we may choose $k = 2$, i.e., two components seem reasonable. We again compute the rotated data matrix as $\mathbf{Z}_{n \times 2} = [\mathbf{Z}_1, \mathbf{Z}_2]$ with $\mathbf{Z}_j = \hat{\mathbf{a}}^T \mathbf{X}$. The covariance matrix of Z is calculated as

$$\mathbf{S_z} = \begin{pmatrix} 9.307 & 0 \\ 0 & 6.105 \end{pmatrix},$$

by the following R code:

```
> Z = X%*%A[,1:2]
> Sz = cov(Z)
```

We can test the same null hypothesis using these two principle components. It is easy to calculate $T^2 = 8.978$ with a corresponding p-value of $p = 0.023$. So at the 95 % significance level, we reject the null and conclude that there is a significant difference in expression of the oxidative phosphorylation pathway between diabetics and normal subjects.

Note that in our setting, we were interested in testing for the pathway effect. There are other applications whose major focus lies in regression, such as principal components regression and factor analysis.

6.5 Factor Analysis

The idea of factor analysis is to describe the covariance relationships among variables in terms of a few, unobservable, underlying random quantities. These quantities usually are called *factors*. Suppose that variables can be grouped by their correlations, i.e., variables within a certain group are highly correlated among themselves, but have relatively small correlations with variables in other groups. Then there may be some unseen constructs or factors that are responsible for such a correlation structure. Factor analysis can be viewed as an extension of PCA, as it also attempts to approximate the covariance structure $\boldsymbol{\Sigma}$. However, factor analysis usually does not have a unique answer, and thus is an "art" in some sense.

6.5.1 Orthogonal Factor Model

Assume that $\mathbf{X}$ is a random vector with mean $\boldsymbol{\mu}$ and covariance $\boldsymbol{\Sigma}$. Further assume that $\mathbf{X}$ is linearly dependent on m random variables, $F_1, \ldots, F_m$. The model can be written as

$$\mathbf{X}_{p \times 1} = \boldsymbol{\mu}_{p \times 1} + \mathbf{L}_{p \times m} \mathbf{F}_{m \times 1} + \boldsymbol{\varepsilon}_{p \times 1}$$

where ε_i is a vector of p sources of errors, and

$$\mu_i = \text{mean of variable } i.$$

$$\varepsilon_i = \text{error for observation } i.$$

$$\mathbf{F}_j = j\text{th common factor}.$$

$$l_{ij} = \text{loading of the } i\text{th variable on the } j\text{th factor}.$$

$\mathbf{F}$ and ε are independent and

$$E(\mathbf{F}) = \mathbf{0} \text{ and } \text{cov}(\mathbf{F}) = \mathbf{I}.$$

$$E(\varepsilon) = \mathbf{0} \text{ and } \text{cov}(\varepsilon) = \mathbf{\Psi} = \text{diag}(\psi_1, \dots, \psi_p).$$

It is straightforward to show that

1. $\text{cov}(\mathbf{X}) = \mathbf{\Sigma} = \mathbf{L}\mathbf{L}^T + \mathbf{\Psi}$, so $\text{var}(\mathbf{X}_i) = l_{i1}^2 + \cdots + l_{im}^2 + \psi_i$ and $\text{cov}(\mathbf{X}_i, \mathbf{X}_k) = l_{i1}l_{k1} + \cdots + l_{im}l_{km}$.
2. $\text{cov}(\mathbf{X}, \mathbf{F}) = \mathbf{L}$, so $\text{cov}(\mathbf{X}_i, \mathbf{F}_j) = l_{ij}$.

This assumes that the model is *linear* in the common factors.

It follows that we can decompose the variance of the ith variable as

$$\underbrace{\sigma_{ii}^2}_{\text{var}(\mathbf{X}_i)} = \underbrace{l_{i1}^2 + \cdots + l_{im}^2}_{\text{communality}} + \underbrace{\psi_i}_{\text{specific variance}}.$$

Under this model, the $p(p+1)/2$ elements in $\mathbf{\Sigma}$ can be reproduced by $m \times p$ factor loadings and p specific variances. If $m = p$, any matrix can be written as exactly $\mathbf{L}\mathbf{L}^T$ with $\mathbf{\Psi} = \mathbf{0}$, so factor analysis is only interesting when m is smaller than p. Typically, each column of $\mathbf{X}$ is standardized to have mean 0 and variance 1. If the sample correlation matrix is close to identity, factor analysis is not helpful, since no specific factors dominate.

Factor analysis often proceeds as follows:

1. Impose conditions that allow $\mathbf{L}$ and $\mathbf{\Psi}$ to be estimated.
2. Rotate the loading matrix $\mathbf{L}$ using an orthogonal matrix designed to facilitate interpretation.
3. After loadings and specific variances are determined, identify and estimate factors.

6.5.2 Estimating the Parameters

Two popular methods are often used for parameter estimation: principal component analysis method and maximum likelihood.

6.5.2.1 The Principal Component Method

Suppose $\boldsymbol{\Sigma}$ is a covariance matrix and has eigenvectors $\mathbf{a}_i$ with the corresponding eigenvalues λ_i, where $\lambda_1 \geq \lambda_2 \geq \cdots \geq \lambda_p \geq 0$. The eigen-analysis of $\boldsymbol{\Sigma}$ gives

$$\boldsymbol{\Sigma} = \lambda_1 \mathbf{a}_1 \mathbf{a}_1^T + \cdots + \lambda_p \mathbf{a}_p \mathbf{a}_p^T$$

$$= \left[\sqrt{\lambda_1} \mathbf{a}_1, \cdots, \sqrt{\lambda_p} \mathbf{a}_p \right] \begin{bmatrix} \sqrt{\lambda_1} \mathbf{a}_1 \\ \vdots \\ \sqrt{\lambda_p} \mathbf{a}_p \end{bmatrix}$$

$$= \mathbf{L} \mathbf{L}^T.$$

The above eigen-analysis provides as many factors as the number of variables in the original data set. Hence, $\psi_i = 0$, and no dimension reduction is provided. Our goal is to find a few common factors. Suppose $\mathbf{S}$ is the sample covariance matrix. One way is to use the eigen-analysis to select only the first m principal components where m is chosen such that the last $p - m$ components contribute relatively little variability. It follows that $\mathbf{L}$ and $\boldsymbol{\Psi}$ can be estimated by

$$\hat{\mathbf{L}} = [\sqrt{\hat{\lambda}_1} \hat{\mathbf{a}}_1, \ldots, \sqrt{\hat{\lambda}_m} \hat{\mathbf{a}}_m]$$

$$\hat{\boldsymbol{\Psi}} = \begin{bmatrix} \hat{\psi}_1 & 0 & \cdots & 0 \\ 0 & \hat{\psi}_2 & \cdots & 0 \\ \vdots & \vdots & \ddots & \vdots \\ 0 & 0 & \cdots & \hat{\psi}_p \end{bmatrix},$$

where $\hat{\psi}_i = s_{ii}^2 - \sum_{j=1}^{m} \hat{l}_{ij}^2$.

6.5.2.2 The Maximum Likelihood Method

Assume that $\mathbf{F}$ and $\boldsymbol{\varepsilon}$ are normally distributed, then $\mathbf{X}_j - \boldsymbol{\mu} = \mathbf{L} \mathbf{F}_j + \boldsymbol{\varepsilon}_j$ is also normal, then the likelihood can be written as

$$l(\boldsymbol{\mu}, \boldsymbol{\Sigma}) = (2\pi)^{np/2} |\boldsymbol{\Sigma}|^{-n/2} \exp \left\{ -\frac{1}{2} \mathrm{tr} \left[\boldsymbol{\Sigma}^{-1} \left(\sum_{j=1}^{n} (\mathbf{x}_j - \bar{\mathbf{x}})(\mathbf{x}_j - \bar{\mathbf{x}})^T + n(\bar{\mathbf{x}} - \boldsymbol{\mu})(\bar{\mathbf{x}} - \boldsymbol{\mu})^T \right) \right] \right\}$$

where $\boldsymbol{\Sigma} = \mathbf{L} \mathbf{L}^T + \boldsymbol{\Psi}$ and $\mathbf{L}$ is an $p \times m$ matrix. We need to impose a uniqueness condition $\mathbf{L}^T \boldsymbol{\Psi}^{-1} \mathbf{L} = \boldsymbol{\Delta}$, where $\boldsymbol{\Delta}$ is a diagonal matrix. Then $\hat{\mathbf{L}}$ and $\hat{\boldsymbol{\Psi}}$ can be obtained by the maximum likelihood method.

Note that the factor loadings, $\mathbf{L}$, are determined only up to an orthogonal matrix $\mathbf{T}$. In fact, we can define $\mathbf{L}^* = \mathbf{L}\mathbf{T}$, and $\mathbf{L}^*$ and $\mathbf{L}$ give the same representation, i.e.,

$$\boldsymbol{\Sigma} = \mathbf{L} \mathbf{L}^T + \boldsymbol{\Psi} = (\mathbf{L}^*)(\mathbf{L}^*)^T + \boldsymbol{\Psi}.$$

Fig. 6.5 A subpart of the
heat map for the correlation
between the proteins

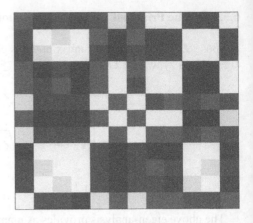

The estimated factor loadings $\hat{\mathbf{L}}$ are often not readily interpretable. As rotating
the loadings does not affect factorization, we can rotate the loadings to achieve a
"simpler" structure such that the loadings of each factor are high on a subset of
variables and low on the other variables. The *varimax* criterion defined below can
be used to determine optimal rotations

$$V = \frac{1}{p} \sum_{j=1}^{m} \left[\sum_{i=1}^{p} (\tilde{l}_{ij}^*)^4 - \frac{\sum_{i=1}^{p} (\tilde{l}_{ij}^*)^2}{p} \right],$$

where $\hat{l}_{ij}^* = \hat{l}_{ij}^* / \sum_{j=1}^{m} \hat{l}_{ij}^*$. Note that maximizing V corresponds to "spread out" the
squares of the loadings on each factor as much as possible.

6.5.3 An Example

We use a proteomic data example to illustrate factor analysis. This data set contains
the intensities of top $p = 17$ proteins measured from blood of $n = 103$
subjects using mass spectrometry. A question of interest is whether we identify any
underlying structures of these proteins, e.g., whether they are related.

Figure 6.5 shows part of a heat map of the correlation structure between all the
proteins. The heatmap is a convenient way to reveal correlation structures among
variables by appropriately arranging the variables based on their correlations. This
map is generated by the following R code

```
> heatmap(cor(X))
```

This plot shows that some variables have good correlations. We estimated the factors
using the PCA method. Figure 6.6 shows the scree plot of the PCA for the proteomic
data, suggesting the data can be reasonably summarized by three common factors,
which explains 95 % of the variability.

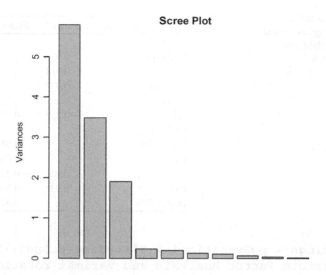

Fig. 6.6 Scree plot for PCA of the proteomic data example

Table 6.3 Factor loadings estimated using the PCA method after varimax rotation

Protein mass/charge	Factor1	Factor2	Factor3
2,854.358		−0.933	
5,281.05	−0.935		
7,572.76	−0.957		
7,944.713	−0.961		
8,944.587		−0.948	
9,546.51	0.243		0.918
10,320.412		−0.949	
11,153.307			0.971
13,400.023			0.976
15,217.792	−0.938		−0.218
15,965.698	−0.922		−0.32
16,771.596			0.917

Table 6.3 provides the loadings of the three factors after varimax rotation, where the loadings that are smaller than 0.2 are shown in blank for the ease of interpretability.

The loadings can also be estimated using the maximum likelihood method. Table 6.4 provides the loadings after varimax rotation, with smaller coefficients (<0.2 in absolute value) shown in blank in the table.

The R code for the principal component method is

```
### Standardize X ###
> X.std = scale(X, center = T, scale =T)
> pc = prcomp(X.std)
> plot(pc, main = "Scree Plot")
### Get Factor Loadings ###
```

Table 6.4 The factor
analysis loadings estimated
from PCA method, after
varimax rotation

Protein mass/charge	Factor1	Factor2	Factor3
2,854.358			0.906
5,281.05	0.932		
7,572.76	0.911		
7,944.713	0.973		
8,944.587			0.917
9,546.51	−0.253	0.875	
10,320.412			0.937
11,153.307		0.95	
13,400.023		0.985	
15,217.792	0.901	−0.225	
15,965.698	0.936	−0.309	
16,771.596		0.93	

```
> loadings = pc$rotation[,1:3]%*%diag(pc$sd[1:3])
### Perform Factor Analysis and varimax rotation ###
> fa = varimax(loadings)
```

The *R* outputs are

```
> fa$loadings
Loadings:
             [,1]      [,2]      [,3]
2854.358     0.181    -0.933
5281.05     -0.935              -0.134
7572.76     -0.957     0.108
7944.713    -0.961     0.141    -0.177
8944.587              -0.948
9546.51      0.243               0.918
10320.412    0.100    -0.949    -0.118
11153.307    0.127               0.971
13400.023    0.159               0.976
15217.792   -0.938     0.102    -0.218
15965.698   -0.922              -0.320
16771.596    0.185     0.170     0.917
             [,1]      [,2]      [,3]
SS loadings    4.626     2.762     3.800
Proportion Var 0.386     0.230     0.317
Cumulative Var 0.386     0.616     0.932
```

The *R*-code for the maximum likelihood method is

```
## Gets the varimax rotated factor loadings
## under normality assumption
> factanal(X,3, rotation = "varimax")
```

The R outputs are

```
> factanal(X,3)$loadings
Loadings:
           Factor1 Factor2 Factor3
2854.358   -0.184           0.906
5281.05     0.932  -0.140  -0.102
7572.76     0.911          -0.117
7944.713    0.973  -0.172  -0.140
8944.587                    0.917
9546.51    -0.253   0.875
10320.412  -0.101  -0.121   0.937
11153.307  -0.141   0.950
13400.023  -0.156   0.985
15217.792   0.901  -0.225  -0.107
15965.698   0.936  -0.309
16771.596  -0.177   0.930  -0.166
               Factor1 Factor2 Factor3
SS loadings     4.523   3.725   2.635
Proportion Var  0.377   0.310   0.220
Cumulative Var  0.377   0.687   0.907
```

Although the primary interest for factor analysis is usually in the factor loadings, estimates for the unobserved random factors $\mathbf{F}_j$ may also be calculated. The jth factor score is given by

$$\hat{\mathbf{f}}_j = \text{estimate of the values } \mathbf{f}_j \text{ attained by } \mathbf{F}_j \text{ (the } j\text{th case).}$$

Assume that $\hat{l}_{ij}$ and $\hat{\psi}_i$ are the true values, then the rotated loadings can be estimated via

1. The weighted least square method

$$\hat{\mathbf{f}}_j = (\hat{\mathbf{L}}^T \hat{\boldsymbol{\psi}}^{-1} \hat{\mathbf{L}})^{-1} \hat{\mathbf{L}}^T \hat{\boldsymbol{\psi}}^{-1} (\mathbf{X}_j - \hat{\boldsymbol{\mu}})$$

2. Or the regression method

$$\hat{\mathbf{f}}_j = \hat{\mathbf{L}}^T \mathbf{S}^{-1} (\mathbf{X}_j - \overline{\mathbf{X}})$$

6.6   Linear Discriminant Analysis

Principal component analysis does not utilize information of class labels and is hence a unsupervised learning method. One example given above shows that sometimes PCA does not provide a good direction to contrast cases and controls.

However, the contrast between cases and controls is usually the major research interest. Linear discriminant analysis (LDA) provides one possible way to achieve this goal by using information of class labels in dimension reduction of variables and is hence a supervised learning method.

Assume we have a training data set, which contains two groups for simplicity. We use X as the group labels. Denote by $X = 1$ the disease group, which contains n_1 subjects, and $X = 2$ the non-disease group, which contains n_2 subjects. Each subject has measurements of p variables (e.g., p genes or proteins), which are denoted in a p-dimensional vector, $\mathbf{Y} = (Y_1, \ldots, Y_p)^T$. The purpose of linear discriminant analysis (LDA) is to identify a *best linear combination* of the p outcomes $Z = \sum_{j=1}^{p} a_j Y_j$ that best separates the two groups, i.e., elucidates the differences between the two groups. In this section, we only deal with the two-group classification problem. Note that the LDA method can also be applied to multiple groups.

In this section, we will use the diabetes microarray data set to illustrate the LDA method. The data set consists of 27 subjects. Each subject belongs to one of the following two groups: 14 subjects with Type 2 diabetes (T2D) and the remaining 13 subjects who are normal (i.e., non-diabetic). Each subject has measurements on seven genes in the mitochondrial activity pathway. We are interested in answering the following questions:

1. What is the overall gene profile difference between diabetics and controls?
2. What is the relative importance of the genes for discriminating diabetes from controls?

6.6.1   Two-Group Linear Discriminant Analysis

In this subsection, we formally define the two-group linear discriminant analysis and apply it to the diabetic data. We first introduce some general notation. Assume that we have two groups of data, each subject has p variables. For group 1, there are n_1 subject and the ith subject data are denoted by

$$\mathbf{Y}_{1i} = (Y_{1i1}, \ldots, Y_{1ip})^T, \quad (i = 1, \ldots, n_1).$$

For group 2, there are n_2 subjects, and the ith subject data are denoted by

$$\mathbf{Y}_{2i} = (Y_{2i1}, \ldots, Y_{2ip})^T, \quad (i = 1, \ldots, n_2).$$

We make the following assumptions:

1. The subjects within each group have the same mean, and different groups have different means. That is,

$$E(\mathbf{Y}_{1i}) = \boldsymbol{\mu}_1, \text{ and}(\mathbf{Y}_{2i}) = \boldsymbol{\mu}_2.$$

2. The two groups share the same covariance structure

$$\text{cov}(\mathbf{Y}_{1i}) = \text{cov}(\mathbf{Y}_{2i}) = \mathbf{\Sigma}.$$

The common covariance between different groups is a relatively strong assumption in multivariate classification problems. We will study the classification under this assumption in this section, and study more general classification methods later.

Under the above assumptions, we can define the LDA method. The idea of LDA is to find a linear combination of $\mathbf{Y}$ that maximizes the standardized distance between the two group means. Define a $p \times 1$ vector $\mathbf{a}$, the transformation of $\mathbf{a}^T \mathbf{Y}$ will transform $\mathbf{Y}$ to a scalar. Thus, we have $Z_{1i} = \mathbf{a}^T \mathbf{Y}_{1i}$ as the transformed scalar for group 1 and $Z_{2i} = \mathbf{a}^T \mathbf{Y}_{2i}$ for group 2. The LDA method finds the optimal $\mathbf{a}$ by maximizing the two-sample t test using the transformed data Z. Specifically, the LDA maximizes the objective function

$$\lambda = \frac{(\hat{Z}_1 - \hat{Z}_2)^2}{s_z^2} = \frac{[\mathbf{a}^T (\overline{\mathbf{Y}}_1 - \overline{\mathbf{Y}}_2)]^2}{\mathbf{a}^T \mathbf{S}_p \mathbf{a}},$$

where $\mathbf{S}_p = \hat{\mathbf{\Sigma}}$ is the pooled sample covariance matrix. Note that λ defined above is proportional to the T_z^2, the two-sample T^2 statistic using the Z data: $\lambda = (n_1^{-1} + n_2^{-1})T_z^2$. Some simple calculations given below show that the LDA solution is

$$\hat{\mathbf{a}} \propto \mathbf{S}_p^{-1}(\overline{\mathbf{Y}}_1 - \overline{\mathbf{Y}}_2).$$

We would like to make several remarks about the above LDA solution.

1. The solution $\hat{\mathbf{a}}$ is not unique but the direction is unique.
2. The magnitude of $\hat{\mathbf{a}}$ measures the relative importance of the individual variables $(Y_1, \ldots, Y_p)$ for discriminating the two groups.
3. In order to make $\mathbf{S}_p$ non-singular, one requirement is $n_1 + n_2 - 2 > p$, i.e., standard LDA does not work for high dimensional data (e.g., p is larger than the number of subjects, $n_1 + n_2$).
4. If p is larger than n, the standard LDA method can be modified by using a regularization technique. For example, we can replace $\mathbf{S}_p$ by $\mathbf{S}_p + \delta \mathbf{I}$, where δ is a small constant. This is very similar to ridge regression in linear regression.
5. Note that $\max_{\mathbf{a}} \lambda(\mathbf{a}) = \hat{\lambda}$, where $\hat{\lambda}$ is the eigenvalue of $\mathbf{S}_p^{-1}\mathbf{D}$, where $\mathbf{D} = (\overline{\mathbf{Y}}_1 - \overline{\mathbf{Y}}_2)(\overline{\mathbf{Y}}_1 - \overline{\mathbf{Y}}_2)^T$.
6. The LDA coefficient vector $\hat{\mathbf{a}}$ is the eigenvector associated with $\hat{\lambda}$.

Both PCA and LDA are to find a linear combination of p outcome variables. PCA tries to find the direction that maximizes the variability of the p variables, while LDA tries to find the direction by maximizing the weighted group difference. Due to the different objectives, the resulting direction usually is not the same. Figure 6.7 shows two simulated two-gene microarray data sets with two groups. The left panel (a) in the figure shows that the PCA direction (the blue line) is almost the same as

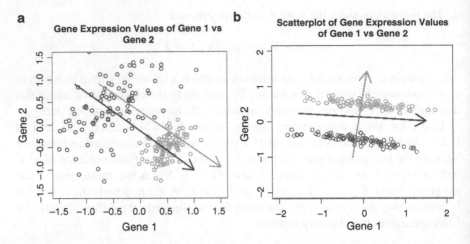

Fig. 6.7 The LDA and PCA comparisons for two simulated two-gene microarray data sets. The *blue line* is the first PC direction, and the *orange line* corresponds to the LDA direction

the LDA direction (the orange line), as this direction maximizes both the variability and best separates the two groups. The right panel (b) in the figure is an example that the PCA direction (the blue line) is almost orthogonal to the LDA direction (the orange line).

6.6.1.1 Derivation of the LDA Solution

In this subsection, we will derive the LDA solution. Note that

$$\{\mathbf{a}^T(\overline{\mathbf{Y}}_1 - \overline{\mathbf{Y}}_2)\}^2 = [\{\mathbf{a}^T\mathbf{S}_p^{1/2}\}\{\mathbf{S}_p^{-1/2}(\overline{\mathbf{Y}}_1 - \overline{\mathbf{Y}}_2)\}]^2.$$

Using the Cauchy-Schwarz inequality, we have

$$\{\mathbf{a}^T(\overline{\mathbf{Y}}_1 - \overline{\mathbf{Y}}_2)\}^2 \le \{\mathbf{a}^T\mathbf{S}_p^{1/2}\mathbf{S}_p^{1/2}\mathbf{a}\}\{(\overline{\mathbf{Y}}_1 - \overline{\mathbf{Y}}_2)^T\mathbf{S}_p^{-1/2}\mathbf{S}_p^{-1/2}(\overline{\mathbf{Y}}_1 - \overline{\mathbf{Y}}_2)\}.$$

Thus, we have

$$\lambda = \frac{[\mathbf{a}^T(\overline{\mathbf{Y}}_1 - \overline{\mathbf{Y}}_2)]^2}{\mathbf{a}^T\mathbf{S}_p\mathbf{a}} \le (\overline{Y}_1 - \overline{Y}_2)^T\mathbf{S}_p^{-1}(\overline{Y}_1 - \overline{Y}_2).$$

The equality is satisfied subject to a multiplicative factor when $\mathbf{S}_p^{1/2}\mathbf{a} = \mathbf{S}_p^{-1/2}(\overline{\mathbf{Y}}_1 - \overline{\mathbf{Y}}_2)$, i.e.,

$$\mathbf{a} = \mathbf{S}_p^{-1}(\overline{\mathbf{Y}}_1 - \overline{\mathbf{Y}}_2).$$

6.6.1.2 Relation Between the LDA Method and the Two-Sample Hotelling's T^2 Test

The LDA method has a strong relation with the two-sample Hotelling's T^2 test. Note that

$$\max_{\mathbf{a}} = (\overline{\mathbf{Y}}_1 - \overline{\mathbf{Y}}_2)^T \mathbf{S}_p^{-1} (\overline{\mathbf{Y}}_1 - \overline{\mathbf{Y}}_2),$$

which is proportional to the two-sample Hotelling's T^2 test for testing $H_0 : \mu_1 = \mu_2$. In fact, the two-sample Hotelling's T^2 test has the omnibus property, that is, exactly the LDA objective function.

By using the LDA method, we can test a null hypothesis that the LDA coefficient vector H_0: $\mathbf{a} = \boldsymbol{\Sigma}^{-1}(\mu_1 - \mu_2) = 0$. This is equivalent to the null hypothesis H_0: $\mu_1 = \mu_2$. Thus, we can use the Hotelling's T^2 statistic to perform the above hypothesis test. Equivalently, one can use the Wilks Λ test

$$\Lambda = \frac{1}{1 + \hat{\lambda}},$$

where has a chi-square distribution with p DFs asymptotically.

6.6.1.3 Relation Between LDA and Multiple Regression

The LDA can also be viewed as a solution of a multiple regression problem. Specifically, define a new "outcome" W as follows:

$$W_i = \begin{cases} \frac{n_2}{n_1 + n_2} \mathbf{Y}_{1i} \in G_1 \\ -\frac{n_1}{n_1 + n_2} \mathbf{Y}_{2i} \in G_2 \end{cases},$$

where G_i denotes group i ($i = 1, 2$). A linear regression model of the "outcome" W on $\mathbf{Y}$ can be fitted by treating $\mathbf{Y}_i = (Y_{i1}, \ldots, Y_{ip})$ as p independent covariates.

$$W_i = b_0 + b_1 Y_{i1} + b_2 Y_{i2} + \cdots + b_p Y_{ip} + \varepsilon_i, \tag{6.6}$$

where $E(\varepsilon_i) = 0$ and $\text{var}(\varepsilon_i) = \sigma^2$.

It can be shown easily that the least squares estimator $\hat{\mathbf{b}}$ of (6.6) is proportional to the LDA coefficient $\hat{\mathbf{a}}$ as

$$\hat{\mathbf{b}} = \frac{n_1 n_2}{(n_1 + n_2)(n_1 + n_2 - 2 + T^2)} \hat{\mathbf{a}},$$

where T^2 is the Hotelling's T^2 statistic.

The R^2 from fitting the linear regression model (6.6) is closely related to the maximized LDA objective function $\hat{\lambda}$ and the Hotelling's T^2 as

$$R^2 = (\overline{\mathbf{Y}}_1 - \overline{\mathbf{Y}}_2)^T \hat{\mathbf{b}} = \frac{n_1 n_2}{(n_1 + n_2)(n_1 + n_2 - 2 + T^2)} \hat{\lambda} = \frac{T^2}{(n_1 + n_2 - 2 + T^2)}.$$

This connection between LDA and linear regression was first discovered by R.A. Fisher.

6.6.1.4   Test for Redundant Variables in LDA

One important questions in linear discrimination analysis is to identify whether a subgroup of variables are redundant (in classification). Consider we have $p + q$ outcome variables, denoted as $Y_1, Y_2, \ldots, Y_p, Y_{p+1}, \ldots, Y_{p+q}$. The problem we are interested in is to test the last q variables $Y_{p+1}, \ldots, Y_{p+q}$ are redundant for discriminating the two groups (e.g., disease/no disease).

Note that the LDA dimension reduction is a linear combination of the outcome variables:

$$Z = \mathbf{a}^T \mathbf{Y} = a_1 Y_1 + a_2 Y_2 + \cdots + a_p Y_p + a_{p+1} Y_{p+1} + \cdots + a_{p+q} Y_{p+q}.$$

To test whether the last q variables are redundant is to test whether $H_0: a_{p+1} = \cdots = a_{p+q} = 0$ is true. By using the connections between LDA and multiple linear regression, this test can be done by a F test:

$$F = \frac{n_1 + n_2 - p - q - 1}{q} \frac{R_{p+q}^2 - R_p^2}{1 - R_{p+q}^2},$$

where R_{p+q}^2 and R_p^2 are the correlation coefficients from the multiple regression of W on $\mathbf{Y}$ with $p + q$ or p variables, respectively.

6.6.2   An Example

Recall the earlier example in this section, that we have 27 subjects, where 14 of them are with Type 2 diabetes (T2D) and the remaining 13 are normal patients. There are seven genes related to mitochondrial activity measured. The questions we want to answer are:

1. What is the gene profile that best discriminates diabetics from controls?
2. What is the relative importance of the genes for discriminating diabetes from controls?

The sample means within each group are

$$
\overline{\mathbf{Y}}_1 = \begin{bmatrix} -0.253 \\ -0.357 \\ -0.368 \\ -0.402 \\ 0.096 \\ 0.049 \\ -0.156 \end{bmatrix} \text{ and } \overline{\mathbf{Y}}_2 = \begin{bmatrix} 0.273 \\ 0.384 \\ 0.397 \\ 0.433 \\ -0.103 \\ -0.053 \\ 0.168 \end{bmatrix}.
$$

The pooled sample covariance matrix is

$$
\mathbf{S}_p = \begin{bmatrix}
0.965 & 0.412 & -0.056 & 0.084 & -0.015 & -0.101 & -0.033 \\
0.412 & 0.892 & -0.088 & 0.063 & -0.026 & -0.049 & -0.192 \\
-0.056 & -0.088 & 0.882 & 0.008 & 0.447 & 0.054 & -0.065 \\
0.084 & 0.063 & 0.008 & 0.852 & -0.297 & 0.204 & 0.243 \\
-0.015 & -0.026 & 0.447 & -0.297 & 1.029 & 0.125 & -0.172 \\
-0.101 & -0.049 & 0.054 & 0.204 & 0.125 & 1.037 & 0.421 \\
-0.033 & -0.192 & -0.065 & 0.243 & -0.172 & 0.421 & 1.012
\end{bmatrix}.
$$

From the LDA solution, $\hat{\mathbf{a}} = \mathbf{S}_p^{-1}(\overline{\mathbf{Y}}_1 - \overline{\mathbf{Y}}_2)$, the LDA estimated coefficient is

$$
\hat{\mathbf{a}} = (-0.13, -0.917, -1.198, -0.722, 0.344, 0.416, -0.517)^T.
$$

The above results can be achieved from the following R codes:

```
Y.std = scale(Y, center = T, scale = T)
Y1 = Y.std[X==1,]
Y2 = Y.std[X==0,]
Ybar1 = apply(Y1,2, mean)
Ybar2 = apply(Y2,2, mean)
S1 = cov(Y1)
S2 = cov(Y2)
Sp = S1*(n1-1)/(n1+n2-2)+S2*(n2-1)/(n1+n2-2)
a = solve(Sp)%*%(Ybar1 - Ybar2)
```

Alternatively, we can directly use the lda function from the MASS package:

```
a = lda(Y.std, X)$scaling
```

Then we find:

$$
\hat{\mathbf{a}}_* = (-0.081, -0.574, -0.751, -0.453, 0.215, 0.261, -0.324)^T.
$$

Note that $\hat{\mathbf{a}}_* = 0.63\hat{\mathbf{a}}$. The magnitude is different, but the direction is the same. These results suggest that genes 2, 3, and 4 are more important in discriminating T2D from normal patients.

To compare, individual gene t test results are summarized as follows:

Gene	a	a^*	p-value
Gene 1	−0.13	−0.081	0.177
Gene 2	−0.917	−0.574	0.052
Gene 3	−1.198	−0.751	0.045
Gene 4	−0.722	−0.453	0.027
Gene 5	0.344	0.215	0.615
Gene 6	0.416	0.261	0.797
Gene 7	−0.517	−0.324	0.411

At the significance level $\alpha = 0.05$, these individual tests also suggest that the genes 3 and 4 are individually differentially expressed, while gene 2 is marginally differentially expressed. These genes have higher LDA weights.

6.7   Classification Methods

6.7.1   Introduction of Classification Methods

The LDA method is a special classification method in multivariate analysis. It requires that two groups (for a two-group classification problem) share the same covariance structure. There are many other classification methods in multivariate analysis, which require different assumptions. We will study some of them in this section. See, e.g., Hastie et al. [4] for an introduction of these methods and others. For simplicity, we will focus on binary classification problems as well in this section.

Assume we have a binary outcome, denoted by Y, where $Y = 1$ if a subject has disease and 0 if a subject does not have disease. Denote the p covariates by $\mathbf{X} = (X_1, \ldots, X_p)^T$, i.e., $\mathbf{X}$ is a $p \times 1$ random vector. For example, these covariates can be p genes for each subject. For n subjects, the layout of the data set is

Subject	Predictors	Group (binary clinical outcome)
1	$X_{11}, X_{12}, \ldots, X_{1p}$	Y_1
2	$X_{21}, X_{22}, \ldots, X_{2p}$	Y_2
$\vdots$	$\vdots$	$\vdots$
n	$X_{n1}, X_{n2}, \ldots, X_{np}$	Y_n

The goal of a classification method is to use the covariates $\mathbf{X}$ to predict (classify) the disease status of a new (incoming) subject.

The way to construct a decision rule is to use the training data set $(\mathbf{X}_i, Y_i)$ to construct a "best" classification rule to predict the class membership of Y_i. Take a linear decision rule as an example, the decision rule can be

$$\hat{\boldsymbol{\alpha}}^T \mathbf{X}_i > c \Rightarrow \hat{Y}_i = 1,$$

$$\hat{\boldsymbol{\alpha}}^T \mathbf{X}_i \leq c \Rightarrow \hat{Y}_i = 0.$$

Here, $\hat{\boldsymbol{\alpha}}$ is estimated from the training data set $(\mathbf{X}_i, Y_i)$. When there is a new subject coming in, we can classify the subject based on $\mathbf{X}_{\text{new}}$ (i.e., there is no Y_{new}). The decision rule will be if $\boldsymbol{\alpha}^T \mathbf{X}_{\text{new}} > c$, we will predict $\hat{Y}_{\text{new}} = 1$.

The performance of the decision rule can be investigated as follows. Use an independent testing data $(\mathbf{X}_i^*, Y_i^*)$ to estimate the misclassification rate, e.g., pretend Y_i^* are unknown and use $\mathbf{X}_i^*$ to predict Y_i^*. The misclassification rate can be calculated by comparing $\hat{Y}_i^*$ with Y_i^*, where $\hat{Y}_i^* = I[\hat{\boldsymbol{\alpha}}^T \mathbf{X}_i^* > c]$.

We use a leukemia microarray data set to illustrate several classification methods. The training data set includes 27 patients with Acute Myelogenous Leukemia (AML), which is coded as $Y = 0$, and 16 patients with acute lymphoblastic leukemia (ALL), which is coded as $Y = 1$. Gene expression profiling was performed, and the eight most differentially expressed genes were found and used for classification (i.e., $p = 8$). The testing (new) data includes 29 leukemia patients within known leukemia status. Expressions of the same eight genes were measured. The goal is to predict the type of leukemia for new patients in the testing data and check how the classification rules work.

6.7.1.1 Error Rates

In order to compare the performance of different classifiers, or even just want to evaluate the performance of a particular classification method, we need to have a definition of (misclassification) error rate. There are several different types of error rates.

True Error Rate

The true error rate is defined as follows:

$$\text{TER} = \Pr(T - |Y = 1) + \Pr(T + |Y = 0) = \text{FNR} + \text{FPR},$$

where $T+$ means test positive, e.g., $T+ = \boldsymbol{\alpha}^T \mathbf{X} > c$ and $T- = \boldsymbol{\alpha}^T \mathbf{X} < c$, $(\mathbf{X}, Y)$ is a new observation, and FNR = false-negative rate and FPR = false-positive rate.

Apparent Error Rate

The apparent error rate uses the training data to estimate $\hat{\alpha}$ and estimate the error rate also using the training data by comparing $\hat{Y}_i$ with Y_i. The apparent error rate often underestimates the true error rate. Specifically, using the observed and predicted data, we can construct a 2×2 table as follows:

	Predicted		
Observed	1	2	Total
1	n_{11}	n_{12}	n_1
2	n_{21}	n_{22}	n_2

Apparent error rate is defined as $(n_{12} + n_{21})/(n_1 + n_2)$, where all the calculations are based on the training data.

Other Unbiased Estimation of the True Error Rate

Here, we provide two unbiased estimation methods of the true error rate:

1. We can use separate test data to estimate the true error rate: Use the training data to estimate $\hat{\alpha}$ and estimate the error rate using the testing data by comparing $\hat{Y}_i^*$ with Y_i^*.
2. We can also use k-fold cross validation to estimate the true error rate. This is usually used when there is no testing data available. The idea is to randomly divide the training data into k equal size groups. Use $(k - 1)$ groups of data to estimate α and estimate the error rate using the remaining kth group by comparing $\hat{Y}_i$ with Y_i. This procedure should be done for a large number of times and take an average to estimate the error rate.

There are many classification methods, such as the K-nearest neighbors method, linear (quadratic) discriminant analysis, logistic regression, support vector machine (SVM), tree and boosting, and neural network. The first four will be discussed in this section.

6.7.2   k-Nearest Neighbor Method

The idea of the k-nearest neighbor method is to use the observations in the training set T that are closest to a new patient's data $\mathbf{X}$ to predict the new patient's class label $\hat{Y}$. Specifically, for a new observation $\mathbf{X}$, define $N_k(\mathbf{X}) =$ neighborhood of $\mathbf{X} = k$ points in T "closest" to $\mathbf{X}$. The decision rule is that if the majority of the observations in the neighborhood of $\mathbf{X}$ have disease (i.e., $Y = 1$), then we will classify the new subject as a disease case ($\hat{Y} = 1$), i.e.,

$$\hat{Y}(\mathbf{X}) = I\left(\frac{1}{k} \sum_{\mathbf{X}_i \in N_k(\mathbf{X})} Y_i \geq \frac{1}{2}\right).$$

Note that closeness here implies some sort of distance metric. The most commonly used distance is the Euclidean distance or the Mahalanobis distance.

6.7.2.1   An Example

Figure 6.8 provides a simulated training data set and a simulated test data set. The left panel in Fig. 6.8 shows all data points in both the training data and the test data, where the green dots show the simulated cases in the training data set (i.e., $Y = 1$), the red dots are the simulated controls with class label $Y = 0$, and the blue dots show those observations in the test data set, which do not have a class label. The KNN decision regions, along with the actual class labels for the test data set, are shown in the right panel. The red region is the predicted control region (i.e., $\hat{Y} = 0$), and the green regions correspond to the predicted region for the cases (i.e., $\hat{Y} = 1$). This plot shows that KNN is not a linear decision rule, and under the current parameter setting (k), there are several misclassified subjects in the test data set.

We further illustrate the KNN method using the leukemia data set. The training data set of the leukemia data contains 27 AML patients ($Y_i = 0$) and 16 ALL patients ($Y_i = 1$). The data contain $p = 8$ genes. In addition to the training data set, there are 29 leukemia patients with unknown leukemia status and eight genes measured. In R, the training data set forms a $(27+16)\times 8$ matrix, $\mathbf{X}$, and a class label vector 44×1, Y. The test data only provide the $\mathbf{X}_{new}$, which is a 29×8 matrix of genes. For illustration, we used a KNN method with $k = 3$, i.e., three observations per neighborhood to classify the above training data set, and predict the class labels for the test data.

The R code is

```
>library(class)
>knnclass = knn(X,Xnew,y,k=3)
```

Note that KNN is provided in the library of class. One needs to load the library into R first. The results are given in the following table.

	True Class		
Predicted	AML	ALL	Total
AML	8	1	9
ALL	1	20	21
	9	21	30

Note that for the testing data set, only two observations were misclassified. If we let $k = 5$, the KNN method will have only a single observation being misclassified. The choice of k is not always clear, although data-driven methods such as cross validation can be used.

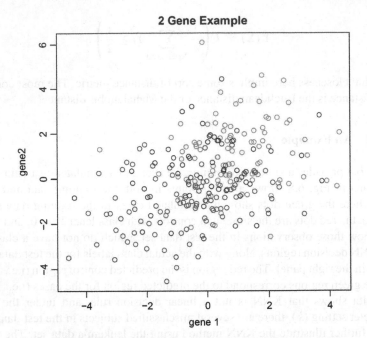

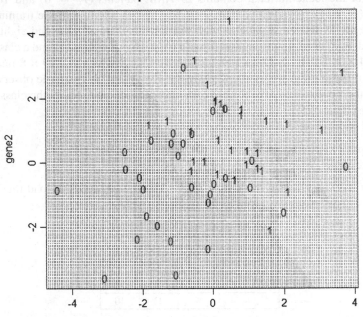

New Data plotted in Blue (true class label), background is decision boundary

6.7.3 Density-Based Classification Decision Rule

We discuss in this section classification methods using density-based methods.

6.7.3.1 Classification Using LDA Method

The LDA method has been discussed in detail in the last section for hypothesis testing. LDA can also be used for classification. We first briefly summarize the LDA method. LDA finds a linear combination of $\mathbf{X}$ as $Z = \boldsymbol{\alpha}^T \mathbf{X}$ that can best separate group 1 ($Y = 1$) from group 0 ($Y = 0$). Note that the LDA method assumes that the two groups share the same covariance, i.e., $\text{cov}(\mathbf{X}_{1i}) = \text{cov}(\mathbf{X}_{2i}) = \boldsymbol{\Sigma}$. The LDA method leads to the LDA direction $\hat{\boldsymbol{\alpha}} = (\overline{\mathbf{X}}_1 - \overline{\mathbf{X}}_0)^T \mathbf{S}_p^{-1}$. Using the transformed data $Z_i = \hat{\boldsymbol{\alpha}}^T \mathbf{X}_i$, the average score of each group is $\overline{Z}_1 = \hat{\boldsymbol{\alpha}}^T \overline{\mathbf{X}}_1$ for group 1 and $\overline{Z}_0 = \hat{\boldsymbol{\alpha}}^T \overline{\mathbf{X}}_0$ for group 2.

One can construct a classification decision rule based on LDA as follows. For a new subject with $\mathbf{X}$, calculate the score $Z = \hat{\boldsymbol{\alpha}}^T \mathbf{X}$. If $Z > (\overline{Z}_1 + \overline{Z}_0)/2$, then $\hat{Y} = 1$; if $Z \le (\overline{Z}_1 + \overline{Z}_0)/2$, then $\hat{Y} = 0$.

Note that there are two equivalent LDA decision rules. The first one is to calculate the Mahalanobis distance: $D_j(\mathbf{X}) = (\mathbf{X} - \overline{\mathbf{X}}_j)^T \mathbf{S}_p^{-1}(\mathbf{X} - \overline{\mathbf{X}}_j)$ ($j = 0, 1$). The decision rule is as follows: If $D_1(\mathbf{X}) < D_0(\mathbf{X})$, $\hat{Y} = 1$; if $D_1(\mathbf{X}) \ge D_0(\mathbf{X})$, $\hat{Y} = 0$. The second equivalent LDA decision rule is to calculate the linear classification function

$$T_j(\mathbf{X}) = \overline{\mathbf{X}}_j^T \mathbf{S}_p^{-1} \mathbf{X} - \frac{1}{2}\overline{\mathbf{X}}_j^T \mathbf{S}_p^{-1} \overline{\mathbf{X}}_j.$$

Assign Y to the group for which $T_j(\mathbf{X})$ is maximized.

6.7.3.2 General Density-Based Classification Decision Rule

LDA is a special case of the density-based classification decision rules. In this subsection, we will provide more general discussions on classification methods based on densities. Define the population disease prevalence (incidence) as

$$\Pr(Y = 1) = \pi_1, \quad \Pr(Y = 0) = \pi_0.$$

Fig. 6.8 The *left panel* shows a simulated training data set (with class labels) and a simulated test data set (without class label) where there are two genes. The *green dots* correspond to class label 1, and the *red dots* are for class label 0. The test data set is shown as *blue dots* in the figure. The *right panel* shows the decision boundary estimated using the KNN method, where two classes are shaded with different colors (*red* corresponds to class label 0, and *green* is for class label 1). The actual class labels for the test data set are shown as 0 or 1 for the test data set

Assume that the density of $\mathbf{X}$ for each population is $f_1(\mathbf{X})$ and $f_0(\mathbf{X})$, respectively. The goal of a density-based classification method is to classify a subject with $\mathbf{X}$ to 1 or 0 based on the probability $\Pr(Y = j|\mathbf{X})$ $(j = 0, 1)$.

By the Bayes rule, we get

$$P(Y = j|\mathbf{X}) = \frac{f_j(\mathbf{X})\pi_j}{f_1(\mathbf{X})\pi_1 + f_0(\mathbf{X})\pi_0}.$$

Thus,

$$\frac{\Pr(Y = 1|\mathbf{X})}{\Pr(Y = 0|\mathbf{X})} = \frac{f_1(\mathbf{X})}{f_0(\mathbf{X})}\frac{\pi_1}{\pi_0}.$$

The optimal decision rule is as follows: if $\Pr(Y = 1|\mathbf{X}) > \Pr(Y = 0|\mathbf{X})$, i.e., $\Pr(Y = 1|\mathbf{X}) > 0.5$, then $\hat{Y} = 1$.

Assume $\mathbf{X} \sim N(\boldsymbol{\mu}_1, \boldsymbol{\Sigma})$ for $Y = 1$ and $\mathbf{X} \sim N(\boldsymbol{\mu}_0, \boldsymbol{\Sigma})$ for $Y = 0$. Note that we assume they have the same covariance structure. Then,

$$\log\left[\frac{P(Y = 1|\mathbf{X})}{P(Y = 0|\mathbf{X})}\right] = \log\left[\frac{f_1(\mathbf{X})}{f_0(\mathbf{X})}\right] + \log\left(\frac{\pi_1}{\pi_0}\right)$$

$$= \mathbf{X}^T \boldsymbol{\Sigma}^{-1}(\boldsymbol{\mu}_1 - \boldsymbol{\mu}_0) - \frac{1}{2}(\boldsymbol{\mu}_1^T \boldsymbol{\Sigma}^{-1}\boldsymbol{\mu}_1 - \boldsymbol{\mu}_0^T \boldsymbol{\Sigma}^{-1}\boldsymbol{\mu}_0) + n\left(\frac{\pi_1}{\pi_0}\right).$$

The optimal decision rule leads to the linear discriminant function

$$T_j(\mathbf{X}) = \mathbf{X}^T \boldsymbol{\Sigma}^{-1}\boldsymbol{\mu}_j - \frac{1}{2}\boldsymbol{\mu}_j^T \boldsymbol{\Sigma}^{-1}\boldsymbol{\mu}_j.$$

When we apply it to the data, the estimated linear discriminant function will be

$$T_j(\mathbf{X}) = \mathbf{X}^T \mathbf{S}_p^{-1}\hat{\boldsymbol{\mu}}_j - \frac{1}{2}\hat{\boldsymbol{\mu}}_j \mathbf{S}_p^{-1}\hat{\boldsymbol{\mu}}_j.$$

The decision rule then is to assign Y to j $(j = 1, 0)$ if $T_j(\mathbf{X})$ is maximized. Note that from the above derivation, the LDA is an optimal (density-based) decision rule when $\mathbf{X}$ is normal and has a common covariance.

6.7.3.3 An Example

Figure 6.9 shows the decision regions based on the simulated training data set discussed in the previous section, where the red region is the region to be predicted as controls, $\hat{Y} = 0$, and the green region corresponds to the region to be predicted as cases, $\hat{Y} = 1$. The actual class labels for the test data are shown as 0 and 1 in the figure. It clearly shows that LDA provides a linear decision boundary. The number of misclassified observations is still large in this simulation setting.

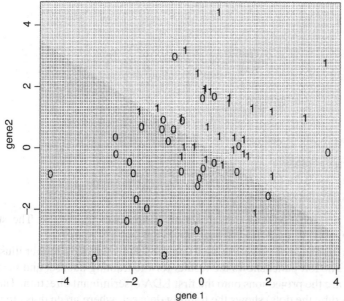

2 Gene Example - LDA - Test Data Classification

gene 1

New Data plotted in Blue (true class label), background is decision boundary

Fig. 6.9 The LDA decision regions based on a simulated training data set, along with the actual class label for a simulated test data set

We also apply the LDA method to the leukemia data set. The following table shows the estimated probabilities for the testing data.

Sample	$P(\widehat{Y} = 0|\mathbf{x})$	$P(\widehat{Y} = 1|\mathbf{x})$	Sample	$P(\widehat{Y} = 0|\mathbf{x})$	$P(\widehat{Y} = 1|\mathbf{x})$
1	0.013	0.987	2	0	1
3	0.003	0.997	4	0	1
5	0.007	1	6	0.002	0.998
7	0	1	8	0	1
9	0	0.999	10	0.008	0.992
11	0.001	1	12	0	1
13	0	1	14	0	1
15	0	0.994	16	0.003	0.997
17	0.006	1	18	0	1
19	0	0	20	0.002	0.998
21	1	0	22	0.999	0.001
23	0.36	64	24	1	0
25	0.895	0.105	26	0	1
27	1	0	28	0.995	0.005
29	0.996	0.004			

The classification results for the test data set are summarized in the following table:

Predicted	True Class		Total
	AML	ALL	
AML	7	0	7
ALL	2	20	22
	9	20	29

These were calculated using the following *R* code:

```
>library(MASS)
>model = lda(X,y)
>predict(model,Xn)
```

Note that the function `lda` is a function in the package `MASS`. The statement `library(MASS)` loads the package.

The performance of classification using the LDA method is further illustrated in Fig. 6.10. The left panel shows the results for the training data set and the test data set. These are the projections onto the first LDA discriminant direction. The bottom line (formed by the dots) shows the training data set, where green dots are the cases and red dots are the controls. The new test data are shown as the top line (formed by blue dots) in the left panel. The right panel shows the estimated densities for the projection of each group. The red region corresponds to the controls, and the blue region is for the cases. There are four lines formed by dots in the right panel. The top line is the same as the bottom line in the left panel, i.e., the dot plot for the training data set. The test data is projected onto the discriminant direction and is plotted as the second line (from the top), without any class labels. The third line in the subplot shows the predicted labels for the training data set. The actual class labels are shown in the bottom line. This plot suggests that the classification performance using LDA for this leukemia data set is reasonably good.

6.7.4   Quadratic Discriminant Analysis

Note that the LDA requires a common covariance structure, i.e., $\Sigma_1 = \Sigma_0 = \Sigma$. There is some indication in the leukemia data that the variances in the two groups might not be the same. The ALL variance seems a little smaller. An alternative to LDA is quadratic discriminant analysis (QDA) which relaxes the common covariance assumption. Assume X is from a multivariate normal distribution but with $\Sigma_1 \neq \Sigma_0$. Similar calculations to show in the last subsection show that

Training/New Data plotted on First Discriminant Direction

Results

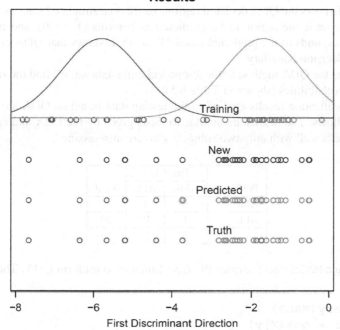

Fig. 6.10 The *left panel* shows the projections of the training data set and the test data set onto the first LDA discriminant direction. The *right panel* shows the densities of individual groups for these projections, the actual class labels and the predicted class labels for the test data. The *green region* is for the cases, and the *red region* corresponds to the controls

$$\log\left[\frac{P(Y=1|\mathbf{X})}{P(Y=0|\mathbf{X})}\right] = \log\left[\frac{f_1(\mathbf{X})}{f_0(\mathbf{X})}\right] + \log\left(\frac{\pi_1}{\pi_0}\right) = -\frac{1}{2}\log\left(\frac{|\boldsymbol{\Sigma}_1|}{|\boldsymbol{\Sigma}_0|}\right)$$

$$-\frac{1}{2}\left[(\mathbf{X}^T\boldsymbol{\Sigma}_1^{-1}\mathbf{X} - 2\mathbf{X}^T\boldsymbol{\Sigma}_1^{-1}\boldsymbol{\mu}_1 + \boldsymbol{\mu}_1^T\boldsymbol{\Sigma}_1^{-1}\boldsymbol{\mu}_1)\right.$$

$$\left.-(\mathbf{X}^T\boldsymbol{\Sigma}_0^{-1}\mathbf{X} - 2\mathbf{X}^T\boldsymbol{\Sigma}_0^{-1}\boldsymbol{\mu}_0 + \boldsymbol{\mu}_0^T\boldsymbol{\Sigma}_0^{-1}\boldsymbol{\mu}_0)\right] + n\left(\frac{\pi_1}{\pi_0}\right),$$

which leads to the quadratic discriminant function

$$T_j(\mathbf{X}) = -\frac{1}{2}\log|\boldsymbol{\Sigma}_j| - \frac{1}{2}(\mathbf{X} - \boldsymbol{\mu}_j)^T\boldsymbol{\Sigma}_1^{-1}(\mathbf{X} - \boldsymbol{\mu}_j) + \log\left(\frac{\pi_1}{\pi_0}\right), \quad i = 0, 1.$$

The decision will be to assign Y to group j if $T_j(\mathbf{X})$ is maximized.

An advantage of QDA over LDA is that it does not require the common variance assumption. Its drawback is that we need to estimate much more parameters, e.g., both $\boldsymbol{\Sigma}_1$ and $\boldsymbol{\Sigma}_0$ need to be estimated. The estimates can be unstable and require a larger sample size to estimate a stable decision rule.

6.7.4.1 An Example

Figure 6.11 shows the QDA decision regions based on a simulated training data set. The red region is the region to be predicted as controls ($\hat{Y} = 0$), and the green region corresponds to the predicted cases ($\hat{Y} = 1$). It shows that QDA provides a nonlinear decision boundary.

Applying the QDA method to the above leukemia data set, we find the following predicted probabilities (shown in Table 6.5):

The classification results applied to the testing data based on QDA are summarized as follows. The results are similar to those given by the LDA, indicating the method works well with only two subjects who are misclassified.

	True Class		
Predicted	AML	ALL	Total
AML	8	1	9
ALL	1	19	20
	9	20	29

R package MASS also provides the qda function to preform QDA. The code is as follows:

```
>library(MASS)
>model = qda(X,y)
>preds = predict(model,Xn)
```

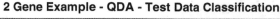

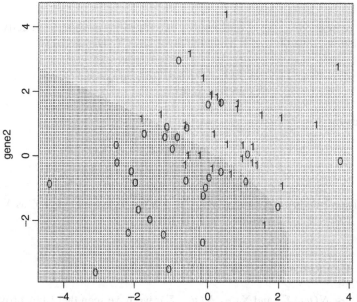

Fig. 6.11 The QDA decision regions based on a simulated training data set (*red* is for controls, and *green* is for cases) and the actual class labels (0 and 1) for a simulated test data set

Table 6.5 The predicted probabilities based on QDA for the leukemia data set

Sample	$P(\hat{Y} = 0\|\mathbf{x})$	$P(\hat{Y} = 1\|\mathbf{x})$	Sample	$P(\hat{Y} = 0\|\mathbf{x})$	$P(\hat{Y} = 1\|\mathbf{x})$
1	1	0	2	0	1
3	0.001	0.999	4	0	1
5	0.014	0.986	6	0.051	0.949
7	0	1	8	0.006	0.994
9	0	1	10	0.101	0.899
11	0	1	12	0	1
13	0	1	14	0	1
15	0	1	16	0.287	0.713
17	0.086	0.914	18	0	1
19	0	1	20	0	1
21	1	0	22	1	0
23	1	0	24	1	0
25	1	0	26	0	1
27	1	0	28	1	0
29	1	0			

6.7.5   Logistic Regression

Logistic regression is also a well-known classification method. It directly models the prediction probability of Y using the covariates $\mathbf{X}$. The model is

$$\log\left\{\frac{P(Y=1|\mathbf{X})}{P(Y=0|\mathbf{X})}\right\} = \beta_0 + \boldsymbol{\beta}_1^T\mathbf{X}.$$

This leads to

$$P(Y=1|\mathbf{X}) = \frac{\exp(\beta_0 + \boldsymbol{\beta}_1^T\mathbf{X})}{1 + \exp(\beta_0 + \boldsymbol{\beta}_1^T\mathbf{X})}.$$

The decision rule is as follows: if $P(Y=1|\mathbf{X}) > 0.5$, we predict Y as $\hat{Y} = 1$; if $P(Y=1|\mathbf{X}) \le 0.5$, we predict Y as $\hat{Y} = 0$.

6.7.5.1   Connections Between Logistic Regression and LDA

When $\mathbf{X}_1 \sim N(\boldsymbol{\mu}_1, \boldsymbol{\Sigma})$ and $\mathbf{X}_0 \sim N(\boldsymbol{\mu}_0, \boldsymbol{\Sigma})$, we have seen that LDA provides the optimal decision rule:

$$\frac{P(Y=1|\mathbf{X})}{P(Y=0|\mathbf{X})} = \log\frac{f_1(\mathbf{X})}{f_0(\mathbf{X})} + \log\frac{\pi_1}{\pi_0}$$

$$= \log\frac{\pi_1}{\pi_0} - \frac{1}{2}(\boldsymbol{\mu}_1 + \boldsymbol{\mu}_0)^T\boldsymbol{\Sigma}^{-1}(\boldsymbol{\mu}_1 - \boldsymbol{\mu}_0) + \mathbf{X}^T\boldsymbol{\Sigma}^{-1}(\boldsymbol{\mu}_1 - \boldsymbol{\mu}_0).$$

The right hand side is a linear function of $\mathbf{X}$ and can be rewritten as

$$\frac{P(Y=1|\mathbf{X})}{P(Y=0|\mathbf{X})} = \alpha_0 + \boldsymbol{\alpha}_1^T\mathbf{X}.$$

Note that the linearity is a result of the normality and common covariance matrix assumptions. The above model takes the same form as the classical logistic regression model.

However, logistic regression is directly maximizing the *conditional likelihood of* $Y|X$, which ignores the parameters in the distribution of $\mathbf{X}$. The LDA method fits the parameters by maximizing the full log-likelihood based on the joint density:

$$L(\mathbf{X}, Y) = \{\phi(\mathbf{X}; \boldsymbol{\mu}_1, \boldsymbol{\Sigma})\pi_1\}^Y\{\phi(\mathbf{X}; \boldsymbol{\mu}_0, \boldsymbol{\Sigma})\pi_0\}^{1-Y},$$

and thus, LDA method uses the marginal density $\mathbf{X}$. If the normality assumption assumed by the LDA assumption is true, the logistic regression method is

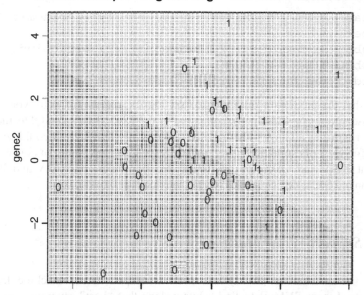

Fig. 6.12 The decision regions of logistic regression based on the simulated training data set, along with the actual class labels of the simulated test data set

asymptotically 30 % less efficient in coefficient estimation than LDA. However, if the normality assumption fails, logistic regression tends to be more robust, because of fewer assumptions.

6.7.5.2   An Example

Figure 6.12 shows the decision regions based on the above simulated training data set, where the red region is for the controls $\hat{Y} = 0$ and the green region corresponds to the cases, $\hat{Y} = 1$. The actual class labels are plotted as 0 and 1. Note that the logistic regression decision regions also have a linear boundary and are very similar to the LDA decision regions as shown in Fig. 6.9.

We also applied the logistic regression method to the leukemia data set; however, it failed to converge. The classification results are:

	True Class		
Predicted	AML	ALL	Total
AML	8	6	14
ALL	1	14	15
	9	20	29

The following R code can be used for logistic regression-based classification.

```
>betas = glm(y~as.matrix(X)+0, family = "binomial")
 $coef
>probs = exp(cbind(1,Xn)%*%betas)/(1+ exp(cbind(1,Xn)
 %*%betas))
>preds = as.numeric(probs>=0.5)
```

Note that the prediction results based on logistic regression were worse (with seven observations misclassified), compared to the KNN, LDA, and QDA methods shown earlier (they all have only two observations misclassified). It could be due to loss of efficiency of estimation using logistic regression and failure of convergence. Note that KNN and QDA are not linear decision rules, but LDA and logistic regression are linear decision rules.

6.7.6 Support Vector Machine

Support vector machine (SVM) was first developed in the machine learning community. Linear SVM is to find a hyperplane that "best" separates groups by maximizing the margins. We will first introduce the linear SVM in this section.

6.7.6.1 Linear Support Vector Machine

Suppose there is a training data set $(\mathbf{X}_i, Y_i)$ as discussed earlier. In this section, the class label Y_i is set up differently as a convention. To distinguish the notation, we assume $y_i \in \{+1, -1\}$ (e.g., ALL and AML in the leukemia data set). The target here is to find a hyperplane

$$\{\mathbf{x} : f(\mathbf{x}) = \mathbf{x}^T \boldsymbol{\beta} + \beta_0 = 0\}$$

where $\|\boldsymbol{\beta}\|_2 = \sqrt{\sum_{j=1}^{p} \beta_j^2} = 1$ is a unit vector. Then the decision of a new data point x is

$$\hat{Y}(\mathbf{x}) = \text{sign}(\mathbf{x}^T \boldsymbol{\beta} + \beta_0).$$

Here, $f(\mathbf{x})$ gives the signed distance from a point $\mathbf{x}$ to the hyperplane.

We first consider the simple situation, where the two classes are separable (i.e., there exists a separating hyperplane that the observations in one class lie on one side of the hyperplane, while the observations in the other class are on the other side of the hyperplane). In this simple case, there exist functions $f(\mathbf{x})$ such that $y_i f(\mathbf{x}_i) > 0$ for any i. The linear SVM tries to make $y f(\mathbf{X})$ as large as possible.

Define $yf(\mathbf{x})$ as the distance to the hyperplane and C the smallest distance. SVM tries to maximize C. The optimization problem is

$$\max_{\beta,\beta_0:\|\beta\|_2=1} C$$

subject to $y_i(\mathbf{x}_i^T \beta + \beta_0) \geq C, \quad i = 1,\dots,n.$

Here, C is called the *margin*, which provides (half of) the length of a band region separating the two groups. The above optimization problem can be rewritten as the following (dual) problem

$$\min_{\beta,\beta_0} \|\beta\|_2^2$$

subject to $y_i(\mathbf{x}_i^T \beta + \beta_0) \geq 1, \quad i = 1,\dots,n.$

This is one of the standard ways of formulating the problem, which can be solved via quadratic programming.

In practice, data are often non-separable (i.e., there does not exist a hyperplane that one class all lies on one side, and the other class all lie on the other side). In this case, we still want to maximize the margin C, but we need to allow some data points to be on the wrong side of the margin. We can introduce "slack" variables $\xi = (\xi_1,\dots,x_n)$ and modify the problem as

$$\max_{\beta,\beta_0,\|\beta\|_2=1} C$$

subject to $y_i(\mathbf{x}_i^T \beta + \beta_0) \geq C - \xi_i, \quad i = 1,\dots,n,$ and $\sum_{i=1}^{n} \xi_i \leq \kappa,$

where κ is a constant, which controls the relative amount of misclassification, i.e., the amount by which predictions fall on the wrong side of the margin for the training data set.

The dual problem of this non-separable case (which is sometimes called the standard definition of the linear SVM) can be rewritten as follows:

$$\min_{\beta,\beta_0} \|\beta\|_2^2$$

subject to $y_i(\mathbf{x}_i^T \beta + \beta_0) \geq 1 - \xi_i, \quad i = 1,\dots,n,$ and $\xi > 0, \sum_{i=1}^{n} \xi_i \leq \kappa.$

This standard SVM corresponds to the following Lagrange form using three parameters (γ, α, μ): $\min_{(\beta,\beta_0,\xi)} L_p$, where

$$L_p = \frac{1}{2}\|\beta\|_2^2 + \gamma \sum_{i=1}^{n} \xi_i - \sum_{i=1}^{n} \alpha_i [y_i(\mathbf{x}_i^T \beta + \beta_0) - (1 - \xi_i)] - \sum_{i=1}^{n} \mu_i \xi_i.$$

This leads to the dual objective function

$$L_D = \sum_{i=1}^{n} \alpha_i - \frac{1}{2} \sum_{i=1}^{n} \sum_{i'=1}^{n} \alpha_i \alpha_{i'} y_i y_{i'} \mathbf{x}_i^T \mathbf{x}_{i'},$$

which is maximized subject to some constraints under which the solution for L_p will be equivalent to L_D. It can be shown that

$$\hat{\beta} = \sum_{i=1}^{n} \hat{\alpha}_i y_i \mathbf{x}_i,$$

where $\hat{\alpha}$ will be nonzeros for only some observations. These observations are called the *support vectors*. Note that the support vectors are essentially those observations on the boundary of the margin-band and there are often a small number of them. Note that the SVM solution has sparsity in observations (i.e., only a small number of support vectors are used) but does not impose any sparsity in the variables $\mathbf{x}$, i.e., all genes are used in creating the hyperplane. In other words, standard SVM does not perform variable selection.

6.7.6.2 Nonlinear SVM

Although a linear hyperplane is convenient in classification, there are many cases that a linear hyperplane is not appropriate. The power of SVM also lies in its ability to be extended for nonlinear classification. The idea is to expand the basis from the original variables to a much larger (or infinite) feature space. The feature space is defined by the basis functions $h(\mathbf{x}_i) = (h_1(\mathbf{x}_i), h_2(\mathbf{x}_i), \ldots, h_1(\mathbf{x}_i))$. We can construct a linear hyperplane in the feature space and then translate the linear boundaries back to the original variable space. The decision boundary usually is nonlinear in the original variable space.

The above idea can be directly defined using the dual objective function of the linear SVM. The (nonlinear) SVM is to maximize the following dual function

$$L_D = \sum_{i=1}^{n} \alpha_i - \frac{1}{2} \sum_{i=1}^{n} \sum_{i'=1}^{n} \alpha_i \alpha_{i'} y_i y_{i'} < h(\mathbf{x}_i), h(\mathbf{x}_{i'}) >$$

where an inner product for the transformed feature vectors is used. After the maximizer for the above L_D is found, the decision function is

$$f(\mathbf{x}) = \sum_{i=1}^{n} \alpha_i y_i < h(\mathbf{x}), h(\mathbf{x}) > + \beta_0.$$

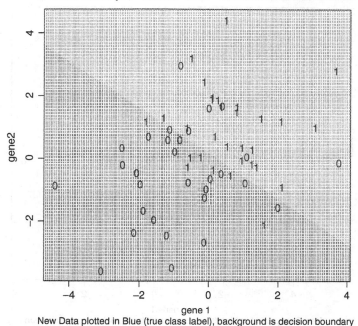

2 Gene Example - Linear SVM - Test Data Classification

New Data plotted in Blue (true class label), background is decision boundary

Fig. 6.13 The decision regions of the linear SVM based on the simulated training data set, along with the actual class labels of the simulated test data set

Note that the transformation from the original variable space to the feature space only uses the inner product in the feature space, we can extend the inner product to a more general kernel function. In other words, we do not need to specify the individual h; we only need to specify the inner product, e.g.,

$$K(\mathbf{x}, \mathbf{x}') = < h(\mathbf{x}), h(\mathbf{x}') >.$$

Here we list two examples of commonly used kernel function K:

1. dth degree polynomial kernel: $K(\mathbf{x}, \mathbf{x}') = (1 + < \mathbf{x}, \mathbf{x}' >)^d$
2. Gaussian kernel: $K(\mathbf{x}, \mathbf{x}') = \exp(-\|\mathbf{x} - \mathbf{x}'\|_2^2 / c)$

The dth degree polynomial kernel corresponds to the functional space expanded by up to the dth polynomials and their interactions. The Gaussian kernel corresponds to the functional space expanded by radial basis.

6.7.6.3 An Example

We here apply linear SVM to the simulated training data set and examine its performance. Figure 6.13 shows the predicted regions and the actual class labels

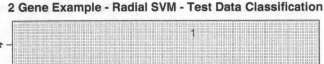

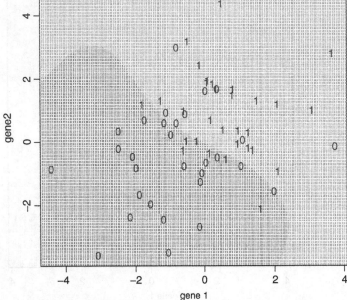

Fig. 6.14 The decision regions of the nonlinear SVM based on the simulated training data set, along with the actual class labels of the simulated test data set

for the test data set. The tuning parameter κ is found by five-fold cross validation. The green region is for the predicted cases ($\hat{Y} = 1$), and the red region corresponds to the predicted controls ($\hat{Y} = 0$). The 0–1 are the actual class labels for the observations of the testing data. Those 0s in the green regions and 1s in the red regions are misclassified.

Figure 6.14 shows the predicted regions based on a nonlinear SVM using the Gaussian kernel. The parameter is also chosen by five-fold cross validation. The linear SVM is also applied to the leukemia data set. The predicted results using the testing data set are summarized in the following table:

	True Class		
Predicted	AML	ALL	Total
AML	8	0	8
ALL	1	20	21
	9	20	29

The *R* package e1701 provides the code for SVM. The *R* code to get the above linear SVM results is

```
>library(e1071)
>##select the best cost parameters via grid search and
5 fold CV
>tune(svm, train.x=as.matrix(X), train.y=as.factor(y),
kernel="linear",
ranges=list(cost=2(-5:5)), control=tune.control
(sampling="cross", cross=5))
>model = svm(X,as.factor(y),kernel="linear",cost=
0.0625)
>preds = predict(model,Xn)
```

The nonlinear SVM classification results using the Gaussian kernel are summarized as

Predicted	True class AML	True class ALL	Total
AML	8	0	8
ALL	1	20	21
	9	20	29

The corresponding *R* code is

```
>library(e1071)
>##select the best cost parameters via grid search and
5 fold CV
>tune(svm, train.x=as.matrix(X), train.y=as.factor(y),
kernel="radial",
ranges=list(cost=2(-5:5), gamma=2(-5,5)),
control=tune.control(sampling="cross", cross=5))
>model= svm(X,as.factor(y),kernel=" radial",cost=0.25,
gamma=0.25)
>preds = predict(model,Xn)
```

Note that both the linear and nonlinear SVM methods misclassify only one observation in the testing data. They improve the performance of the classifiers discussed earlier. In this situation a linear boundary is already very good, and expanding it using nonlinear basis does not improve the results.

6.8 Variable Selection

With the advance of biotechnology, high-throughput omic data, such as microarray gene expression data or SNPs in genome-wide association studies, have become rapidly available, where hundreds of thousands of variables are collected. Statistical analysis of such large dimensional data is very challenging. One is often interested in selecting a subset of variables that are true signals. If we know in advance some variables are pure noises, or only a small portion is important, we can pre-select some of them for a later stage analysis. However, this pre-screening step usually needs opinions from experts and is difficult to do in practice. An automatic or data-driven variable selection procedure is hence desirable.

In this section, we will discuss several variable selection methods in regression settings. Assume that we have a data set with n observations. Let $i = 1, \ldots, n$, Y_i be a continuous scalar outcome and $\mathbf{X}_i$ be a $p \times 1$ covariate vector. Here we assume p is likely to be large. The major goal in this section is to jointly study the association between $\mathbf{X}$ and Y, i.e., we want to construct a prediction model of Y based on the covariates $\mathbf{X}$. Meanwhile, we do not want to use all of the covariates as many are noises. We want to build the prediction model only based on a subset of the p covariates that are true signals.

In this section, we will use a prostate cancer data as an example to illustrate different variable selection techniques. The data set contains 54 prostate cancer patients, i.e., $n = 54$. The outcome is the prostate-specific antigen (PSA), and the covariates are 33 microarray gene expressions (i.e., $p = 33$). We want to identify a subset of genes that best predict PSA. Note that p is large relative to n, which provides a challenge compared to standard linear regression settings.

6.8.1 Linear Regression Model

Here we focus on a simple prediction model using linear regression. Assume that Y follows a linear model as follows:

$$Y_i = \beta_0 + \mathbf{X}_i^T \beta + \varepsilon_i = \beta_0 + \sum_{j=1}^{p} \beta_j X_{ij} + \varepsilon_i,$$

where $\varepsilon_i \sim N(0, \sigma^2)$. After obtaining the regression coefficient estimates $\hat{\beta}_0$ and $\hat{\beta}$, the prediction model using a new $\mathbf{X}_{\text{new}}$ is

$$\hat{Y}_{\text{new}} = \hat{\beta}_0 + \sum_{j=1}^{p} \hat{\beta}_j X_{j,\text{new}}.$$

A standard estimator for β is the least square (LS) estimator $\hat{\beta} = (\hat{\beta}_1, \ldots, \hat{\beta}_p)$. It is well known that $\hat{\beta}$ is a consistent estimator of β.

However, if the underlying $\beta_j = 0$, the usual LS estimator $\hat{\beta}_j$ is not zero. In this situation, the prediction model that includes $\hat{\beta}_j$ will introduce noises and increase the prediction error. In fact, the standard LS estimators of $\hat{\beta}_j$ are all nonzeros; thus, it does not help with selecting the true parsimonious model. Assume that $\hat{\beta} = \hat{\beta}(\mathbf{Y}, \mathbf{X})$, which is some estimator of β using the data. Assume the true model only contains $X_1, \ldots, X_q$ and can be written as

$$Y = \beta_0^0 + \mathbf{X}^T \beta^0 = \beta_0^0 + X_1\beta_1^0 + \cdots + X_q\beta_q^0,$$

i.e., the true value β^0 $(p \times 1)$ has the last $(q - p)$ βs to be zero:

$$\beta_1^0, \ldots, \beta_q^0 \neq 0, \beta_{q+1}^0 = \cdots = \beta_p^0 = 0.$$

To measure the performance of a selected model, we often minimize some types of errors. Here we define two types of errors:

1. The prediction error (PE) is defined as

$$PE(\beta) = E_{\mathbf{X},Y}(Y_{new} - \mathbf{X}_{new}^T\beta)^2,$$

where $(Y_{new}, \mathbf{X}_{new})$ are new data.
2. The model error (ME) is defined as

$$ME(\beta) = E_{\mathbf{X},Y}(\mu_{new} - \mu_{new}^0)^2 = E_{\mathbf{X},Y}(\mathbf{X}_{new}^T\beta - \mathbf{X}_{new}^T\beta^0)^2,$$

where $\mu_{new} = E(Y_{new}|\mathbf{X}_{new}, \beta)$ is the mean of Y_{new} under an assumed model and $\mu_{new}^0 = E(Y_{new}|\mathbf{X}_{new}, \beta^0)$ is the mean of Y_{new} under the true model.

The relationship between PE and ME is that

$$PE(\beta) = ME(\beta) + \sigma^2.$$

The usual model (variable) selection criterion is to find the best model, which is the one that minimizes the prediction error (or model error).

The least squares estimator gives a consistent estimator of β, i.e., if $\beta_j = 0$, $\hat{\beta}_j$ will be 0 asymptotically. A natural question is that, since we have a consistent estimator, i.e., a consistent model is chosen asymptotically, why are we interested in variable selection?

6.8.2 Motivation for Variable Selection

To answer this question, consider the p-covariate full model as follows:

$$Y = \beta_0 + \beta_1 X_1 + \cdots + \beta_p X_p + \varepsilon. \tag{6.7}$$

The true model (i.e., with the first q covariates) $(q < p)$ is defined as

$$Y = \beta_0 + \beta_1 X_1 + \cdots + \beta_q X_q + \varepsilon, \tag{6.8}$$

i.e., $\beta_{q+1} = \cdots + \beta_p = 0$.

The least square method tries to minimize the RSS with respect to $\boldsymbol{\beta}$: $\sum_{i=1}^{n}$ $(Y_i - \mathbf{X}_i^T \boldsymbol{\beta})^2$. If we have an LS estimator from the full model, does it minimize the PE(ME)? That is, is it the best model in terms of prediction?

The LS estimator using the full model (6.7) is

$$\hat{\boldsymbol{\beta}}_F = \left[\sum_{i=1}^{n} \mathbf{x}_i \mathbf{x}_i^T \right]^{-1} \left[\sum_{i=1}^{n} \mathbf{x}_i^T Y_i \right],$$

and the LS estimator using the true model (6.8) is

$$\hat{\boldsymbol{\beta}}_{(q)} = \left[\sum_{i=1}^{n} (\mathbf{x}_{i(q)} \mathbf{x}_{i(q)}^T) \right]^{-1} \left[\sum_{i=1}^{n} \mathbf{x}_{i(q)}^T Y_i \right],$$

where $\mathbf{X}_{i(q)} = (X_{i1}, \ldots, X_{iq})^T$. Here we assume that the last $(p - q)$ βs are zeros and only fit the regression using the first q Xs, i.e., $\hat{\boldsymbol{\beta}}_T = (\hat{\boldsymbol{\beta}}_{(q)}, 0_{p-q})$. However, $\hat{\boldsymbol{\beta}}_T$ cannot be obtained from the data.

The errors of different models are summarized in the following table:

Model	Prediction error	Model error
Full model ($\hat{\boldsymbol{\beta}}_F$)	$\sigma^2 + p/n$	p/n
True model ($\hat{\boldsymbol{\beta}}_T$)	$\sigma^2 + q/n$	q/n
True value ($\boldsymbol{\beta}^0 = (\boldsymbol{\beta}_{(q)}^0, 0_{p-q})$)	σ^2	0

It suggests that the least squares estimator obtained from fitting the full model gives a consistent estimator of the model but has a larger PE and ME. If p is comparable to n and n is small or moderate,

$$\text{PE}(\hat{\boldsymbol{\beta}}_F) \gg \text{PE}(\hat{\boldsymbol{\beta}}_T) \text{ and } \text{ME}(\hat{\boldsymbol{\beta}}_F) \gg \text{ME}(\hat{\boldsymbol{\beta}}_T).$$

The least squares estimator from the full model $\hat{\boldsymbol{\beta}}_F$ is a consistent estimator of $\boldsymbol{\beta}$ but has a much larger variance than $\hat{\boldsymbol{\beta}}_T$ if p is large and q is small. The purpose of variable selection is to select the best model to minimize the PE or the ME.

6.8.3 Traditional Variable Selection Methods

Variable selection is a classical topic in statistics. There are many classical variable selection methods. For example, the traditional variable selection criteria include

1. Mallow's C_p criteria:

$$C_p = \frac{\text{RSS}}{\hat{\sigma}^2} - (n - 2p).$$

2. AIC criteria:

$$\text{AIC} = n \log\left(\frac{\text{RSS}}{n}\right) + 2p.$$

3. BIC criteria:

$$\text{BIC} = n \log\left(\frac{\text{RSS}}{n}\right) + p \log n.$$

One chooses the best model by minimizing one of these criteria. As there are $2^p - 1$ sub-models to fit, we hence need a search strategy. Here we list three traditional strategies:

1. *Best subset selection:* This procedure enumerates all possible subsets of the p covariates. At each setting, the model is fitted, and one of the above criteria is calculated. The optimal model corresponds to the model with the smallest value of the criterion function.
2. *Forward selection:* This procedure starts from the most significant single variable to be included in the model, and then add one variable a time to the model. The criteria value is calculated as well. The procedure stops when the value of the criterion function stops decreasing.
3. *Backward selection:* This procedure starts from the full model and then removes one variable a time from the model. The value of the criterion function is calculated as well. The procedure stops when the value of the criterion function is minimized.

See more discussion on these traditional variable selection methods in classical regression textbooks.

The limitation of the best subset selection strategy is that all sub-models need to be fit and computational is intensive if p is moderate and large. The limitation of the forward and backward selection procedures is that they are discrete processes and often exhibits higher variances. New techniques such as shrinkage methods usually are continuous, and thus do not suffer from high variability.

6.8.4 Regularization and Variable Selection

Several modern variable selection procedures have been proposed in the last decade. Here we list a few of them: nonnegative (NN) Garrote, boosting, least absolute shrinkage and selection method (LASSO), penalized regression with smoothly clipped absolute deviation (SCAD) penalty, least angle regression (LAR, Efron et al. [5]), adaptive LASSO, and Dantzig selector.

In this section, we will mainly focus on variable selection methods using penalized (or regularized) regression techniques, which can simultaneously perform variable selection and parameter estimation. Consider the linear regression model (6.7). The penalized least square method is defined to minimize

$$\sum_{i=1}^{n} \left(Y_i - \beta_0 - \sum_{j=1}^{p} \beta_j X_{ij} \right)^2 + P_\lambda(\boldsymbol{\beta}),$$

where $P_\lambda(\boldsymbol{\beta})$ is a penalty and λ is a tuning parameter that controls the goodness of fit and the model complexity. Two commonly employed penalty functions are:

1. The L_2 penalty function: $P_\lambda(\boldsymbol{\beta}) = \lambda \sum_{j=1}^{n} \beta_j^2$. This penalty corresponds to ridge regression. Ridge regression is usually used to handle high collinearity between covariates in a regression problem and when the number of variables p is large relative to the sample size n. However, it does not perform variable selection.
2. The LASSO penalty or the L_1 penalty: $P_\lambda(\boldsymbol{\beta}) = \lambda \sum_{j=1}^{n} |\beta_j|$. This LASSO penalty was proposed in Tibshirani (1996) for variable selection in regression settings.

The tuning parameter λ can be estimated using cross validation, GCV or AIC, and BIC. We will not discuss the details in this section. Note that the penalized (regularized) regression coefficient estimator $\hat{\boldsymbol{\beta}}$ is a shrinkage estimator and usually is continuous with respect to different values of λ. If $\lambda = 0$, this is no penalty at all, and $\hat{\boldsymbol{\beta}}$ is the ordinary least squares estimator. If $\lambda = \infty$, then $\hat{\boldsymbol{\beta}} = 0$.

6.8.4.1 Ridge Regression

Ridge regression is a popular penalized least square method, which minimizes the penalized RSS with a L_2 penalty of $\boldsymbol{\beta}$ as

$$\sum_{i=1}^{n} \left(Y_i - \beta_0 - \sum_{j=1}^{p} X_{ij} \beta_j \right)^2 + \lambda \sum_{j=1}^{p} \beta_j^2.$$

It can be derived that the estimated ridge regression coefficients are

$$\hat{\boldsymbol{\beta}}_{\text{Ridge}} = (\mathbf{X}^T \mathbf{X} + \lambda \mathbf{I})^{-1} \mathbf{X}^T \mathbf{Y}.$$

Note that the ridge regression coefficients $\hat{\beta}_{\text{Ridge}}$ shrink the least squares estimator $\hat{\beta}_{LS}$ toward 0. Ridge regression is useful for obtaining stable prediction of Y. However, it does not perform variable selection since it uses all variables for prediction and does not produce sparse solutions.

6.8.4.2 Nonnegative Garrote

Breiman (1995) proposed the nonnegative garrote method, which performs variable selection. The method over-parameterizes the model as

$$Y_i = \beta_0 + \sum_{j=1}^{p} \gamma_j \beta_j X_{ij} + \varepsilon_{ij}.$$

The estimation procedure is the as follows:

1. Obtain an initial estimator of $\hat{\beta}$, which is denoted by $\hat{\beta}^{\text{init}}$; e.g., this can be the ordinary least squares estimator or the ridge regression estimator.
2. Calculate γ to minimize

$$\sum_{i=1}^{n} \left(Y_i - \sum_{j=1}^{p} X_{ij} \hat{\beta}_j^{\text{init}} \gamma_j \right)^2 + \lambda \sum_{j=1}^{p} \gamma_j,$$

where $\gamma_j \geq 0 \ (j = 1, \ldots, p)$. The final estimate is

$$\hat{\beta}_j = \hat{\gamma}_j \hat{\beta}_j^{\text{init}}.$$

Note that the second step here is equivalent to minimizing square loss function under the constraints $0 \leq \sum_{j=1}^{p} \gamma_j \leq s$. When s is small (i.e., λ large), some coefficients will be exactly zero. Thus, this leads to a simultaneously shrinkage and variable selection.

6.8.4.3 LASSO Regression

LASSO penalty is proposed by Tibshirani (1996). It provides a simultaneous procedure to estimate the coefficients and perform variable selection, by using the penalized least square method with the L_1 penalty to minimize

$$\sum_{i=1}^{n} \left(Y_1 - \beta_0 - \sum_{j=1}^{p} X_{ij} \beta_j \right)^2 + \lambda \sum_{j=1}^{p} |\beta_j|.$$

Similar to the above nonnegative Garrote, the LASSO estimator is equivalent to minimizing the RSS

$$\sum_{i=1}^{n} \left(Y_1 - \beta_0 - \sum_{j=1}^{p} X_{ij} \beta_j \right)^2$$

subject to the constraints $\sum_{j=1}^{p} |\beta_j| \leq c$. When c is sufficiently small (i.e., λ is large), some $\hat{\beta}_j$ will be exactly 0. Hence, LASSO achieves the shrinkage and variable selection simultaneously.

Note that the L_1 constraint makes the solutions nonlinear with respect to Y. A quadratic programming algorithm can be used to find the LASSO estimator for a given λ. The new method, least angle regression (LAR), provides an efficient algorithm to calculate the LASSO estimator for all possible values of λ quickly. When the X_js are orthogonal, it can been shown that the LASSO estimator is the same as soft thresholding.

There are several limitations of LASSO regression. When the number of variables p is much larger than the sample size n, the LASSO selects at most n variables before it saturates. The connection between the LASSO method and the soft thresholding method suggests that the LASSO method over-penalizes large β_js, and hence produces biased estimators if the true coefficient β_js are large. Finally, under some regularity conditions, the LASSO has been shown to be not consistent for model selection, i.e., it does not have oracle property (i.e., cannot perform as well as you know the truth, even we have a relatively large sample size). See some discussions on this issue in Fan and Li [6] and Zhao and Bin [7].

6.8.4.4   SCAD Method

Knight and Fu [8] proposed an alternative penalty function to overcome the limitations of LASSO that is termed the smoothed clipped absolute deviation (SCAD) penalty, which leaves large values of $\boldsymbol{\beta}$ not excessively penalized. The SCAD penalty is defined as:

$$P'_\lambda(\beta) = \lambda \left\{ I(\beta \leq \lambda) + \frac{(a\lambda - \beta)_+}{(a-1)} \lambda I(\beta > \lambda) \right\},$$

for some tuning parameter $a > 2$ and $\lambda > 0$. The SCAD penalty is proposed to overcome the above limitations of LASSO. Knight and Fu [8] showed the corresponding penalized least square and penalized likelihood method enjoy the oracle property.

Computation of the penalized least square with the SCAD penalty is more complicated than the LASSO method. It can be done by the local quadratic Newton-Raphson method and is often much slower than computing the LASSO estimator.

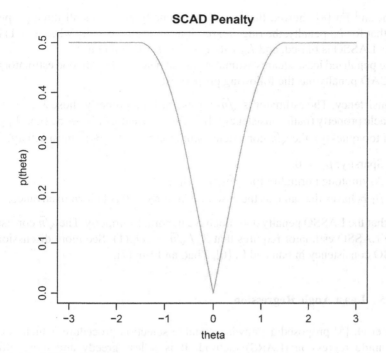

Fig. 6.15 The SCAD penalty function diagram

The one-step SCAD method provides a faster way to approximate the estimate, and builds a connection between the SCAD and LASSO-type estimator.

Note that the SCAD penalty corresponds to a quadratic spline with knots at λ and $a\lambda$:

$$P_\lambda(\beta) = \begin{cases} \lambda|\beta| & \text{if} |\beta| \leq \lambda \\ -\frac{|\beta|^2 - 2a\lambda|\beta| + \lambda^2}{2(a-1)} & \text{if} \lambda < |\beta| \leq a\lambda \\ \frac{(a+1)\lambda^2}{2} & \text{if} |\beta| > a\lambda \end{cases}$$

Figure 6.15 shows the SCAD penalty. The SCAD function has a similar form as the L_1 penalty for small coefficients. For larger coefficients, SCAD applies a constant penalty in contrast to the LASSO penalty increases linearly with the coefficient.

In Knight and Fu [8], they argued that a "good" penalty function will produce an estimator that satisfies the following properties:

1. Unbiasness: The estimator is nearly unbiased when the true unknown parameter is large. This property avoids model bias.
2. Sparseness: Small coefficients are automatically set to zero, and thus, the model complexity will be reduced.
3. Continuity: The estimator is continuous in data to avoid instability in model prediction.

Knight and Fu [8] showed that the SCAD penalty possesses all three properties. Note that for L_q penalty, the ridge regression (or any L_q penalty when $q > 1$) is not sparse, LASSO is biased, and L_q with $q < 1$ is not continuous.

The penalized least squares estimator (or the penalized likelihood estimator) with the SCAD penalty has the following properties:

1. Consistency: The estimator is $\sqrt{n}$-consistent for a properly chose λ_n.
2. Oracle property (path consistency): If $\lambda_n/n \to 0$ and $\lambda/\sqrt{n} \to \infty$ (i.e., λ_n grows not too quickly), the $\sqrt{n}$-consistent estimator $\hat{\beta} = (\hat{\beta}_1, \hat{\beta}_2)^T$ must satisfy:

   (a) Sparsity: $\hat{\beta}_2 = 0$.
   (b) Asymptotic normality for $\sqrt{n}(\hat{\beta}_1 - \beta_{10})$.
   (c) $\hat{\beta}_1$ behaves the same as the case in which $\beta_2 = 0$ is known in advance.

Note that the LASSO penalty does not have the oracle property. The $\sqrt{n}$ consistency of the LASSO estimator requires that $\lambda_n/\sqrt{n} = O_p(1)$. See more discussions on LASSO consistency in Fan and Li [6], Zhao and Bin [7].

6.8.4.5 Least Angle Regression

Efron et al. [5] proposed a stepwise variable selection procedure, which is called least angle regression (LARS) method. It is a less greedy and more efficient (computationally) version of forward selection procedures.

First of all, assume all variables Xs are normalized to have mean 0 and standard deviation 1. The forward stagewise regression (FSR) procedure is defined as follows:

1. Initialize $\hat{\mu} = X\hat{\beta} = 0$ (i.e., $\hat{\beta} = 0$, start from all coefficients as zeros).
2. Find the current vector of correlations: $\hat{c} = c(\hat{\mu}) = X^T(y - \hat{\mu})$.
3. Set $\hat{j} = \text{argmax}|\hat{C}_j|$ and take a small step in the direction of greatest correlation, i.e.,

$$\hat{\mu} = \hat{\mu} + \varepsilon * \text{sign}(\hat{c}_j)X_j,$$

   where ε is some small constant.
4. Repeat step 2 till all variables are included.

The LARS is a stylized version of the forward stagewise regression method. It speeds up computation via a relatively simple mathematical formula. The procedure of LARS is as follows:

1. Start with all of the coefficients zeros. This is the same as in the FSR procedure.
2. Find the variable most correlated with the response, e.g., X_{j_1}.
3. Take the largest possible step in the direction of this predictor until another predictor, e.g., X_{j_2}, becomes equally correlated with the residual.
4. Instead of moving toward X_{j_2} as FSR procedure does, the LARS procedure moves equiangularly between X_{j_1} and X_{j_2} until a third covariate, X_{j_3}, becomes equally correlated.

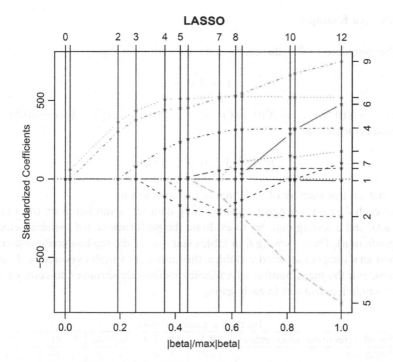

Fig. 6.16 One-sample LASSO solution path

5. The LARS again moves equiangularly between X_{j_1}, X_{j_2}, and X_{j_3} (along the "least angle direction") until the next covariate is equally correlated.
6. Continue the procedure to the full model.

Note that the LARS method uses only p steps to obtain a full set of solutions, instead of taking thousands of steps.

Because of the similarity of the LARS method and the FSR method, a slight modification of the LARS algorithm yields the FSR solutions. Another minor modification of the LARS algorithm allowing variables to leave the active set (i.e., the index set of the nonzero variables) generates the entire LASSO solution path. The LASSO solution and the LARS solution are piecewise linear with change in slope for each variable only when:

1. A new variable enters the active set.
2. A variable is removed from the active set.

The LARS algorithm thus makes it possible to obtain the entire LASSO solution path in the same order of time as OLS. Other variable selection methods also developed LARS-like algorithms to obtain computational efficiency, because this piecewise linearity speeds up computation immensely. Figure 6.16 shows one solution path, which clearly visualizes the piecewise linearity and the slope change when a new variable enters or is removed from the active set.

6.8.4.6   An Example

Here we simulate 1,000 data sets from the following settings,

$$y = \mathbf{X}\boldsymbol{\beta} + \varepsilon$$

where $\varepsilon \sim N(0, 9)$, $X_j \sim N(0, 1)$ ($j = 1, \ldots, 8$), and $\text{cor}(X_j, X_{j'}) = (0.5)^{|j - j'|}$. The true coefficient is

$$\boldsymbol{\beta} = (3, 1.5, 0, 0, 2, 0, 0, 0)^T.$$

Note that the true number of variables is 3 for this setting.

For each data set, we generated a training data set, a validation set (or a tuning data set), and a testing data set to evaluate the performance (of variable selection and prediction). The following four tables summarize the median prediction errors, the median number of selected variables, the number of variables correctly shrunken to zeros, and the mean number of variables incorrectly shrunken to zero, of these 1,000 simulation data sets in each setting.

The median prediction errors

Sample sizes (training/tuning/test)	OLS	LASSO	Garrote	LARS	SCAD
20/20/500	15.81	12.45	12.87	12.46	12.31
40/40/500	11.49	10.50	10.56	10.51	10.42
100/100/500	9.82	9.56	9.49	9.56	9.46

The median number of variables in final model

Sample sizes (training/tuning/test)	OLS	LASSO	Garrote	LARS	SCAD
20/20/500	8	5.35	4.51	5.39	3.81
40/40/500	8	5.75	4.85	5.77	3.80
100/100/500	8	5.82	4.68	5.82	3.98

The median number of variables correctly shrunken to zeros

Sample sizes (training/tuning/test)	OLS	LASSO	Garrote	LARS	SCAD
20/20/500	0	2.42	3.00	2.41	3.81
40/40/500	0	2.22	3.03	2.20	3.80
100/100/500	0	2.18	3.31	2.20	3.98

The median number of variables incorrectly shrunken to zeros

Sample sizes (training/tuning/test)	OLS	LASSO	Garrote	LARS	SCAD
20/20/500	0	0.22	0.49	0.21	0.35
40/40/500	0	0.04	0.12	0.04	0.09
100/100/500	0	0.00	0.01	0.00	0.01

Among the OLS, LASSO, garrote, LARS, and SCAD methods, the SCAD method provides the smallest median prediction errors and the most parsimonious model, which shows the "best" performance under this simulation settings.

The number of variables correctly shrunken to zeros shows that SCAD is very "aggressive" in terms of shrinkage. However, the mean number of variables incorrectly shrunken to zero is close to the highest. The tables show that SCAD and garrote are "aggressive" in terms of shrinkage. The LASSO and LARS are less aggressive than the other two methods. When the sample size is large, all methods have similar prediction performance and keep the signal variables. For the noise variables, the LASSO and LARS methods intend to keep some and are more conservative than SCAD and garrote.

6.8.5   Summary

In summary, in this subsection, we have studied several variable selection techniques and regularization methods. Ridge regression produces models with good prediction error but does not produce sparse solutions. Nonnegative garrote simultaneously shrinks estimators but relies on possibly unstable initial estimates. LASSO also performs simultaneous shrinkage and variable selection but could produce biased estimators for large coefficients and does not perform well in the presence of highly correlated predictors. The SCAD penalty has attractive theoretical properties and performs well in simulations but is computationally expensive.

In general, when the true model is sparse, all of the presented methods tend to select too many variables. When the true model contains many variables with small effects, none of the methods producing sparse solutions work very well.

References

1. Anderson TW (1984) An introduction to multivariate statistical analysis. Wiley, New York
2. Jolliffe, I.T. Principal Component Analysis, Series: Springer Series in Statistics, 2nd ed., Springer, NY, 2002, XXIX, 487 p. 28 illus. ISBN 978-0-387-95442-4
3. Johnson RA, Wichern DW(2002) Applied multivariate statistical analysis, vol 5. Prentice Hall, Upper Saddle River
4. Hastie T, Tibshirani R, Friedman JH (2001) The elements of statistical learning, vol 1. Springer, New York
5. Efron B, Hastie T, Johnstone I, Tibshirani R (2004) Least angle regression. Ann Stat 32(2):407–499
6. Fan J, Li R (2001) Variable selection via nonconcave penalized likelihood and its oracle properties. J Am Stat Assoc 96(456):1348–1360
7. Zhao P, Bin Y (2006) On model selection consistency of Lasso. J Machine Learn Res 7:2541–2563
8. Knight K, Fu W (2000) Asymptotics for lasso-type estimators. Ann Stat 28:1356–1378

Chapter 7
Association Analysis for Human Diseases: Methods and Examples

Jurg Ott and Qingrun Zhang

7.1 Why Do We Need Statistics?

Many biologists see statistics and statistical analysis of their data more as a nuisance than a necessary tool. After all, they are convinced their data are correct and demonstrate what the experiment was supposed to show. But this view is precisely why we need statistics—researchers tend to see their results in too positive a light. Statistical analysis, if done properly, provides an unbiased assessment of the outcome of an experiment. Here are some additional reasons for using statistics.

Some statistical facts are not intuitive so that intuition alone can lead us astray. For example, consider coin tossings [1]. You toss a coin, which comes up either head or tail. If you do this several times, whenever head (or tail) shows at time i and tail (or head) shows at time $i + 1$, then we say a change has occurred. In 10,000 coin tossings, how many changes do you expect to occur? Most people will say "several thousand." However, statistical analysis shows that the median number of changes is equal to $0.337\sqrt{n}$. With $n - 10,000$, this leads to a median number of changes of only 34. As an application of this principle, consider two men who play a game at which they are equally strong. Sometimes Peter will win and sometimes Xingyu. One might think that in a game that lasts twice as long, Peter should lead about twice as often. But the expected increase in lead times is only $\sqrt{n}$, where n is the number of times the game is played.

At the Bar Harbor course of Medical Genetics in the summer of 2001, Craig Venter [2] mentioned in his presentation that "it was not the powerful machines that gave us the edge in sequencing the genome—the decisive factor was the math, the mathematical methods for analyzing the data."

J. Ott (✉) • Q. Zhang
Chinese Academy of Sciences, Beijing Institute of Genomics, Beijing, China
e-mail: ottjurg@psych.ac.cn

R. Jiang et al. (eds.), *Basics of Bioinformatics: Lecture Notes of the Graduate Summer*   233
School on Bioinformatics of China, DOI 10.1007/978-3-642-38951-1_7,
© Tsinghua University Press, Beijing and Springer-Verlag Berlin Heidelberg 2013

At one of those Bar Harbor courses, a local researcher pointed out that "We don't need statistics. If a result is unclear we just breed more mice." Fair enough, but in human genetics, we usually don't have the luxury of essentially unlimited data, so we want to extract as much information from the limited data at hand.

7.2 Basic Concepts in Population and Quantitative Genetics

In this brief review of Mendelian inheritance, current concepts of genetic modeling are introduced (genotype, haplotype, penetrance, heritability, etc.). An example of risk calculation for a Mendelian trait (cystic fibrosis) will show that this "simple" inheritance model can still pose challenging problems.

We first need to learn some basic terms:

- Locus = (genomic position of) a heritable quantity (plural: loci).
- Polymorphism = fact that a locus is polymorphic (has >1 allele); also stands for genetic marker.
- Genetic marker = well-characterized locus. Examples: microsatellites, single-nucleotide polymorphisms (SNPs).
- Allele = one of the variants of a gene or marker.
- Allele frequency (= gene frequency) is the relative proportion of an allele among all alleles at a locus.
- Genotype = set of two alleles at a locus (gene) in an individual. Examples: A/G (marker alleles), N/D (disease alleles). May be homozygous (G/G) or heterozygous (T/G).
- Haplotype = set of alleles, one each at different loci, inherited from one parent (or being located on the same chromosome).
- Diplotype = set of genotypes, one each at different loci. Other expressions for the same thing are "genotype pattern" or "genotype array."
- Phenotype = "what you see," expression of a genotype. Examples: A/G or AG (marker), "affected" (disease).

The relationship between (underlying) genotypes and (observed) phenotypes is formulated in terms of penetrance = conditional probability of observing phenotypes given a genotype. Table 7.1 shows penetrances for the ABO blood type locus. In Table 7.1, penetrances sum to 1 in a given column. In a given row, for example, for the A blood type, we see that two genotypes can give rise to the same phenotype. We also see that the A allele is dominant over the O allele (A shows whether or not O is present) while O is recessive relative to A. On the other hand, in the AB phenotype, the A and B alleles are codominant with respect to each other.

An important concept is the Hardy-Weinberg equilibrium (HWE). Discovered independently by Hardy [3] and Weinberg [4], if alleles are inherited independently in a genotype, then genotype frequencies depend only on allele frequencies. For example, a SNP has two alleles that combine into three genotypes. Usually, these alleles are denoted by A and B with $p = P(A)$ being the A allele frequency. Under HWE the genotype frequencies are predicted to be p^2, $2p(1-p)$, and $(1-p)^2$, for

Table 7.1 Penetrances
at ABO blood type locus

Phenotype	Genotype					
	A/A	A/B	A/O	B/B	B/O	O/O
A	1	0	1	0	0	0
B	0	0	0	1	1	0
AB	0	1	0	0	0	0
O	0	0	0	0	0	1
Sum	1	1	1	1	1	1

the respective genotypes *AA*, *AB*, and *BB*. The Hardy-Weinberg law generally holds for genetic marker loci but may be violated by genotyping errors or ascertainment of specific genotypes, for example, in affected individuals. With n being the number of alleles, the total number of genotypes at a locus is $n(n + 1)/2$. Thus, it is easier to work with alleles than genotypes.

A locus is called polymorphic if its most common allele has frequency <0.99. The degree of polymorphism is often measured by the heterozygosity

$$H = 1 - \sum_{i=1}^{n} p_i^{2},$$

which is the probability that an individual is heterozygous.

Each human cell contains 23 pairs of chromosomes, with genes and loci arranged along the chromosomes as portions of the DNA. Among these 23 chromosome pairs, 22 pairs (the so-called autosomes) have equally shaped member chromosomes while the last pair consists of an X chromosome and a Y chromosome, the so-called sex chromosomes, where females have two X chromosomes and males have one X chromosome and one Y chromosome. Most of what has been said so far refers to autosomal loci. Loci on the X and Y chromosomes exhibit very specific modes of inheritance that are different from autosomal inheritance.

Heritable diseases may be divided into rare and common traits. The former (e.g., cystic fibrosis, Huntington disease) generally follow a Mendelian mode of inheritance while the latter (e.g., diabetes, heart disease) do not and generally represent a high burden for public health. In Mendelian traits, generally one parent of an affected individual is affected while the other parent is unaffected. These traits tend to occur in large family pedigrees. On the other hand, for an individual affected with a recessive trait, parents are generally unaffected and such traits tend to occur only in one sibship and not elsewhere among close relatives.

In generalized Mendelian inheritance, each genotype has its own penetrance. For example, genotypes *AA*, *AB*, and *BB* have the respective associated penetrances f_1, f_2, and f_3, with $f_1 \leq f_2 \leq f_3$, and genotype probabilities $(1 - p)^2$, $2p(1 - p)$, and p^2. The incidence (trait frequency among newborns) predicted by this model is given by $(1 - p)^2 f_1 + 2p(1 - p)f_2 + p^2 f_3$. With equal mortality of all three genotypes, this is also the prevalence (trait frequency in population). For example, for cystic fibrosis (CF), the disease allele frequency in many Western populations is $p = 0.025$ so that the incidence of carriers (heterozygotes, unaffected) is 0.0488 or about 1 in 20 while the incidence of affected individuals is 1 in 1,600.

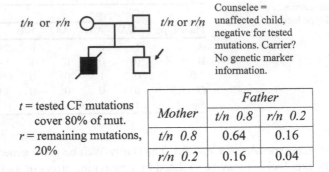

	Father	
Mother	*t/n* 0.8	*r/n* 0.2
t/n 0.8	0.64	0.16
r/n 0.2	0.16	0.04

t = tested CF mutations cover 80% of mut.

r = remaining mutations, 20%

Fig. 7.1 Genetic counseling for CF family

Fig. 7.2 Offspring genotypes for given mating types in genetic counseling of CF family

	Counselee's genotype			
Mating types	¼	¼	¼	¼
t/n × t/n 0.64	t/t 0.16	t/n 0.16	t/n 0.16	n/n 0.16
t/n × r/n 0.32	t/r 0.08	t/n 0.08	r/n 0.08	n/n 0.08
r/n × r/n 0.04	r/r 0.01	r/n 0.01	r/n 0.01	n/n 0.01

At the CF locus, a large number of disease variants exist. Molecular genetics labs have probes for many of them so that they can be detected directly by an assay. However, only about 80 % of these variants can be detected in this manner. Many family relatives of an affected individual are seeking genetic counseling, which often must be carried out by statistical genetics calculations. For example, consider the situation depicted in Fig. 7.1. The counselee is unaffected and tested negative for known CF mutations but his brother died of CF. He wants to know his probability of being a carrier of a CF variant. The parents are unaffected and each must be heterozygous. Formally we distinguish two types of CF variants: detectable variants (t) and the relatively small proportion of undetectable variants (r). Thus, each parent has one of two possible genotypes, t/n or r/n, where n stands for the normal allele. Disregarding order of the parents, the four mating types in Fig. 7.1 may be represented as three mating types, and each of these leads to four possible offspring genotypes with probability ¼ each (Fig. 7.2). Only the gray underlined genotypes (r/n and n/n) are compatible with the counselee's phenotype, and among these, the r/n genotype corresponds to carrier status. Thus, as shown in Fig. 7.2, the desired conditional probability of carrier status is given by $2/7 = 29$ %. Unfortunately, such calculations are often carried out incorrectly leading to wrong results.

7.3 Genetic Linkage Analysis

Linkage analysis examines for two loci whether they are inherited independently when alleles are passed from parents to offspring. If two genes are on the same chromosome in close proximity to each other, then the two alleles on the

Fig. 7.3 Simple assumed example showing how a crossover leads to recombinant (*R*) and nonrecombinant (*N*) offspring

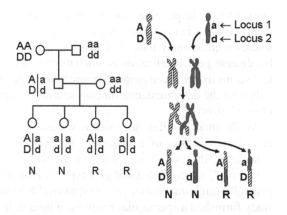

same chromosome (at the two gene loci) will travel together as one "package" (one haplotype, or in one gamete) from a parent to a child. Specifically, assume two alleles *D* and *d* at locus 1 and alleles *A* and *a* at locus 2. Also assume we know from the genotypes of grandparents that the *D* and *A* alleles are on one chromosome while the *d* and *a* alleles are on the other (i.e., we know phase), which may be written as the genotype *DA/da*. If there is only a small distance between the two loci, then usually *DA* or *da* will be inherited by a child, but when the loci are on different chromosomes, then their alleles will be inherited independently. Figure 7.3 shows these principles on a simple assumed example.

Genetic linkage analysis is carried out in family data and generally requires working under an assumed disease inheritance model. One usually postulates a disease locus and then tests for linkage between this locus and one after the other of the marker loci. Linkage between loci as far apart as 10–20 cM (1 cM $\approx$ 1 MB) may be detected. Since the 1980s, interest in this type of genetic investigation has gradually decreased and is being replaced by association analysis, but linkage analysis is still an important tool for finding genes underlying Mendelian traits. For details please refer to relevant textbooks [5].

7.4 Genetic Case-Control Association Analysis

A new mutation on a given chromosome initially is in coupling with all alleles on that same chromosome. As crossovers occur, this association between alleles along a chromosome tends to be broken up, but for loci very close to the position of the original mutation, no crossover may have occurred between them so that the original complete association tends to persist over many generations. This is the rationale for the current genetic association studies: When alleles or genotypes at a marker loci exhibit different frequencies between people with and without a heritable trait, then we conclude that the gene underlying the trait is in close proximity of the marker locus. In practice, association between two marker loci tends to exist only if the distance between the two loci is less than about 0.1 cM. Thus, association studies require many more loci distributed over the genome than linkage analysis but disease gene localization is much more precise. Association as a gene-mapping tool seems to have been proposed more than 80 years ago [6] but became fashionable only with the documentation that association is more powerful than linkage for loci of small effects [7].

While small families are suitable for association studies [8], the most common data design is that of a case-control study. A set of individuals affected with a heritable disease ("cases") and a number of individuals without the disease ("controls") are collected and genotyped for large numbers of genetic marker loci, generally single-nucleotide polymorphism (SNP) markers. The first chip-based such study furnished a spectacular result even though it was based on only 96 cases and 50 controls: A functional SNP for age-related macular degeneration was identified [9]. Generally, fairly large numbers of case and control individuals are used for such studies [10].

7.4.1 Basic Steps in an Association Study

Assume you have collected respective numbers n_A and n_U of case and control individuals and had them genotyped for, say, $m = 500$ K SNPs. Companies like Affymetrix and Illumina furnish their genotyping results in the form of large text files, for example, with rows corresponding to individuals and columns referring to SNPs. The body of such a large array will contain genotypes AA, AB, and BB, with some code like NN denoting "missing." Given such data, basic analysis steps are as follows:

- Check each SNP whether it contains only one genotype for everybody. If everyone in your data is homozygous AA or BB, that SNP is just not polymorphic in your dataset. It also happens that everybody is heterozygous AB, which reflects a genotyping problem. Any such SNP must be discarded. Also discarded should be SNPs with genotype numbers $m_{AA} = m - 1$ and $m_{AB} = 1$. Such markers with a single minor allele have very extreme allele frequencies. Also, that one minor allele may just reflect an error.
- For each SNP determine the proportion of missing observations and reject any SNP with call rates lower than, say, 90 %.
- Do an analogous analysis for individuals. Of course, one does not want to delete individuals lightly, so the rules for rejection should not be too stringent. For example, reject an individual with missing rates exceeding 40 or 50 % in cases or controls.
- The above tests led to the deletion of several hundred SNPs in the 100 K AMD study [9]. One more test is now generally done: Genotype frequencies are checked for agreement with Hardy-Weinberg equilibrium (HWE) proportions.

Strong deviations from HWE are often caused by genotyping errors, for example, when homozygotes AA tend to be misinterpreted as heterozygotes AB. On the other hand, ascertainment of affected individuals from the general population leads to a subset of individuals within which HWE may not hold even though it holds in the population. The reason for this is that an SNP in close proximity to a disease locus may be associated with the disease so that selection of affected individuals also selects for some SNP genotypes. Therefore, one does not want to reject an SNP lightly just because of deviations from HWE. The test of HWE results in a chi-square with 1 df and an associated p-value. It is prudent to reject HWE only when $mp < 0.05$ or $mp < 0.01$.

After these QC (quality control) steps, you are ready to proceed to association analysis. The basic tests generally carried out are an allele test and a genotype test for each SNP. That is, chi-square is computed for a 2×2 table of alleles (each individual furnishes two entries to this table) with rows corresponding to cases and controls and columns referring to the two SNP alleles. Analogously, chi-square is computed for a 2×3 table with columns representing the three SNP genotypes (here, of course, each individual contributes one entry). The allele test is valid only under HWE. Otherwise the two alleles in a genotype are not independent [11] and the allele test tends to have an inflated type 1 error (rate of false-positive results).

Various additional association tests have been proposed. A powerful test is MAX2, which is obtained as follows [12]. For each 2×3 table of genotypes, you form two sub-tables: one with columns $(AA + AB, BB)$ and another with columns $(AA, AB + BB)$. That is, you assume dominant and recessive inheritance of the SNP. Then chi-square is computed for each of the 2×2 sub-tables and the larger of the two chi-squares is retained as the relevant test statistic. Of course, if chi-square tables are consulted for determining p-values, this procedure leads to an increased rate of false-positive results. Therefore, correct p-values must be obtained by appropriate methods (randomization, see below). Alternatively, you may want to apply the FP test, which could be used as the only test procedure [13]. It tests for differences in allele frequency and inbreeding coefficient between case and control individuals and is more powerful than the genotype test for recessive-like traits while being equal in power to the genotype test for dominant traits.

7.4.2   Multiple Testing Corrections

Assume now that for each SNP, one statistical association test is carried out. The test result is called significant if $p \leq 0.05$, for example. This means that one in 20 tests will be significant just by chance without there being an association. While this risk of a false-positive result is generally considered acceptable when a single test is carried out, it is unacceptable when 1,000s of tests are done. If these tests are independent, then the probability of at least one of them becoming significant is given by $p_e = 1 - (1 - p)^m$. For small p, this is very nearly equal to $p_e \approx mp$, where p is often called the pointwise significance level and p_e the experimentwise significance level. Thus, to keep p_e below 0.05, one must keep p below $0.05/m$. This adjustment is called the Bonferroni correction for multiple testing. It is generally rather conservative; alternative, more powerful corrections have been developed (see below). For example, with 100,000 SNPs, a single SNP is only then called significantly disease associated if its significance level is smaller than $0.05/100,000 = 5 \times 10^{-7}$ or, equivalently, if $100,000p$ is smaller than 0.05.

At this point it is good to sit back and *look at the results*—do they make sense? Is there anything unusual about them? For example, it is a good thing to make a histogram of all the 100,000s of p-values. In the absence of any association (under the null hypothesis), p-values should have a uniform distribution from 0 to 1. If the histogram consists of 20 bars, then on average each bar should have a height of 0.05, that is, each bar comprises 5 % of the observations. You may see a trend toward an excess of small p-values, which would indicate potentially significant results. If you have strong excesses toward 1, for example, then something is not right and you should investigate to find the reason for this abnormal situation. It may also be useful to rank the p-values so that the p-value ranked 1 is the smallest. Then plot $-\log(p)$ against ranks, which may show you something similar as what you find on slide #11: You see a smooth curve rising toward small p-values (large values of $-\log(p)$) with an abrupt change at some point. The values beyond the abrupt change are likely to represent outliers, that is, are likely to be significant.

A new concept of significance is now often used in association analysis. While the p-value is the conditional probability of a significant result given the null hypothesis, $P(\text{sig}|H_0)$, the so-called false discovery rate (FDR) is the conditional probability of a false positive among all significant results, $P(H_0|\text{sig})$. The FDR is a more intuitive concept than the significance level (p-value). On the other hand, while p-values are well defined, it is not immediately clear how to determine among significant results which one is a false positive. Various procedures have been developed for identifying the FDR associated with a number of test results. Probably the simplest procedure is the Benjamini-Hochberg method [14, 15]. A more powerful approach are the so-called q-values [16]: All those test results with a q-value of at most 0.05 are associated with an FDR of no more than 0.05. A computer program is available at (http://genomics.princeton.edu/storeylab/qvalue/) for determining q-values based on p-values.

The above procedures for multiple testing correction do not specifically take into account that dense sets of SNPs on the human genome yield somewhat correlated results. Allowing for this dependency can potentially gain power. The most reliable way of allowing for the dependence structure among SNPs is randomization [17]. Recall that the p-value is the conditional probability of a significant result given H_0. For case-control data we have an easy way to generate data under the null hypothesis of no association: We randomly permute the labels "case" and "control" but leave everything else intact. Such a dataset with permuted disease status assignments clearly has no association between disease and genetic marker data. We now compute the same test statistic in the randomized dataset as in the observed dataset, for example, the largest chi-square among all SNPs. We repeat the randomization and computation of max(chi-square) many times and determine the proportion of randomized datasets in which max(chi-square) is at least as large as in the observed data. This proportion is an unbiased estimate for the p-value associated with the observed maximum chi-square. The *sumstat* program is an example of software carrying out randomization (http://www.genemapping.cn/sumstat.html).

7.4.3 Multi-locus Approaches

Because complex traits are thought to be due to multiple possibly interacting underlying genes and environmental triggers, approaches have been developed that look for joint association of multiple markers [18, 19]. This area is currently very active and new methods are being developed by various statistical geneticists.

7.5 Discussion

You may want to learn more about statistical gene-mapping methods, for example, by taking courses and reading books. However, while it is useful to become better acquainted with genetic linkage and association analysis methods, it is recommended to leave a comprehensive data analysis in the hands of people trained in statistics. There are all kinds of pitfalls that an unsuspecting analyst can fall into. Many analysis procedures have been implemented in computer programs, so it may appear that nothing is easier than to "run my data through some programs." Unfortunately, not all published programs are trustworthy, and it is often better to write your own programs unless you can trust the person who wrote a published program.

Although strong genetic determinants exist for various traits, for example, Alzheimer's disease [20], one must not forget that environmental effects can also be very strong. For example, individuals who eat fish at least once a week have an at least 50 % lower risk of developing Alzheimer's disease [21]. Geneticists tend to overlook such environmental risk factors, partly because relevant research is published in journals not usually read by geneticists.

References

1. Feller W (1967) An introduction to probability theory and its applications, 3rd edn. Wiley, New York
2. Venter JC (2007) A life decoded: my genome, my life. Viking, New York
3. Hardy GH (1908) Mendelian proportions in a mixed population. Science 28(706):49–50
4. Weinberg W (1908) Über den Nachweis der Vererbung beim Menschen. Jahresh Ver vaterl Naturkunde Württemberg 64:369–382

5. Ott J (1999) Analysis of human genetic linkage, 3rd edn. Johns Hopkins University Press, Baltimore
6. Sax K (1923) The association of size differences with seed-coat pattern and pigmentation in *Phaseolus vulgaris*. Genetics 8(6):552–560
7. Risch N, Merikangas K (1996) The future of genetic studies of complex human diseases. Science 273(5281):1516–1517
8. Horvath S, Xu X, Laird NM (2001) The family based association test method: strategies for studying general genotype-phenotype associations. Eur J Hum Genet 9(4):301–306
9. Klein RJ, Zeiss C, Chew EY et al (2005) Complement factor H polymorphism in age-related macular degeneration. Science 308(5720):385–389
10. Bentley D, Brown MA, Cardon LA et al (2007) Genome-wide association study of 14,000 cases of seven common diseases and 3,000 shared controls. Nature 447(7145):661–678
11. Sasieni PD (1997) From genotypes to genes: doubling the sample size. Biometrics 53(4): 1253–1261
12. Zheng G, Freidlin B, Gastwirth JL (2006) Comparison of robust tests for genetic association using case-control studies. IMS Lect Notes Monogr Ser 49:253–265
13. Zhang Q, Wang S, Ott J (2008) Combining identity by descent and association in genetic case-control studies. BMC Genet 9(1):42
14. Benjamini Y, Hochberg Y (1995) Controlling the false discovery rate: a practical and powerful approach to multiple testing. J R Stat Soc Ser B (Methodol) 57(1):289–300
15. Benjamini Y, Drai D, Elmer G et al (2001) Controlling the false discovery rate in behavior genetics research. Behav Brain Res 125(1–2):279–284
16. Storey JD, Tibshirani R (2003) Statistical significance for genomewide studies. Proc Natl Acad Sci USA 100(16):9440–9445
17. Manly BFJ (2006) Randomization, Bootstrap and Monte Carlo methods in biology, 3rd edn. Chapman & Hall/CRC Press, New York
18. Hoh J, Wille A, Ott J (2001) Trimming, weighting, and grouping SNPs in human case-control association studies. Genome Res 11(12):2115–2119
19. Ritchie MD, Hahn LW, Moore JH (2003) Power of multifactor dimensionality reduction for detecting gene-gene interactions in the presence of genotyping error, missing data, phenocopy, and genetic heterogeneity. Genet Epidemiol 24(2):150–157
20. Strittmatter WJ, Saunders AM, Schmechel D et al (1993) Apolipoprotein E: high-avidity binding to beta-amyloid and increased frequency of type 4 allele in late-onset familial Alzheimer disease. Proc Natl Acad Sci USA 90(5):1977–1981
21. Friedland RP (2003) Fish consumption and the risk of Alzheimer disease: is it time to make dietary recommendations? Arch Neurol 60(7):923–924

Chapter 8
Data Mining and Knowledge Discovery Methods with Case Examples

S. Bandyopadhyay and U. Maulik

8.1   Introduction

This chapter deals with the area of knowledge discovery and data mining that has emerged as an important research direction for extracting useful information from vast repositories of data of various types. The basic concepts, problems, and challenges are first briefly discussed. Some of the major data mining tasks like classification, clustering, and association rule mining are then described in some detail. This is followed by a description of some tools that are frequently used for data mining. Two case examples of supervised and unsupervised classification for satellite image analysis are presented. Finally, an extensive bibliography is provided.

Huge amount of data is generated routinely in banks, telephones, supermarkets, credit card companies, insurance, and other business transactions as well as in scientific domains. For example, AT&T handles billions of calls per day and Google searches more than four billion pages in a day which results in several terabyte orders of data. Similarly, astronomical data of the order of gigabytes per second, as well as large amount of biological data, data from e-commerce transactions, etc., are generated regularly. These data sets are not only huge but also complex and sometimes even unstructured.

Traditionally, manual methods were employed to turn data into knowledge. However, analyzing these data manually and making sense out of it is slow, expensive, subjective, and prone to errors. Hence, the need to automate the process arose, thereby leading to research in the fields of data mining and knowledge discovery. Knowledge discovery from databases (KDD) evolved as a research direction that appears at the intersection of research in databases, machine learning, pattern

S. Bandyopadhyay (✉) • U. Maulik
Indian Statistical Institute, Kolkata, India
e-mail: sanghami@isical.ac.in

R. Jiang et al. (eds.), *Basics of Bioinformatics: Lecture Notes of the Graduate Summer School on Bioinformatics of China*, DOI 10.1007/978-3-642-38951-1_8,
© Tsinghua University Press, Beijing and Springer-Verlag Berlin Heidelberg 2013

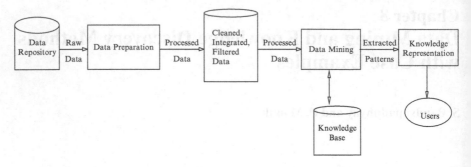

Fig. 8.1 The knowledge discovery process [15]

recognition, statistics, artificial intelligence, reasoning with uncertainty, expert systems, information retrieval, signal processing, high-performance computing, and networking [19, 30].

Data mining and knowledge discovery is the nontrivial process of extraction of valid, previously unknown, potentially useful, and ultimately understandable patterns from data. It is the science of extracting useful information from large data sets or databases. Data mining techniques attempt to use raw data for:

- Increasing business, e.g., focused marketing, inventory logistics
- Improving productivity, e.g., analysis of weather data
- Reducing costs, e.g., fraud detection
- Scientific discovery, e.g., biological applications, drug discovery

The application areas of data mining are very large. Some of these are as follows:

- Science and technology – astronomy, bioinformatics, medicine, drug discovery, etc.
- Business – customer relationship management (CRM), fraud detection, e-commerce, manufacturing, sports/entertainment, telecom, targeted marketing, health care, retail sales, etc.
- Web – search engines, advertising, web and text mining, etc.
- Government – surveillance, crime detection, profiling tax cheaters, etc.
- Natural resource study and estimation – agriculture, forestry, geology, environment, etc.
- Astronomy – mining large astronomical databases
- Image mining – content-based image retrieval from large databases

The task of knowledge discovery can generally be classified into data preparation, data mining, and knowledge presentation. Data mining is the core step where the algorithms for extracting the useful and interesting patterns are applied. In this sense, data preparation and knowledge representation can be considered, respectively, to be preprocessing and postprocessing steps of data mining. Figure 8.1 presents a schematic view of the steps involved in the process of knowledge discovery.

8.2   Different Tasks in Data Mining

Data mining tasks can be classified as descriptive and predictive [30]. While the descriptive techniques provide a summary of the data, the predictive techniques learn from the current data in order to make predictions about the behavior of new data sets. The commonly used tasks in data mining are as follows:

- Classification: predicting an item class
- Clustering: finding groups in data
- Associations: finding associations between different items
- Visualization: proper depiction of the results so as to facilitate knowledge discovery
- Summarization: describing a group of related items
- Deviation detection: finding unexpected changes in the data
- Estimation: predicting a continuous value of a variable
- Link analysis: finding relationships between items/events

The data mining tasks of classification, clustering, and association mining are described in detail in the following subsections.

8.2.1   Classification

The problem of classification involves taking an input pattern that is characterized by a set of features and making a decision about its belongingness to one (or more) of the class (classes). In case the classifier is designed using a set of labelled patterns, it is called a supervised classifier. The classification problem can be modelled in a variety of ways, e.g., by generating a set of rules, learning decision trees, generating class boundaries capable of distinguishing among the different classes. Some well-known classification methods are described below.

8.2.1.1   Nearest Neighbor Rule

A simple and well-known approach of classification is the nearest neighbor rule.

Let us consider a set of n pattern (or points) of known classification $\{x_1, x_2, \ldots, x_n\}$, where it is assumed that each pattern belongs to one of the classes $C_1, C_2, \ldots, C_k$. The NN classification rule then assigns a pattern x of unknown classification to the class of its nearest neighbor, where $x_i \in \{x_1, x_2, \ldots, x_n\}$ is defined to be the nearest neighbor of x if

$$D(x_i, x) = \min_l \{D(x_l, x)\}, \quad l = 1, 2, \ldots, n \tag{8.1}$$

where D is any distance measure definable over the pattern space.

Since the above algorithm uses the class information of only the nearest neighbor to $\mathbf{x}$, it is known as the 1-NN rule. If K neighbors are considered for classification, then the scheme is termed as the K-NN rule. The K-NN rule assigns a pattern $\mathbf{x}$ of unknown classification to class C_i if the majority of the K nearest neighbors belongs to class C_i. The details of the K-NN rule along with the probability of error are available in Duda and Hart [23], Fukunaga [29], and Tou and Gonzalez [56].

8.2.1.2   Bayes Maximum Likelihood Classifier

Bayes maximum likelihood classifier [3, 56] is another well-known and widely used classifier. In most of the real-life problems, the features are usually noisy and the classes in the feature space are overlapping. In order to model such systems, the feature values $x_1, x_2, \ldots, x_j, \ldots, x_N$ are considered as random values in the probabilistic approach. The most commonly used classifier in such probabilistic systems is the Bayes maximum likelihood classifier, which is now described.

Let P_i denote the a priori probability and $p_i(\mathbf{x})$ denote the class conditional density corresponding to the class C_i $(i = 1, 2, \ldots, k)$. If the classifier decides $\mathbf{x}$ to be from the class C_i, when it actually comes from C_l, it incurs a loss equal to L_{li}. The expected loss (also called the conditional average loss or risk) incurred in assigning an observation $\mathbf{x}$ to the class C_i is given by

$$r_i(\mathbf{x}) = \sum_{l=1}^{k} L_{li} \; p\left(\frac{C_l}{\mathbf{x}}\right), \tag{8.2}$$

where $p(C_l/\mathbf{x})$ represents the probability that $\mathbf{x}$ is from C_l. Using Bayes formula, Eq. (8.2) can be written as

$$r_i(\mathbf{x}) = \frac{1}{p(\mathbf{x})} \sum_{l=1}^{k} L_{li} \; p_l(\mathbf{x}) P_l, \tag{8.3}$$

where

$$p(\mathbf{x}) = \sum_{l=1}^{k} p_l(\mathbf{x}) P_l.$$

The pattern $\mathbf{x}$ is assigned to the class with the smallest expected loss. The classifier which minimizes the total expected loss is called the *Bayes classifier*.

Let us assume that the loss (L_{li}) is zero for correct decision and greater than zero but the same for all erroneous decisions. In such situations, the expected loss, Eq. (8.3), becomes

$$r_i(\mathbf{x}) = 1 - \frac{P_i \, p_i(\mathbf{x})}{p(\mathbf{x})}. \tag{8.4}$$

Since $p(\mathbf{x})$ is not dependent upon the class, the Bayes decision rule is nothing but the implementation of the decision functions

$$D_i(\mathbf{x}) = P_i \, p_i(\mathbf{x}), \qquad i = 1, 2, \ldots, k, \qquad (8.5)$$

where a pattern $\mathbf{x}$ is assigned to class C_i if $D_i(\mathbf{x}) > D_l(\mathbf{x})$, $\forall l \neq i$. This decision rule provides the minimum probability of error. The naive Bayes classifier assumes that each feature x_i is conditionally independent of every other feature x_j for $j \neq i$. Therefore, the decision function in Eq. 8.5 is written as

$$D_i(\mathbf{x}) = P_i \prod_{j=1}^{N} p_i(x_j), \qquad i = 1, 2, \ldots, k. \qquad (8.6)$$

It is to be noted that if the a priori probabilities and the class conditional densities are estimated from a given data set, and the Bayes decision rule is implemented using these estimated values (which may be different from the actual values), then the resulting classifier is called the *Bayes maximum likelihood classifier*.

8.2.1.3 Support Vector Machines

Support vector machine (SVM) is considered to be the state-of-the-art classifier. The underlying principle of SVMs is to map the objects to a high dimensional space where the classes become linearly separable. The task then is to estimate the hyperplane that optimally separates the two classes in the high dimensional space. An interesting feature of the SVM is that the mapping to the higher dimensional space is not explicit. Rather, it is done implicitly when the inner product between two vectors $\phi(\mathbf{x}_1)$ and $\phi(\mathbf{x}_2)$ in the higher dimensional space is computed as a kernel function defined over the input feature space. Here $\phi(.)$ is the mapping that transforms the vectors $\mathbf{x}_1$ and $\mathbf{x}_2$ in the input feature space to the higher dimensional space. In other words,

$$\phi^T(\mathbf{x}_1)\phi(\mathbf{x}_2) = K(\mathbf{x}_1, \mathbf{x}_2). \qquad (8.7)$$

When used for pattern classification, the SVM basically uses two mathematical operations [32]:

1. Nonlinear mapping of an input vector into a high dimensional feature space. This is in accordance with Cover's theorem that states that a complex pattern-classification problem cast in a high dimensional space nonlinearly is more likely to be linearly separable than in a low dimensional space [20].
2. Construction of an optimal hyperplane for separating the transformed patterns computed in the feature space. Construction of the hyperplane is performed in accordance with the principle of structural risk minimization that has its root in the Vapnik–Chervonenkis (VC) dimension theory.

Note that here the input space refers to the original feature space while feature space refers to the transformed higher dimensional space.

Let the training set be denoted by $(\mathbf{x}_i, d_i)_{i=1}^n$, where $\mathbf{x}_i$ is the ith input pattern and d_i is its class. The SVM first learns n parameters, $\alpha_1, \alpha_2, \ldots, \alpha_n$, the Lagrange multipliers, by maximizing the objective

$$\sum_{i=1}^n \alpha_i - \frac{1}{2} \sum_{i=1}^n \sum_{j=1}^n \alpha_i \alpha_j d_i d_j K(\mathbf{x}_i, \mathbf{x}_j) \qquad (8.8)$$

subject to the constraints

$$\sum_{i=1}^n \alpha_i d_i = 0 \qquad (8.9)$$

and

$$0 \le \alpha_i \le C, \text{ for } i = 1, 2, \ldots, n, \qquad (8.10)$$

where C is a user-specified positive parameter. Thereafter, given an unlabelled vector $\mathbf{x}$, the SVM classifies it based on the decision function

$$f(\mathbf{x}) = \text{sgn}(\sum_{i=1}^n \alpha_i d_i K(\mathbf{x}, \mathbf{x}_i) + b), \qquad (8.11)$$

where b is a bias term computed from the already-learned Lagrange multipliers and the support vectors. Here, the support vectors are the vectors that are closest to the optimal hyperplane, and hence the most difficult to classify.

8.2.2 Clustering

Clustering [2, 22, 31, 34, 56] is an important unsupervised classification technique where a set of patterns, usually vectors in a multidimensional space, are grouped into clusters in such a way that patterns in the same cluster are similar in some sense and patterns in different clusters are dissimilar in the same sense. For this purpose, a measure of similarity which will establish a rule for assigning patterns to a particular cluster is defined. One such measure of similarity may be the Euclidean distance $\mathbf{D}$ between two patterns $\mathbf{x}$ and $\mathbf{z}$ defined by $\mathbf{D} = \|\mathbf{x} - \mathbf{z}\|$. The smaller the distance between $\mathbf{x}$ and $\mathbf{z}$, the greater is the similarity between the two and vice versa.

An alternative measure of similarity is the dot product between $\mathbf{x}$ and $\mathbf{z}$, which, physically, is a measure of the cosine of the angle between these two vectors. Formally,

$$d(\mathbf{x}, \mathbf{z}) = \mathbf{x}^T \mathbf{z}. \qquad (8.12)$$

Clearly, the smaller the Euclidean distance between these two vectors, the more similar they are, and therefore, the larger will be the inner product between them. It can be shown that minimization of the Euclidean distance corresponds to maximization of the inner product.

Let us assume that the n points $\{x_1, x_2, \ldots, x_n\}$, represented by the set P, are grouped into K clusters $C_1, C_2, \ldots, C_K$. Then, in general, for crisp clustering,

$$C_i \neq \emptyset \qquad for \ i = 1, \ldots, K,$$
$$C_i \cap C_j = \emptyset \ for \ i = 1, \ldots, K, \ j = 1, \ldots, K \ and \ i \neq j, \ and$$
$$\bigcup_{i=1}^{K} C_i = P.$$

Clustering techniques may be partitional or hierarchical [2]. Among the partitional clustering techniques, where a partitioning of the data is obtained only on termination of the algorithm, the K-means technique [56] has been one of the more widely used ones. In hierarchical clustering, the clusters are generated in a hierarchy, where every level of the hierarchy provides a particular clustering of the data, ranging from a single cluster to n clusters. A clustering technique is said to be crisp if it assigns a point to exactly one cluster. An alternate clustering strategy is fuzzy clustering where a point can have nonzero membership to more than one cluster simultaneously. Fuzzy c-means is a well-known partitional clustering technique that belongs to this category. The basic steps of the K-means, fuzzy c-means, and single-linkage hierarchical clustering algorithms are described below.

8.2.2.1 K-Means Algorithm

The K-means clustering algorithm essentially consists of an alternating sequence of cluster assignment followed by center update. The steps of the algorithm are as follows:

1. Choose K initial cluster centers $z_1, z_2, \ldots, z_K$ randomly from the n points $\{x_1, x_2, \ldots, x_n\}$.
2. Assign point x_m, $m = 1, 2, \ldots, n$ to cluster C_j, $j \in \{1, 2, \ldots, K\}$ iff

$$\|x_m - z_j\| < \|x_m - z_p\|, \ \ p = 1, 2, \ldots, K, \ and \ j \neq p.$$

   Ties are resolved arbitrarily.
3. Compute new cluster centers $z_1^*, z_2^*, \ldots, z_K^*$ as follows:

$$z_i^* = \frac{1}{n_i} \Sigma_{x_j \in C_i} x_j \ \ i = 1, 2, \ldots, K,$$

   where n_i is the number of elements belonging to cluster C_i.
4. If $z_i^* = z_i$, $i = 1, 2, \ldots, K$ then terminate. Otherwise, continue from Step 2.

Note that in case the K-means algorithm does not terminate normally, it is executed for a predefined maximum number of iterations.

Sometimes, the K-means algorithm may converge to some local optima [54]. Moreover, global solutions of large problems cannot be found within a reasonable amount of computation effort [55]. As a result, several approximate methods, including genetic algorithms and simulated annealing [8, 14, 41], are developed to solve the underlying optimization problem. These methods have also been extended to the case where the number of clusters is variable [7, 42], and to fuzzy clustering [43].

The K-means algorithm is known to be sensitive to outliers, since such points can significantly affect the computation of the centroids, and hence the resultant partitioning. K-medoid attempts to alleviate this problem by using the medoid, the most centrally located object, as the representative of the cluster. Partitioning around medoid (PAM) [36] was one of the earliest K-medoid algorithms introduced. PAM finds K clusters by first finding a representative object for each cluster, the medoid. The algorithm then repeatedly tries to make a better choice of medoids by analyzing all possible pairs of objects such that one object is a medoid and the other is not. PAM is computationally quite inefficient for large data sets and a large number of clusters. The Clustering LARge Applications (CLARA) algorithm was proposed in Kaufman and Rousseeuw [36] to tackle this problem. CLARA is based on data sampling, where only a small portion of the real data is chosen as a representative of the data, and medoids are chosen from this sample using PAM. CLARA draws multiple samples and outputs the best clustering from these samples. As expected, CLARA can deal with larger data sets than PAM. However, if the best set of medoids is never chosen in any of the data samples, CLARA will never find the best clustering. The CLARANS algorithm proposed in Ng and Han [44] tries to mix both PAM and CLARA by searching only a subset of the data set. However, unlike CLARA, CLARANS does not confine itself to any sample at any given time, but draws it randomly at each step of the algorithm. Based upon CLARANS, two spatial data mining algorithms, the spatial dominant approach, SD(CLARANS), and the nonspatial dominant approach, NSD(CLARANS), were developed. In order to make CLARANS applicable to large data sets, use of efficient spatial access methods, such as R^*-tree, was proposed [24]. CLARANS has a limitation that it can provide good clustering only when the clusters are mostly equisized and convex. DBSCAN [25], another popularly used density clustering technique that was proposed by Ester et al., could handle nonconvex and nonuniformly sized clusters. Balanced Iterative Reducing and Clustering using Hierarchies (BIRCH), proposed in Zhang et al. [61], is another algorithm for clustering large data sets. It uses two concepts, the clustering feature and the clustering feature tree, to summarize cluster representations which help the method achieve good speed and scalability in large databases. Discussion on several other clustering algorithms may be found in Han and Kamber [30].

8.2.2.2   Fuzzy c-Means Clustering Algorithm

Fuzzy c-means (FCM) [17] is a widely used technique that uses the principles of fuzzy sets to evolve a fuzzy partition matrix for a given data set. The set of all $c \times n$, where c is equal to the number of clusters, nondegenerate constrained fuzzy partition matrices, denoted by M_{fcn}, is defined as

$$M_{fcn} = \{U \in R^{c \times n} \mid \sum_{i=1}^{c} u_{ik} = 1, \ \sum_{k=1}^{n} u_{ik} > 0, \ \forall i \ \text{ and}$$
$$u_{ik} \in [0,1]; 1 \leq i \leq c; 1 \leq k \leq n\}. \tag{8.13}$$

Here, u_{ik} is the membership of the kth point to the ith cluster. The minimizing criterion used to define good clusters for fuzzy c-means partitions is the FCM function defined as

$$J_\mu(U, Z) = \sum_{i=1}^{c} \sum_{k=1}^{n} (u_{ik})^\mu D_{ik}^2. \tag{8.14}$$

Here, $U \in M_{fcn}$ is a fuzzy partition matrix; $\mu \in [1, \infty]$ is the weighting exponent on each fuzzy membership; $Z = [\mathbf{z}_1, \dots, \mathbf{z}_c]$ represents c cluster centers; $\mathbf{z}_i \in IR^N$; and D_{ik} is the distance of $\mathbf{x}_k$ from the ith cluster center. The fuzzy c-means theorem [17] states that if $D_{ik} > 0$, for all i and k, then (U, Z) may minimize J_μ, only if when $\mu > 1$

$$u_{ik} = \frac{1}{\sum_{j=1}^{c} (\frac{D_{ik}}{D_{jk}})^{\frac{2}{\mu-1}}}, \quad \text{for } 1 \leq i \leq c, \ 1 \leq k \leq n, \tag{8.15}$$

and

$$\mathbf{z}_i = \frac{\sum_{k=1}^{n} (u_{ik})^\mu \mathbf{x}_k}{\sum_{k=1}^{n} (u_{ik})^\mu}, \quad 1 \leq i \leq c. \tag{8.16}$$

A common strategy for generating the approximate solutions of the minimization problem in Eq. (8.14) is by iterating through Eqs. (8.15) and (8.16) (also known as the Picard iteration technique). A detailed description of the FCM algorithm may be found in Bezdek [17].

Note that in fuzzy clustering, although the final output is generally a crisp clustering, the users are free to still utilize the information contained in the partition matrix. The FCM algorithm shares the problems of the K-means algorithm in that it also gets stuck at local optima depending on the choice of the initial clusters, and requires the number of clusters to be specified a priori.

8.2.2.3 Single Linkage Hierarchical Clustering Technique

The single-linkage clustering scheme is a noniterative method based on a local connectivity criterion [34]. Instead of an object data set X, single linkage processes sets of n^2 numerical relationships, say $\{r_{jk}\}$, between pairs of objects represented by the data. The value r_{jk} represents the extent to which object j and k are related in the sense of some binary relation ρ. It starts by considering each point in a cluster of its own. The single-linkage algorithm computes the distance between two clusters C_i and C_j as

$$\delta_{SL}(C_i, C_j) = \min_{x \in C_i, y \in C_j} \{d(x, y)\}, \tag{8.17}$$

where $d(x, y)$ is some distance measure defined between objects x and y. Based on these distances, it merges the two closest clusters, replacing them by the merged cluster. The distance of the remaining clusters from the merged one is recomputed as above. The process continues until a single cluster, comprising all the points, is formed. The advantages of this algorithm are that (1) it is independent of the shape of the cluster, and (2) it works for any kind of attributes, both categorical and numeric, as long as a similarity of the data objects can be defined. However, the disadvantages of this method are its computational complexity and its inability to handle overlapping classes.

8.2.3 Discovering Associations

The root of the association rule mining (ARM) [30] problem lies in the market basket or transaction data analysis. Association analysis is the discovery of rules showing attribute–value associations that occur frequently.

Let us assume $I = \{i_1, i_2, \ldots, i_n\}$ be a set of n items and X be an itemset where $X \subset I$. A k-itemset is a set of k items. Let $T = \{(t_1, X_1), (t_2, X_2) \ldots, (t_m, X_m)\}$ be a set of m transactions, where t_i and X_i, $i = 1, 2, \ldots, m$, are the transaction identifier and the associated itemset, respectively. The *cover* of an itemset X in T is defined as follows:

$$cover(X, T) = \{t_i | (t_i, X_i) \in T, X \subset X_i\}. \tag{8.18}$$

The *support* of an itemset X in T is

$$support(X, T) = |cover(X, T)|, \tag{8.19}$$

and the *frequency* of an itemset is

$$frequency(X, T) = \frac{support(X, T)}{|T|}. \tag{8.20}$$

In other words, support of an itemset X is the number of transactions where all the items in X appear in each transaction. The frequency of an itemset represents the probability of its occurrence in a transaction in T. An itemset is called frequent if its support in T is greater than some threshold min_sup. The collection of frequent itemsets with respect to a minimum support min_sup in T, denoted by $\mathcal{F}(T, min\_sup)$, is defined as

$$\mathcal{F}(T, min\_sup) = \{X \subset I, support\,(X, T) > min\_sup\}. \qquad (8.21)$$

The objective in association rule mining is to find all rules of the form $X \Rightarrow Y$, $X \cap Y = \emptyset$ with probability $c\%$, indicating that if itemset X occurs in a transaction, the itemset Y also occurs with probability $c\%$. X is called the *antecedent* of the rule and Y is called the *consequent* of the rule. Support of a rule denotes the percentage of transactions in T that contains both X and Y. This is taken to be the probability $P(X \cup Y)$. An association rule is called *frequent* if its support exceeds a minimum value min_sup.

The confidence of a rule $X \Rightarrow Y$ in T denotes the percentage of the transactions in T containing X that also contains Y. It is taken to be the conditional probability $P(X|Y)$. In other words,

$$confidence(X \Rightarrow Y, T) = \frac{support(X \cup Y, T)}{support(X, T)}. \qquad (8.22)$$

A rule is called *confident* if its confidence value exceeds a threshold min_conf. The problem of association rule mining can therefore be formally stated as follows: *Find the set of all rules R of the form $X \Rightarrow Y$ such that*

$$R = \{X \Rightarrow Y | X, Y \subset I, X \cap Y = \emptyset, X \cup Y \in \mathcal{F}(T, min\_sup),$$

$$confidence(X \Rightarrow Y, T) > min\_conf\}. \qquad (8.23)$$

The association rule mining process, in general, consists of the following two steps:

1. Finding all frequent itemsets
2. Generating strong association rules from the frequent itemsets

It may be noted that the number of itemsets grows exponentially with the number of items, $|I|$, and therefore, generating all the frequent itemsets is a challenging problem. *Apriori* algorithm [1] is commonly used for this purpose. The *Apriori* algorithm finds frequent itemsets of length k from frequent itemsets of length $k - 1$. The important concept, utilized for pruning many unnecessary searches, is that if an itemset I of length $k - 1$ is not frequent, then all itemsets I' such that $I \in I'$ cannot be frequent, and hence, this branch of the tree can be effectively pruned from the search space.

Table 8.1 A set of six
transactions defined over
a set of four items
$I = \{i_1, i_2, i_3, i_4\}$

$I = \{i_1, i_2, i_3, i_4\}$

Transaction	Items
t_1	i_1, i_3
t_2	i_1, i_2, i_3
t_3	i_1, i_2
t_4	i_1, i_4
t_5	i_1, i_2, i_3, i_4
t_6	i_2, i_4

Consider the example shown in Table 8.1. Here, there are four items and six transactions. Let the problem be to identify the associations, if any, that can be extracted from this example. Let the $min\_sup = 2$, and $min_con f$ of a rule be 0.9. From the example, it can be seen that

$$cover(i_1, i_3) = \{t_1, t_2, t_5\} \text{ and } cover(i_1, i_2) = \{t_2, t_3, t_5\}.$$

Hence,

$$support(i_1, i_3) = 3 \text{ and } support(i_1, i_2) = 3,$$

both of which exceed the min_sup, and hence, they are frequent itemsets. In fact, it can be verified that only these two are the largest frequent itemsets. Therefore, the rules to be evaluated are

(1) $R_1 : i_1 \Rightarrow i_3$ (2) $R_2 : i_3 \Rightarrow i_1$ (3) $R_3 : i_1 \Rightarrow i_2$ (4) $R_4 : i_2 \Rightarrow i_1$

It can be easily shown that

(1) $confidence(R_1) = \frac{3}{5} = 0.6$ (2) $confidence(R_2) = \frac{3}{3} = 1.0$
(3) $confidence(R_3) = \frac{3}{5} = 0.6$ (4) $confidence(R_4) = \frac{3}{4} = 0.75$.

Hence, only rule R_2, i.e., $i_3 \Rightarrow i_1$, is above the confidence threshold. So ARM will return R_2 as the discovered association rule.

8.2.4   Issues and Challenges in Data Mining

The earlier mining algorithms were usually applied on data that had a fixed structure and were relatively simple. With the advent of new technologies, the data being collected nowadays is increasingly unstructured, complex, and high dimensional data [15]. Typical domains from where such data are routinely collected are:

1. Biometrics data, e.g., fingerprints, iris, face
2. Biological data, e.g., DNA and RNA sequences, 3-D protein structure, gene regulatory networks
3. Web data, e.g., hyperlink structure, user profiles, access profiles

4. Business data, e.g., credit card information, mobile company data, stock market data
5. Meteorological data, e.g., weather patterns, cloud cover

Because of the sheer quantity of data collected, it is often the case that some data might not be collected and/or noise might be inadvertently introduced. For example, the weather information of a particular day might not be collected simply because of unavailability of a technician, or noise might be introduced while sequencing a genome. Performing data mining under such circumstances necessitates the use of sophisticated data cleaning, integration, and estimation methods. The mining algorithms should also be extremely robust to noise and outliers, and scalable and adaptive to new information.

The purpose of knowledge discovery is to identify interesting patterns. The definition of "interestingness" is itself subjective and dependent on the application domain. Thus, automatic identification of interesting patterns becomes extremely difficult. Moreover, if the data and the environment are frequently changing, e.g., stock market time series data and weather data, then designing suitable algorithms that can dynamically adapt their goals is an important issue.

While huge amount of information presents a challenge to data mining in many domains, lack of it throws an equally important challenge to algorithm designers in several application areas. For example, in gene expression data analysis, an important task is to predict the cancer type of an individual from the expression level of thousands of genes. The training data in many such situations might be the expression values of thousands of genes for a few hundreds of patients, or even less. Such applications demand highly sophisticated feature selection methods that can identify only those features that are relevant for the task in hand. Another important issue in data mining arises due to the fact that some events are much rarer than some others making the different classes in the data highly unbalanced. Intrusion detection is one such area where intrusions are relatively rare, and hence, learning their characteristics and identifying them accurately are difficult tasks. Moreover, in many domains, the cost of error is not the same irrespective of the class. For example, in particular cases, the cost of a false-positive prediction might be less than a false-negative prediction. Designing algorithms that can suitably weight the errors is therefore an important issue.

Distributed data mining [35], where the data is distributed over several sites, has become extremely important in recent times. The data might be distributed horizontally in such a way that the schema viewed in every site is the same. Alternatively, the data might be distributed vertically where each site has a different schema and a different view of the data. Though collecting all the data at a central site and then executing the algorithms is a possible way of dealing with such situations, it is evidently highly inefficient. Moreover, privacy and security are nowadays top priority in many organizations, e.g., credit card companies and pharmaceutical organizations, and hence, sharing local data is not a feasible proposition for them. Thus, designing algorithms that can carry out the computation in a distributed, secure, and privacy-preserving manner is of serious concern and a major research thrust worldwide. Other issues of importance are designing mining algorithms over peer-to-peer networks [16], grids, and cluster computers.

8.3 Some Common Tools and Techniques

In this section, some commonly used optimization, learning, and representation tools are discussed in brief.

8.3.1 Artificial Neural Networks

Artificial neural networks (ANN) [21, 32, 49] are a parallel and layered intercon- nected structure of a large number of artificial neurons, each of which constitutes an elementary computational primitive. According to Kohonen [37], ANNs are massively parallel adaptive networks of simple nonlinear computing elements called neurons which are intended to abstract and model some of the functionality of the human nervous system in an attempt to partially capture some of its computational strengths. The distributed representation of the interconnections through massive parallelism achieved out of the inherent network structure bestows upon such networks properties of graceful degradation and fault tolerance.

In the most general form, an ANN is a layered structure of neurons. The neurons can be of three types, namely, input, hidden, and output. The input neurons are designated to accept stimuli from the external world. The output neurons generate the network outputs. The hidden neurons, which are shielded from the external world, are entrusted with the computation of intermediate functions necessary for the operation of the network. A signal function operates within the neurons that generates its output signal based on its activation. In general, these activation functions take an input as an infinite range of activations $(-\infty, +\infty)$ and transform them in the finite range $[0, 1]$ or $[-1, +1]$.

The neurons are connected based on an interconnection topology which basically house the memory of the network. These connections may be excitatory (+), inhibitory (−) or absent (0). Based on the signals received on its input connection and the signal function applicable for the neuron, its output is computed. Neural networks posses the ability to learn from examples. The learning rule provides the basis for modifying the network dynamics with an aim to improve its performance. Learning rules/algorithms define an architecture-dependent procedure to encode pattern information into interneuron interconnections. Learning in a neural network is data driven and proceeds by modifying these connection weights. Some well- known models of ANNs, distinguished by the interconnection topology, activation function, and learning rules, are the multilayer perceptron (MLP), self-organizing map (SOM), and Hopfield network [32].

Fig. 8.2 A typical MLP [10]

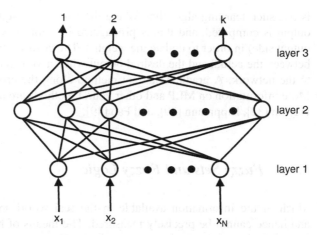

A multilayer perceptron (MLP) consists of several layers of simple neurons with full connectivity existing between neurons of adjacent layers. Figure 8.2 shows an example of a three-layer MLP which consists of an input layer (*layer 1*), one hidden layer (*layers 2*), and an output layer (*layer 3*).

The neurons in the input layer serve the purpose of fanning out the input values to the neurons of *layer 2*. Let

$$w_{ji}^{(l)}, \quad l = 2, 3 \tag{8.24}$$

represent the connection weight on the link from the ith neuron in layer $l - 1$ to the jth neuron in layer l. Let $\theta_j^{(l)}$ represent the threshold or the bias of the jth neuron in layer l. The total input, $x_j^{(l)}$, received by the jth neuron in layer l is given by

$$x_j^{(l)} = \sum_i y_i^{(l-1)} w_{ji}^{(l)} + \theta_j^{(l)}, \quad l = 2, 3, \tag{8.25}$$

where $y_i^{(l-1)}$ is the output of the ith neuron in layer $l - 1$. For the input layer,

$$y_i^{(1)} = x_i, \tag{8.26}$$

where x_i is the ith component of the input vector. For the other layers,

$$y_i^{(l)} = f(x_i^{(l)}) \quad l = 2, 3. \tag{8.27}$$

Several functional forms like threshold logic, hard limiter, and sigmoid can be used for $f(.)$.

There are several algorithms for training the network in order to learn the connection weights and the thresholds from a given training data set. Backpropagation (BP) is one such learning algorithm, where the least mean square error of the network output is computed, and this is propagated in a top-down manner (i.e., from the output side) in order to update the weights. The error is computed as the difference between the actual and the desired output when a known input pattern is presented to the network. A gradient descent method along the error surface is used in BP. More information on MLP and other neural methods are available in Dayhoff [21], Haykin [32], Lippmann [39], and Pao [49].

8.3.2   Fuzzy Sets and Fuzzy Logic

Much of the information available in the real world are not numeric in nature and hence cannot be precisely measured. The means of human communication is inherently vague, imprecise, and uncertain. Fuzzy logic was developed in order to mathematically model this vagueness and imprecision. The fuzzy set theory, introduced by Zadeh [60], explains the varied nature of ambiguity and uncertainty that exist in the real world. This is in sheer contradiction to the concept of crisp sets, where information is more often expressed in quantifying propositions. Fuzzy logic is a superset of conventional (Boolean) logic that has been extended to handle the concept of partial truth, i.e., truth values between completely true and completely false.

A generic fuzzy system comprises the following modules. A fuzzification interface fuzzifies the numeric crisp inputs by assigning grades of membership using fuzzy sets defined for the input variable. A fuzzy rule base/knowledge base comprises a data-derived or heuristic rule base. The data-derived rule base is usually generated by clustering techniques or neural networks using sensor databases. The heuristic rule base, on the other hand, is generated by human experts through some intuitive mechanisms. A fuzzy inference engine infers fuzzy outputs by employing the fuzzy implications and the rules of inference of fuzzy logic. Finally, a defuzzification interface is present that yields a non-fuzzy crisp control action from an inferred fuzzy control action.

Fuzzy set theory has found a lot of applications in data mining [4, 50, 59]. Examples of such applications may be found in clustering [38], association rules [57], time series [18], and image retrieval [28].

8.3.3   Genetic Algorithms

Genetic algorithms (GAs) [33] are randomized search and optimization technique guided by the principles of natural genetic systems. These algorithms are characterized by a population of encoded trial solutions and a collection of operators to act on the population. The basic philosophy behind these algorithms is to encode the parameters of the problems and then parallely search the space of the encoded solutions by the application of the embedded operators so as to arrive at an optimal solution. Generally, two types of operators are used, namely, reproduction and evolution. The reproduction operator is guided by a selection mechanism. The evolution operator includes the crossover and mutation operators. The search technique is implemented through a series of iterations, whereby the different operators are applied in a loop on the initial population. Each iteration is referred to as a generation. Each generation produces a new solution space, which are selectively chosen for participating in the next generation of the optimization procedure. The selection of the participating solutions for the next generation is decided by a figure of merit, often referred to as the fitness function. The essential components of GAs are the following:

- A representation strategy that determines the way in which potential solutions will be coded to form string-like structures called *chromosomes*
- A population of *chromosomes*
- Mechanism for evaluating each chromosome
- Selection/reproduction procedure
- Genetic operators
- Probabilities of performing genetic operations

It operates through a simple cycle of

1. evaluation of each chromosome in the population to get the fitness value,
2. selection of chromosomes, and
3. genetic manipulation to create a new population of chromosomes,

over a number of iterations (or generations) till one or more of the following termination criteria is satisfied:

- The average fitness value of a population becomes more or less constant over a specified number of generations.
- A desired objective function value is attained by at least one string in the population.
- The number of iterations is greater than some predefined threshold.

A schematic diagram of the basic structure of a genetic algorithm is shown in Fig. 8.3.

Applications of genetic algorithms and related techniques in data mining include extraction of association rules [40], predictive rules [26,27], clustering [7,8,14,41–43], program evolution [53], and web mining [45,46,51,52].

8.4   Case Examples

In this section, we provide two case examples of the application of the above-mentioned tools and techniques for solving two real-world problems. They deal with the applications of genetic algorithms for supervised classification and clustering, respectively, both dealing with analysis of remote sensing imagery.

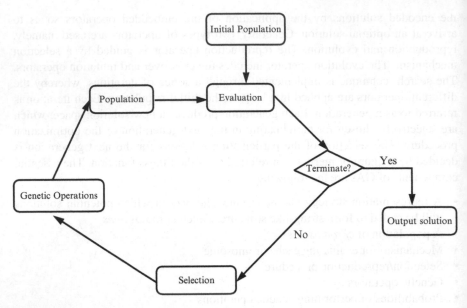

Fig. 8.3 Basic steps of a genetic algorithm [10]

8.4.1 Pixel Classification

The problem of classification can be viewed as one of generating decision boundaries that can successfully distinguish the various classes in the feature space. In real-life problems, the boundaries between the different classes are usually nonlinear. In this section, a classifier, called *GA-classifier*, that utilizes the characteristics of GAs in searching for a number of linear segments which can approximate the nonlinear boundaries while providing minimum misclassification of training sample points is described [11, 48].

The *GA-classifier* attempts to place H hyperplanes in the feature space appropriately such that the number of misclassified training points is minimized. From elementary geometry, it can be derived that a hyperplane in N-dimensional space can be represented by $N - 1$ angle variables and a perpendicular distance variable. These are encoded in a chromosome in the GA. Thus, each chromosome encoding H hyperplanes is of length $l = H((N - 1) * b_1 + b_2)$, where b_1 and b_2 are the numbers of bits used for encoding an angle and the perpendicular distance, respectively.

The computation of the fitness is done for each string in the population. The fitness of a string is characterized by the number of points it misclassifies. A string with the lowest misclassification is therefore considered to be the fittest among the population of strings. If the number of misclassified points for a string is denoted by $miss$, then the fitness of the string is computed as $(n - miss)$, where n is the number of training data points. The best string of each generation or iteration is the one which has the fewest misclassifications. This string is stored after each iteration. If the best string of the previous generation is found to be better than the best string of the current generation, then the previous best string replaces the worst string of the current generation. This implements the *elitist strategy*, where the best string seen up to the current generation is propagated to the next generation.

Since it is difficult to estimate the proper number of hyperplanes a priori, in Bandyopadhyay et al. [13], the concept of variable H and hence variable chromosome length was introduced. This resulted in a classifier called *VGA-classifier*. The chromosomes in the *VGA-classifier* are represented by strings of 1, 0, and # (don't care), encoding the parameters of variable number of hyperplanes. The fitness of string i, encoding H_i hyerplanes, is defined as

$$\text{fit}_i = (n - miss_i) - \alpha H_i, \ 1 \le H_i \le H_{\max}, \qquad (8.28)$$

$$= \ 0, \qquad\qquad \text{otherwise}, \qquad\qquad (8.29)$$

where n = size of the training data set and $\alpha = \frac{1}{H_{\max}}$. Here, $H_{\max}$ is the maximum number of hyperplanes that are considered for approximating the decision boundary. Note that the maximization of the fitness function leads to minimization of the number of misclassified points and also minimization of the number of hyperplanes. The genetic operators of crossover and mutation are redefined appropriately so as to tackle the variable length chromosomes [13].

The *VGA-classifier* was further enhanced in Bandyopadhyay and Pal [9] by the incorporation of the concept of chromosome differentiation proposed in Bandyopadhyay et al. [12]. This resulted in a classifier called *VGACD-classifier*. Here, two classes of chromosomes exist in the population. Crossover (mating) is allowed only between individuals belonging to these categories. A schema analysis of GAs with chromosome differentiation (GACD), vis á vis that of conventional GA (CGA), was conducted in Bandyopadhyay et al. [12] which showed that in certain situations, the lower bound of the number of instances of a schema sampled by GACD is greater than or equal to that of CGA.

The *VGACD-classifier* was used in Bandyopadhyay and Pal [9] to classify a 512×512 *SPOT* image of a part of the city of Calcutta that is available in three bands. The image in the near-infrared band is shown in Fig. 8.4. The design set comprises 932 points belonging to 7 classes that are extracted from the above image. A two-dimensional scatter plot of the training data is shown in Fig. 8.5. The seven classes are *turbid water* (TW), *pond water* (PW), *concrete* (Concr.), *vegetation* (Veg), *habitation* (Hab), *open space* (OS), and *roads* (including bridges) (B/R). The classifier trained using the design set is then utilized for classifying the pixels in the 512×512 image.

Figure 8.6 shows the full Calcutta image classified using the *VGACD-classifier* [9]. As can be seen, most of the landmarks in Calcutta have been properly classified. The optimum number of hyperplanes was found automatically to be equal to 13. The performance of the *VGACD-classifier* was studied in comparison with those of the

Fig. 8.4 *SPOT* image of Calcutta in the near-infrared band

VGA-classifier, K-NN rule, and Bayes maximum likelihood classifier. It showed that the *VGACD-classifier* provided the best performance, and its rate of convergence was also enhanced over the *VGA-classifier* [9].

8.4.2 Clustering of Satellite Images

The purpose of any clustering technique is to evolve a $K \times n$ partition matrix $U(X)$ of a data set X ($X = \{\mathbf{x}_1, \mathbf{x}_2, \ldots, \mathbf{x}_n\}$) in IR^N, representing its partitioning into a number, say K, of clusters $(C_1, C_2, \ldots, C_K)$. The partition matrix $U(X)$ may be represented as $U = [u_{kj}]$, $k = 1, \ldots, K$ and $j = 1, \ldots, n$, where u_{kj} is the membership of pattern $\mathbf{x}_j$ to cluster C_k. Clustering methods are also often categorized as crisp and fuzzy. In crisp clustering, a point belongs to exactly one cluster, while in fuzzy clustering, a point can belong to more than one cluster with varying degrees of membership.

The different clustering methods, in general, try to optimize some measure of goodness of a clustering solution either explicitly or implicitly. The clustering problem can therefore be mapped to one of searching for an appropriate number of suitable partitions such that some goodness measure is optimized. It may be

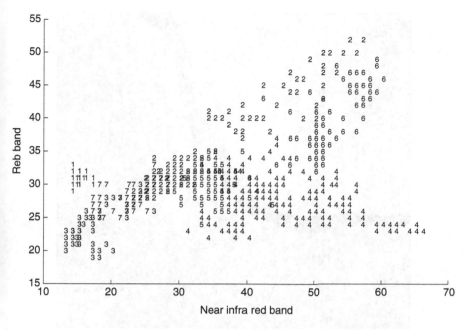

Fig. 8.5 Scatter plot for a training set of *SPOT image* of Calcutta containing seven classes (1,...,7)

noted that searching the exhaustive set of all possible partitions of the data is prohibitively exhaustive since the search space is huge and complex, with numerous local optima. Consequently, heuristic search methods are often employed in this domain, and these often get stuck at local optima. Several attempts have also been made to investigate the effectiveness of GAs and related methods for clustering [5–7, 14, 41, 43, 47].

As earlier, variable length GAs (VGAs) are used for automatically evolving the near-optimal nondegenerate fuzzy partition matrix U^* [43]. The centers of a variable number of clusters are encoded in the chromosomes. Given a set of centers encoded in the chromosome, the fuzzy partition matrix is computed using Eq. (8.15). Thereafter, the centers are updated using Eq. (8.16). For the purpose of optimization, the Xie–Beni [58] cluster validity index is used. The Xie–Beni (XB) index is defined as a function of the ratio of the total variation σ to the minimum separation sep of the clusters. Here, σ and sep can be written as

$$\sigma(U, Z; X) = \sum_{i=1}^{c} \sum_{k=1}^{n} u_{ik}^2 D^2(\mathbf{z}_i, \mathbf{x}_k), \qquad (8.30)$$

and

$$sep(Z) = \min_{i \neq j}\{||\mathbf{z}_i - \mathbf{z}_j||^2\}, \qquad (8.31)$$

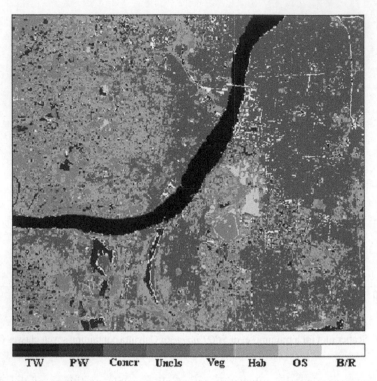

| TW | PW | Concr | Uncls | Veg | Hab | OS | B/R |

Fig. 8.6 Classified *SPOT* image of Calcutta using the *VGACD-classifier* (H_{max} =15, final value of H=13)

where $||.||$ is the Euclidean norm, and $D(\mathbf{z}_i, \mathbf{x}_k)$ is the distance between the pattern $\mathbf{x}_k$ and the cluster center $\mathbf{z}_i$. The XB index is then written as

$$XB(U, Z; X) = \frac{\sigma(U, Z; X)}{n \, sep(Z)} = \frac{\sum_{i=1}^{c}(\sum_{k=1}^{n} u_{ik}^2 D^2(\mathbf{z}_i, \mathbf{x}_k))}{n(\min_{i \neq j}\{||\mathbf{z}_i - \mathbf{z}_j||^2\})}. \tag{8.32}$$

Note that when the partitioning is compact and good, value of σ should be low while sep should be high, thereby yielding lower values of the XB index. The objective is therefore to minimize the XB index for achieving proper clustering. In other words, the best partition matrix U^* is the one such that

$$U^* \in \mathcal{U} \text{ and } XB(U^*, Z^*, X) = \min_{U_i \in \mathcal{U}} XB(U_i, Z_i, X), \tag{8.33}$$

where Z^* represents the set of cluster centers corresponding to U^*. Here, both the number of clusters as well as the appropriate fuzzy clustering of the data are evolved simultaneously using the search capability of genetic algorithms. The chromosome representation and other genetic operators used are described in detail in Maulik and Bandyopadhyay [43].

Fig. 8.7 IRS image of Mumbai in the near-infrared band with histogram equalization

The effectiveness of the genetic fuzzy clustering technique in partitioning the pixels into different landcover types is demonstrated on an Indian remote sensing (IRS) satellite image of a part of the city of Mumbai [43] (Fig. 8.7 shows the image in the near-infrared band). Detailed description is available in Maulik and Bandyopadhyay [43]. The segmented Mumbai image is shown in Fig. 8.8. The method automatically yielded seven clusters, which are labelled concrete (Concr.), open spaces (OS1 and OS2), vegetation (Veg), habitation (Hab), and turbid water (TW1 and TW2), based on the ground information available from earlier studies. The classes Hab, referring to the regions which have concrete structures and buildings, but with relatively lower density than the class concrete, and Concr. share common properties. Figure 8.8 shows that the large water body of the Arabian Sea has been distinguished into two classes which are named TW1 and TW2. The islands, dockyard, and several road structures have mostly been correctly identified in the image. Within the islands, as expected, there is a predominance of open space and vegetation. The southern part of the city, which is heavily industrialized, has been classified as primarily belonging to habitation and concrete. Some confusion within these two classes, namely, Hab and Concr, is observed (as reflected in the corresponding label); this is expected since these two classes are somewhat similar.

Figure 8.9 demonstrates the Mumbai image partitioned into seven clusters using the FCM technique. As can be seen, the water of the Arabian Sea has been

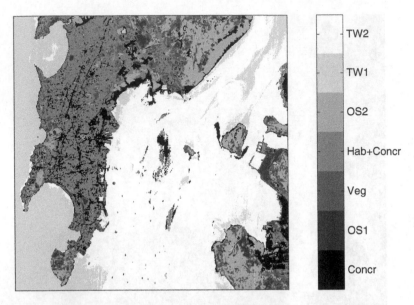

Fig. 8.8 Clustered image of Mumbai using genetic fuzzy clustering

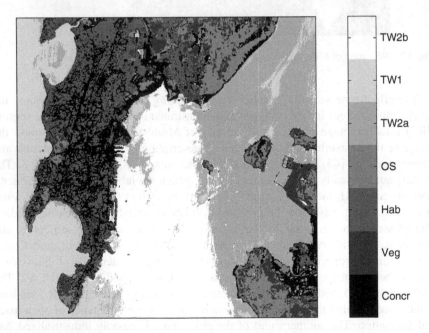

Fig. 8.9 Clustered image of Mumbai using FCM clustering

partitioned into three regions rather than two as obtained earlier. The other regions appear to be classified more or less correctly for this data. It was observed that the FCM algorithm gets trapped at local optima often enough, and the best value of the XB index was worse than that obtained using the genetic fuzzy clustering scheme.

8.5 Discussion and Conclusions

The basic concepts and issues in data mining and knowledge discovery have been discussed in this chapter. The challenges being faced by data miners, namely, very high dimensional and extremely large data sets, unstructured and semi-structured data, temporal and spatial patterns, and heterogeneous data, are mentioned. Some major data mining tasks are discussed with emphasis on algorithms developed for solving them. These include description of classifiers based on Bayes rule; nearest neighbor rule and support vector machines; clustering algorithms like K-means, fuzzy c-means, and single linkage; and association rule mining methods. The utility of some tools like neural networks, genetic algorithms, and fuzzy sets that are frequently used in data mining is discussed. Finally, two case examples are presented.

Traditional data mining generally involved well-organized database systems such as relational databases. With the advent of sophisticated technology, it is now possible to store and manipulate very large and complex data. The data complexity arises due to several reasons, e.g., high dimensionality, semi- and/or unstructured nature, and heterogeneity. Data related to the World Wide Web, the geoscientific domain, VLSI chip layout and routing, multimedia, financial markets, sensor networks, and genes and proteins constitute some typical examples of complex data. In order to extract knowledge from such complex data, it is necessary to develop advanced methods that can exploit the nature and representation of the data more efficiently. In Bandyopadhyay et al. [15], several such interesting attempts for knowledge discovery from complex data have been described.

References

1. Agrawal R, Srikant R (1994) Fast algorithms for mining association rules. In: Bocca JB, Jarke M, Zaniolo C (eds) Proceedings of the 20th international conference on very large data bases, VLDB, Morgan Kaufmann, Bases, VLDB. Morgan Kaufmann, pp 487–499. citeseer.ist.psu. edu/agrawal94fast.html
2. Anderberg MR (1973) Cluster analysis for application. Academic, New York
3. Anderson TW (1958) An introduction to multivariate statistical analysis. Wiley, New York
4. Baldwin JF (1996) Knowledge from data using fuzzy methods. Pattern Recognit Lett 17: 593–600

5. Bandyopadhyay S (2005) Satellite image classification using genetically guided fuzzy clustering with spatial information. Int J Remote Sens 26(3):579–593
6. Bandyopadhyay S (2005) Simulated annealing using reversible jump Markov chain Monte Carlo algorithm for fuzzy clustering. IEEE Trans Knowl Data Eng 17(4):479–490
7. Bandyopadhyay S, Maulik U (2001) Non-parametric genetic clustering: comparison of validity indices. IEEE Trans Syst Man Cybern C 31(1):120–125
8. Bandyopadhyay S, Maulik U (2002) An evolutionary technique based on k-means algorithm for optimal clustering in R^N. Inf Sci 146:221–237
9. Bandyopadhyay S, Pal SK (2001) Pixel classification using variable string genetic algorithms with chromosome differentiation. IEEE Trans Geosci Remote Sens 39(2):303–308
10. Bandyopadhyay S, Pal SK (2007) Classification and learning using genetic algorithms: applications in bioinformatics and web intelligence. Springer, Hiedelberg
11. Bandyopadhyay S, Murthy CA, Pal SK (1995) Pattern classification using genetic algorithms. Pattern Recognit Lett 16:801–808
12. Bandyopadhyay S, Pal SK, Maulik U (1998) Incorporating chromosome differentiation in genetic algorithms. Inf Sci 104(3/4):293–319
13. Bandyopadhyay S, Murthy CA, Pal SK (2000) *VGA-classifier*: design and application. IEEE Trans Syst Man Cybern B 30(6):890–895
14. Bandyopadhyay S, Maulik U, Pakhira MK (2001) Clustering using simulated annealing with probabilistic redistribution. Int J Pattern Recognit Artif Intell 15(2):269–285
15. Bandyopadhyay S, Maulik U, Holder L, Cook DJ (eds) (2005) Advanced methods for knowledge discovery from complex data. Springer, Berlin/Heidelberg/New York
16. Bandyopadhyay S, Giannella C, Maulik U, Kargupta H, Liu K, Datta S (2006) Clustering distributed data streams in peer-to-peer environments. Inf Sci 176(14):1952–1985
17. Bezdek JC (1981) Pattern recognition with fuzzy objective function algorithms. Plenum, New York
18. Chiang DA, Chow LR, Wang YF (2000) Mining time series data by a fuzzy linguistic summary system. Fuzzy Sets Syst 112:419–432
19. Cios KJ, WPedrycz, Swiniarski RW, Kurgan LA (2007) Data mining: a knowledge discovery approach. Springer, New York
20. Cover TM (1965) Geometrical and statistical properties of systems of linear inequalities with applications in pattern recognition. IEEE Trans Electron Comput 14:8326–334
21. Dayhoff JE (1990) Neural network architectures an introduction. Van Nostrand Reinhold, New York
22. Devijver PA, Kittler J (1982) Pattern recognition: a statistical approach. Prentice-Hall, London
23. Duda RO, Hart PE (1973) Pattern classification and scene analysis. Wiley, New York
24. Ester M, Kriegel H-P, Xu X (1995) Knowledge discovery in large spatial databases: focusing techniques for efficient class identification. In: Advances in spatial databases. Springer, Berlin/Heidelberg
25. Ester M, Kriegel H-P, Sander J, Xu X (1996) A density-based algorithm for discovering clusters in large spatial databases with noise. In: Proceedings of 2nd international conference on knowledge discovery and data mining, vol 96. AAAI Press, Portland, pp 226–231
26. Flockhart IW (1995) GA-MINER: parallel data mining with hierarchical genetic algorithms – final report. Technical report EPCC-AIKMS-GA-MINER-REPORT 1.0, University of Edinburgh, Edinburgh. citeseer.ist.psu.edu/flockhart95gaminer.html
27. Flockhart IW, Radcliffe NJ (1996) A genetic algorithm-based approach to data mining. In: Simoudis E, Han JW, Fayyad U (eds) Proceedings of the second international conference on knowledge discovery and data mining (KDD-96). AAAI Press, Portland, pp 299–302. citeseer. nj.nec.com/44319.html
28. Frigui H, Krishnapuram R (1999) A robust competitive clustering algorithm with application in computer vision. IEEE Trans Pattern Anal Mach Intell 21(1):450–465

29. Fukunaga K (1972) Introduction to statistical pattern recognition. Academic, New York
30. Han J, Kamber M (2000) Data mining: concepts and techniques. Morgan Kaufmann Publishers, San Francisco
31. Hartigan JA (1975) Clustering algorithms. Wiley, New York
32. Haykin S (1994) Neural networks, a comprehensive foundation. Macmillan College Publishing Company, New York
33. Holland JH (1975) Adaptation in natural and artificial systems. The University of Michigan Press, Ann Arbor
34. Jain AK, Dubes RC (1988) Algorithms for clustering data. Prentice-Hall, Englewood Cliffs
35. Kargupta H, Chan P (eds) (1999) Advances in distributed and parallel knowledge discovery. MIT/AAAI Press, Cambridge
36. Kaufman L, Rousseeuw PJ (1990) Finding groups in data: an introduction to cluster analysis. Wiley, New York
37. Kohonen T (2001) Self-organization and associative memory. Springer, Heidelberg
38. Krishnapuram R, Joshi A, Nasraoui O, Yi L (2001) Low complexity fuzzy relational clustering algorithms for web mining. IEEE Trans Fuzzy Syst 9:595–607
39. Lippmann RP (1987) An introduction to computing with neural nets. IEEE ASSP Mag 4(2):4–22
40. Lopes C et al (1999) Rule-evolver: an evolutionary approach for data mining. In: New directions in rough sets, data mining, and granular-soft computing. Springer, Berlin/Heidelberg, pp 458–462
41. Maulik U, Bandyopadhyay S (2000) Genetic algorithm based clustering technique. Pattern Recognit 33:1455–1465
42. Maulik U, Bandyopadhyay S (2002) Performance evaluation of some clustering algorithms and validity indices. IEEE Trans Pattern Analy Mach Intell 24(12):1650–1654
43. Maulik U, Bandyopadhyay S (2003) Fuzzy partitioning using a real-coded variable-length genetic algorithm for pixel classification. IEEE Trans Geosci Remote Sens 41(5):1075–1081
44. Ng R, Han J (1994) Efficient and effective clustering method for spatial data mining. In: Proceedings of the 1994 international conference on very large data bases, Santiago, Chile, pp 144–155
45. Oliver A, Monmarché N, Venturini G (2002) Interactive design of web sites with a genetic algorithm. In: Proceedings of the IADIS international conference www/internet, Lisbon
46. Oliver A, Regragui O, Monmarché N, Venturini G (2002) Genetic and interactive optimization of web sites. In: Eleventh international world wide web conference, Honolulu, Hawaii
47. Pakhira MK, Bandyopadhyay S, Maulik U (2004) Validity index for crisp and fuzzy clusters. Pattern Recognit 37(3):487–501
48. Pal SK, Bandyopadhyay S, Murthy CA (1998) Genetic algorithms for generation of class boundaries. IEEE Trans Syst Man Cybern 28(6):816–828
49. Pao YH (1989) Adaptive pattern recognition and neural networks. Addison-Wesley, New York
50. Pedrycz W (1998) Fuzzy set technology in knowledge discovery. Fuzzy Sets Syst 98:279–290
51. Picarougne F, Fruchet C, Oliver A, Monmarché N, Venturini G (2002) Web searching considered as a genetic optimization problem. In: Local search two day workshop, London, UK
52. Picarougne F, Monmarché N, Oliver A, Venturini G (2002) Web mining with a genetic algorithm. In: Eleventh international world wide web conference, Honolulu, Hawaii
53. Raymer ML, Punch WF, Goodman ED, Kuhn LA (1996) Genetic programming for improved data mining: an application to the biochemistry of protein interactions. In: Proceedings of first annual conference on genetic programming, MIT Press, Cambridge, pp 375–380
54. Selim SZ, Ismail MA (1984) K-means type algorithms: a generalized convergence theorem and characterization of local optimality. IEEE Trans Pattern Anal Mach Intell 6:81–87
55. Spath H (1989) Cluster analysis algorithms. Ellis Horwood, Chichester
56. Tou JT, Gonzalez RC (1974) Pattern recognition principles. Addison-Wesley, Reading

57. Wei Q, Chen G (1999) Mining generalized association rules with fuzzy taxonomic structures. In: Proceedings of NAFIPS 99, IEEE Press, New York, pp 477–481
58. Xie XL, Beni G (1991) A validity measure for fuzzy clustering. IEEE Trans Pattern Analy Mach Intell 13:841–847
59. Yager RR (1996) Database discovery using fuzzy sets. Int J Intell Syst 11:691–712
60. Zadeh LA (1965) Fuzzy sets. Inf Control 8:338–353
61. Zhang T, Ramakrishnan R, Livny M (1996) Birch: an efficient data clustering method for very large databases. In: Proceedings of the 1996 ACM SIGMOD international conference on management of data. ACM Press, New York, pp 103–114. http://doi.acm.org/10.1145/233269. 233324

Chapter 9
Applied Bioinformatics Tools

Jingchu Luo

9.1 Introduction

A hands-on course mainly for the applications of bioinformatics to biological problems was organized at Peking University. The course materials are from http://abc.cbi.pku.edu.cn. They are divided into individual pages (separated by lines in the text):

- Welcome page
- About this course
- Lectures
- Exercises
- Projects
- Online literature resources
- Bioinformatics databases
- Bioinformatics tools

This chapter lists some of the course materials used in the summer school. The course pages are being updated and new materials will be added (Fig. 9.1).

9.1.1 Welcome

Welcome to ABC – the website of Applied Bioinformatics Course. We'll learn, step by step, the ABCs of:

- How to access various bioinformatics resources on the Internet
- How to query and search biological databases

J. Luo (✉)
Center for Bioinformatics, School of Life Sciences, Peking University, Beijing 100871, China
e-mail: luojc@mail.cbi.pku.edu.cn

R. Jiang et al. (eds.), *Basics of Bioinformatics: Lecture Notes of the Graduate Summer School on Bioinformatics of China*, DOI 10.1007/978-3-642-38951-1_9,
© Tsinghua University Press, Beijing and Springer-Verlag Berlin Heidelberg 2013

Fig. 9.1 The website of the applied bioinformatics course

- How to use bioinformatics tools to analyze your own DNA or protein sequences
- How to predict the three-dimensional structure of your favorite protein
- How to draw a nice tree for the bunch of sequences at your hand
- And lots more!

9.1.1.1 How We Learn

We will run the course in a training room (see the above picture). Each student will have a PC with both Linux and Windows installed. We start with simple Linux commands and the WebLab bioinformatics platform developed by CBI, to get familiar with dozens of bioinformatics tools through hands-on practice. We will do a lot of exercises for sequence alignment, database similarity search, motif finding, gene prediction, as well as phylogenetic tree construction and molecular modeling. Finally, we will focus on several projects to solve real biological problems. You are also encouraged to bring your own problems to discuss and, hopefully, to solve during the course!

9.1.1.2 What You Need Before the Course

- A desktop PC or laptop hooked to the Internet
- Good background of molecular biology
- Ability to read and write in English
- At least another 6 h to stick on the Internet doing homework of the course

9.1.1.3 What You Gain from the Course

Don't expect to become a bioinformatics expert, but you will know:

Half day on the web, saves you half month in the lab!

9.1.2 About This Web Site

This website is an online portal for the teaching of the semester course of computer application to molecular biology since 2000. We now use Applied Bioinformatics Course, or ABC as a simple name for this course. Here, ABC means that it is an entry level introductory course, rather than an advanced one.

9.1.2.1 Students

The aim of this course is mainly for graduate students of biology to learn how to use bioinformatics tools to solve his or her own problems. The students are mainly from:

- College of Life Sciences, Peking University (PKU)
- Graduate school, Chinese Academy of Agricultural Sciences (CAAS)

We also run training courses for graduate students and junior researchers. For example, with the help of colleagues from EMBnet (Dr. Valverde, Spanish node, and Dr. Moulton, University of Manchester, UK), two 3-day courses (30 h each) were organized for:

- Institute of Botany, Chinese Academy of Sciences (CAS)
  And two training courses (15 h each) were given for participants of:
- 2007 Summer School of Bioinformatics, Lab of Bioinformatics, Tsinghua University (TSU)

9.1.2.2 For Course Students

The course will be given mainly in Chinese, mixed with English terms. Some of the lecture slides are in English. You should have a good background of English especially in scientific reading if you plan to study this course. Registration is needed for both PKU and CAAS students taking this as a semester course. Please see the Notice page for more details.

9.1.2.3 For Web Visitors

The web pages of this site are in English. However, the teaching materials, e.g., the lectures given by students, are in Chinese. They are in PDF format and freely available for download. For non-native Chinese speakers, however, you should have a good command of Chinese.

9.1.3 Outline

A. Getting started with applied bioinformatics
    1. An overview of bioinformatics
    2. The basis of molecular biology

B. Bioinformatics resources over the Internet
    1. International bioinformatics centers (NCBI, EBI, ExPASy, RSCB)
    2. Literature references (PubMed, PubMed Central, online books, bioinformatics courses, tutorials)
    3. Databases of molecular biology (genome, sequence, function structure databases)
    4. Bioinformatics tools (programs, packages, online web servers)

C. Database query
    1. Database query with the NCBI Entrez system
    2. Database query with the EBI SRS platform
    3. Database query with the MRS platform

D. DNA and protein sequence analysis
    1. Sequence alignment and dotplot
    2. Protein sequence analysis
    3. DNA sequence analysis
    4. Database similarity search (BLAST, FASTA)
    5. Phylogenic analysis and tree construction

E. Molecular modeling
    1. Molecular graphics and visualization
    2. Molecular modeling and structure prediction

F. Projects

1. Sequence analysis of Pisum sativum post-floral specific gene
2. MDR – gene prediction and dotplot analysis of fugu multidrug resistance gene
3. Sequence, structure, and function analysis of the bar-headed goose hemoglobin
4. CEA – protein engineering of carcinoembryonic antigen
5. Structure comparison of spider toxins and prediction of antifungal peptide
6. Sequence and structure comparison of human metallothioneins
7. Systematic analysis of the Arabidopsis transcription factor family of Squamosa-promoter binding protein

9.1.4   Lectures

Although the main approach of this course is hands-on practice, it is necessary to communicate among each other during the course. Most of the lectures were given by students in the class.

9.1.4.1   Introduction

• My View on Bioinformatics (Luo JC, CBI) [PDF]

9.1.4.2   UNIX and EMBOSS

• Unix – Carver T (EBI) [PDF] | Tian YL (CAAS06)F [PDF]
• EMBOSS – Fan L (CAAS07F) [PDF]

9.1.4.3   BLAST and Database Search

• BLAST – Luo JC (CBI) [PDF] | Tian YL (CAAS06F) [PDF] | Xie C (PKU08S1) [PDF] | Bian Y (PKU08S1) [PDF]
• Database search – Gao G (CBI) [PDF]
• Multiple sequence alignment – Liu K (PKU07S) [PDF]
• Scoring Matrix – Yuan YX (PKU07S) [PDF]

9.1.4.4   Resources

• PubMed – Huang LY (CAAS07F) [PDF] | Luo JC (CBI) [PDF]
• NCBI Databases – Li L (CAAS07F) [PDF]

- ExPASy – Gong JY (CAAS07F) [PDF]
- PDB – Gao HB (CAAS07F) [PDF]
- SRS – Huang BY (CAAS07F) [PDF]

9.1.4.5 Phylogeny

- MEGA – Hu YP (CAAS06F) [PDF]
- Phylogeny – Gao G (CBI) [PDF] | Huang BY (CAAS07F) [PDF] | Li Z (CBI) [PDF] | Shi XL (CBI) [PDF] | Zhu HW (CAAS07F) [PDF] | Li J (PKU08S) [PDF] | Yang Q (PKU08S) [PDF]

9.1.4.6 Molecular Modeling

- 3D modeling – Ye ZQ (CBI) [PDF] | Zhu HW (CAAS07F) [PDF] | Wang JL (CAAS08S) [PDF]
- 3D of Transcription Factors – Luo JC (CBI) [PDF]
- Swiss PDB Viewer – Li W (CAAS07F) [PDF]

9.1.4.7 Projects

- Hemoglobin – Li WQ (PKU07S) [PDF]
- MDR – Xu LM (PKU08S1) [PDF]
- CEA – Wang Q (CAAS07F) [PDF]
- SBP – Guo AY (CBI) [PDF]
- Rice – Zhu QH (CBI) [PDF]
- CA3 – Hou XH (CAAS07F) [PDF]
- PhyB – Wu FQ (CAAS07F) [PDF]
- P53 – Shen Y (CAAS07F) [PDF]
- Text Mining – Wang X (CAAS07F) [PDF]

All the PDF files of the above lectures can be downloaded freely for teaching. The copyright belongs to the original authors.

9.1.5 Exercises

As a hands-on practical course, we will introduce many exercises to be practiced during the course. This page collects a variety of exercises such as literature search, database query, and database search, sequence alignment, motif search, phylogenetic analysis, and molecular modeling.

Entrez – Literature search with PubMed and database query with the NCBI Entrez system.

ExPASy – Find protein sequences from the Swiss-Prot database with the ExPASy system.

SRS – Database query with the EBI Sequence Retrieval System (SRS).

Dotplot – DNA and protein sequence comparison using the dot plot approach.

Align – Pairwise and multiple sequence alignment using both local and global algorithms.

BLAST – Sequence similarity search against DNA and protein sequence databases.

DNA sequence analysis – Analysis of DNA sequences with several bioinformatics tools.

Protein sequence analysis – Analysis of protein sequences with several bioinformatics tools.

Motif – Identification of conserved sequence motifs and domains.

Phylogeny – Phylogenetic analysis and construction of phylogenetic tress with simple examples.

Modeling – Analysis of three-dimensional structure of proteins and prediction of protein structures.

9.2 Entrez

Literature search with PubMed and database query with the NCBI Entrez system.

9.2.1 PubMed Query

Find papers with the following keywords and compare the query results:

- "Hemoglobin," "human hemoglobin," "human hemoglobin AND structure," and "human hemoglobin AND function"
- "Hemoglobin [TI]," "human hemoglobin [AB]", "human hemoglobin [TI] AND structure [TIAB]," and "human hemoglobin [TI] AND structure [TIAB] AND Function [TIAB]"
- "Hemoglobin [TI] OR haemoglobin [TI]" and "hemoglobin [TI] OR haemoglobin [TI] AND structure [TIAB]"

Find papers with the following author names and compare the query results:

- "Sodmergen"
- "Danchin" and "Danchin A"
- "Li," "Li Y," and "Li YX"
- "Smith," "Smith T," and "Smith TT"

Find papers with the following query and compare the query results:

- "Rice," "rice [TI]," "rice [AU]," and "rice [AU] AND rice"
- "Rice," "Oryza sativa," and "rice OR Oryza sativa"
- "Luo J," "Luo JC," "Luo J[AU] AND bioinformatics," and "Luo J[AU] AND Peking University[AD]"
- "Luo J[AU] AND Peking University[AD] AND bioinformatics OR database OR rice"

9.2.2   Entrez Query

- Find the GenBank entry of the post-floral-specific gene (PPF-1) with the following query and compare query results: "Y12618," "PPF-1," and "post-floral-specific gene."
- Find the GenPept entry of the post-floral-specific protein (PPF-1) with the following query and compare query results: "Q9FY06," "PPF-1," and "post-floral-specific protein."
- Find the three-dimensional structure of the bar-headed goose hemoglobin.

9.2.3   My NCBI

- Register in My NCBI and make the following query: "bioinformatics [TI]."
- Save the above query and set the email delivery options.
- Configure the display options to show query results.

9.3   ExPASy

Find protein sequences from the Swiss-Prot database with the ExPASy system.

9.3.1   Swiss-Prot Query

- Find protein sequence of human hemoglobin alpha chain.
- Find protein sequence of mouse and rat hemoglobin alpha chains.
- Find protein sequence of human hemoglobin beta chain.
- Find protein sequence of mouse and rat hemoglobin beta chains.
- Find all entries of hemoglobin alpha chain in Swiss-Prot.
- Find all entries of hemoglobin beta chain in Swiss-Prot.

9.3.2   Explore the Swiss-Prot Entry HBA_HUMAN

- Retrieve the mRNA and coding sequence of human hemoglobin through cross link to GenBank.
- Find how many entries are deposited to Swiss-Prot through the Taxon cross link.
- Find the literatures related to crystal structure of human hemoglobin.
- Make a summary of the annotation of human hemoglobin based on the Comments of this entry.
- Make a summary of mutation of human hemoglobin based on the sequence features.
- Find out the alpha helices of human hemoglobin alpha chain.
- Find out the two Histidines which bind to the heme.
- Retrieve the FASTA sequence from this entry.

9.3.3   Database Query with the EBI SRS

SRS stands for sequence retrieval system. It is actually the main bioinformatics database query platform maintained by EBI and other bioinformatics centers around the world. SRS was originally developed by Etzold and Argos at the European Molecular Biology Laboratory (EMBL) in the early 1990s. It was moved to EBI and being continually developed by a group led by Etzold during the middle of the 1990s. In 1998, SRS became the main bioinformatics product of the biotechnology company LION Biosciences. Although it was a commercial software, SRS had been free for academic use until 2006 when it was acquired by BioWisdom.

9.3.3.1   Query

- Find human hemoglobin alpha chain sequence from Swiss-Prot with ID "HBA_HUMAN" or Accession "P69905," use Quick Search, Standard Query Form and Extended Query Form, compare the difference of search steps, and query results.
- Find human, mouse, and rat hemoglobin alpha chain sequences simultaneously from Swiss-Prot using ID "HBA_HUMAN," "HBA_MOUSE," and "HBA_RAT"; use Standard Query Form.
- Find all hemoglobin alpha chain sequence from Swiss-Prot with ID "HBA_."

9.3.3.2   Display and Save

- Display human hemoglobin alpha chain sequence ("HBA_HUMAN") with different sequence formats.

- Save all hemoglobin alpha chain sequences (HBA_) with FASTA format.
- Choose accession, sequence length, and species to display and save all hemoglobin alpha chain sequences (HBA_).

9.3.3.3 Dotplot

DNA and protein sequence comparison using the dot plot approach:

- Draw a dotplot for a test DNA sequence with a tandem repeat; use different word sizes to compare the results.
- Draw a dotplot for a test protein sequence (SENESCENSE); use different word sizes to compare the results.
- Draw a dotplot for the fugu cosmid sequence (AF164138).
- Retrieve similar regions from (AF164138) and draw a dotplot.
- Find similar domains from human and mouse carcinoembryonic antigens (CEAM5_HUMAN, CEAM1_MOUSE).
- Find out the sequence pattern of the Drosophila slit protein sequence (P24014) with the online Dotlet web server or the EMBOSS dotmatcher program.
- Find the zinc proteinase domain between human MS2 cell surface antigen (P78325) and adamalysin II (P34179) from Crotalus adamanteus (eastern diamondback rattlesnake) venom.
- Find the special sequence feature of serine-repeat antigen protein precursor (P13823) using dotplot.
- Find the special sequence feature of human zinc finger protein (Q9P255).
- Find the special sequence feature of human ubiquitin C (NP_066289) using dotplot.

9.4 Sequence Alignment

Pairwise and multiple sequence alignments using both local and global algorithms.

9.4.1 Pairwise Sequence Alignment

Use Needle and Water to align the following sequence pairs with different scoring matrices and gap penalties; compare the results:

- "AFATCAT" and "AFASTCAT"
- "THEFASTCATCATCHESAFATRAT" and "THEFATCATCATCHESADEAD-RAT"
- "AREALFRIENDISAFRIENDINNEED" and "AFRIENDINNEEDISAFRIEND INDEED"

Use Needle and Water to align the following sequence pairs with different scoring matrices and gap penalties; compare the results:

- Human hemoglobin alpha chain (Swiss-Prot Accession: P69905) and human hemoglobin beta chain (Swiss-Prot Accession: P68871)
- Human hemoglobin alpha chain (Swiss-Prot Accession: P69905) and yellow lupine leghemoglobin (Swiss-Prot Accession: P02240)
- Pisum sativum post-floral specific protein 1 (PPF-1, Swiss-Prot Accession: Q9FY06) and Arabidopsis inner membrane protein (ALBINO3, Swiss-Prot Accession: Q8LBP4)
- The coding sequence of PPF-1, GenBank Accession: Y12618 and ALBINO3, GenBank Accession: U89272
- The full length mRNA sequence of PPF-1, GenBank Accession: Y12618 and ALBINO3, GenBank Accession: U89272
- The rice histidine transporter (GenPept Accession: CAD89802) and the Arabidopsis lysine and histidine transporter (GenPet Accession: AAC49885)

Use Needle to align the following three sequences between each other; make a summary of your analysis results:

- The protein sequence of human, mouse, and rat hemoglobin alpha chains (Swiss-Prot entry name: HBA_HUMAN, HBA_MOUSE, HBA_RAT)
- The coding sequence of human, mouse, and rat hemoglobin alpha chains (GenBank Accession: V00493, V00714, M17083)

9.4.2   Multiple Sequence Alignment

- Make multiple sequence alignment for the protein sequence of hemoglobin alpha chain from 7 vertebrates [FASTA].
- Make multiple sequence alignment for the protein sequence of 12 human globins [FASTA].
- Make multiple sequence alignment for the protein sequence of 15 Arabidopsis SBP transcription factors [FASTA]; use different programs (ClustalW, T-Coffee, and DIALIGN) and compare the results.
- Make multiple sequence alignment for the 9 repeat sequences of human ubiquitin C protein (NP_066289).
- Make multiple sequence alignment for spider toxin peptides [FASTA]; use manual editing to improve the results.

9.4.3   BLAST

Sequence similarity search against DNA and protein sequence databases.

9.4.3.1 Online BLAST

- Search a virtual peptide sequence (ACDEFGHI) against the NCBI NR database.
- Search a virtual random peptide sequence (ADIMWQVRSFCYLGHTKEPN) against the NCBI NR database.
- Search a virtual DNA sequence (ACGTACGTACGTACGTACGT) against the NCBI NR database.
- Search the delta sleep-inducing peptide (Swiss-Prot Accession: P01158) against the NCBI NR database.
- Search PPF-1 protein sequence (Swiss-Prot Accession: Q9FY06) against the Swiss-Prot database.
- Search OsHT01 protein sequence (CAD89802) against the NCBI plants database.
- Search OsHT01 protein sequence (CAD89802) against the NCBI NR database.
- Search protein sequence (Q57997) against the NCBI NR database.
- Search cytokine induced protein sequence (NP_149073) against the NCBI NR database.
- Search olfactory receptor protein sequence (NP_001005182) against the NCBI NR database.

9.4.3.2 Local BLAST

- Construct a local BLAST database with hemoglobin alpha subunit protein sequences retrieved from Swiss-Prot and do BLAST search locally with a query sequence [207hba.fasta].
- Construct a local BLAST database with hemoglobin beta subunit protein sequences retrieved from Swiss-Prot and do BLAST search locally with a query sequence.
- Build local BLAST database for maize transcription factors [zmtf-mrna.fasta, zmtf-pep.fasta], and do BLAST search to find SBP TFs in maize TFs using seed sequence of the SPL3 DNA-binding domain [atsbpd3.fasta].

9.5 DNA Sequence Analysis

Analysis of DNA sequences with several bioinformatics tools.

9.5.1 Gene Structure Analysis and Prediction

- Draw the gene structure of the Drosophila melanogaster homolog of human Down syndrome cell adhesion molecule (DSCAM) (AF260530).
- Predict the potential genes from the fugu cosmid sequence [af164138.fasta].

9.5.2   Sequence Composition

- Find GC content of 2 pairs of rice sequence fragment; draw a plot [rice-pair1.fasta, rice-pair2.fasta].
- Find the CpG island of the fugu cosmid sequence [af164138.fasta].
- Calculate the codon usage of the coding sequence of fugu MDR3 [fugu-mdr3-cds.fasta].
- Calculate the GC content of coding and noncoding sequences of fugu cosmid [af164138-cds.fasta, af164138-ncs.fasta].

9.5.3   Secondary Structure

- Predict the tandem repeat of the human herpes virus 7 gene locus (HH7TETRA).
- Predict the secondary structure of human mitochondria tRNA-Leu (AB026838).

9.6   Protein Sequence Analysis

Analysis of protein sequences with several bioinformatics tools.

9.6.1   Primary Structure

- Use EMBOSS Pepstat to calculate amino acid composition of Pisum sativum post-floral specific protein 1 (PPF-1, Q9FY06).
- Use EMBOSS Pepinfo to create flowcharts of Pisum sativum post-floral specific protein 1 (PPF-1, Q9FY06).
- Use ExPASy Protscale to create flowcharts to display various properties of the Pisum sativum post-floral specific protein 1 (PPF-1, Q9FY06).

9.6.2   Secondary Structure

- Predict the secondary structure of human hemoglobin alpha chain (P69905) and beta chain (P68871).
- Predict the secondary structure of the N-terminal domain of mouse carcinoembryonic antigen (CEAM1_MOUSE).
- Predict the potential coiled-coil region of the yeast amino acid biosynthesis general control protein (GCN4_YEAST).

9.6.3   Transmembrane Helices

- Use EMBOSS TMAP and ExPASy TMHMM, TMPred, TopPred, and SOUSI to predict the transmembrane helices of the Pisum sativum post-floral specific protein 1 (PPF-1, Q9FY06).
- Use EMBOSS TMAP and ExPASy TMHMM, TMPred, TopPred, and SOUSI to predict the transmembrane helices of sheep ovine opsin (OPSD_SHEEP).
- Use EMBOSS TMAP and ExPASy TMHMM, TMPred, TopPred, and SOUSI to predict the transmembrane helices of GCR2 protein (NP_175700).

9.6.4   Helical Wheel

- Draw the four helical wheels for the transmembrane helices of the Pisum sativum post-floral specific protein 1 (PPF-1, Q9FY06).
- Draw the helical wheel for the last two helices of human hemoglobin alpha chain (P69905).

9.7   Motif Search

Identification of conserved sequence motifs and domains.

9.7.1   SMART Search

- Search SMART to find conserved domain for the post-floral-specific protein 1 (PPF-1, Swiss-Prot Accession: Q9FY06).
- Search SMART to find conserved domain for the RNA-binding protein AT1G60000.
- Search SMART to find conserved domain for the transcription factor SBP protein (At1g27370).

9.7.2   MEME Search

- Search MEME to find conserved domain for the protein sequence of 15 Arabidopsis SBP transcription factors [15atsbp.fasta].

9.7.3   HMM Search

- Use HMMER hmmbuild to build HMM model (atsbpd.hmm) for 15 SBP DNA-binding domains (15atsbpd.fasta).
- Use HMMER hmmcalibrate to adjust the above HMM model (atsbpd.hmm).
- Use HMMER hmmsearch using the HMM model (atsbpd.hmm) to identify SBP DNA-binding domains against maize transcription factors using the above HMM model.
- Build HMM model for 15 SBP DNA-binding domains (15atsbpd.fasta).

9.7.4   Sequence Logo

- Create a sequence logo for the SNPs of 20 SARS coronaviruses [20sars.fasta].
- Create a sequence logo for the DNA-binding domain of 15 Arabidopsis SBP proteins [15atsbpd.fasta].

9.8   Phylogeny

Phylogenetic analysis and construction of phylogenetic tress with simple examples.

9.8.1   Protein

- Construct phylogenetic tree with the maximum parsimony method for the hemoglobin alpha chain from 7 vertebrates [FASTA].
- Construct phylogenetic tree with the maximum parsimony method for the hemoglobin alpha chain using different datasets from 209 species [209hba.fasta] based on the taxonomy table [209HBA.PDF].
- Construct phylogenetic tree with the distance method for 12 human globulins [FASTA].
- Construct phylogenetic tree with the maximum parsimony method for a segment of lysine/histidine transporter from 10 plants [FASTA].
- Construct phylogenetic tree with both distance and maximum parsimony methods for the DNA-binding domain of 15 Arabidopsis SBP proteins [FASTA].

9.8.2   DNA

• Construct phylogenetic tree with distance and maximum parsimony methods for a set of 6 test DNA sequences [FASTA].
• Construct phylogenetic tree with the distance method for the SNPs of 20 SARS coronaviruses [FASTA].

9.9   Projects

As the name of this course implies, we focus on the application of bioinformatics tools to solve real problems in biological research. We chose several samples as working projects to learn how to find the literature, how to obtain sequence and structure data, how to do the analysis step by step, and how to make a summary from the analysis results. You are most encouraged to work on your own projects during the course:

PPF – sequence analysis of Pisum sativum post-floral specific gene
MDR – gene prediction and dotplot analysis of fugu multidrug resistance gene
BGH – sequence, structure, and function analysis of the bar-headed goose hemoglobin
CEA – protein engineering of carcinoembryonic antigen
AFP – structure comparison of spider toxins and prediction of antifungal peptide
HMT – sequence and structure comparison of human metallothioneins
SBP – systematic analysis of the Arabidopsis transcription factor family of Squamosa-promoter binding protein

9.9.1   Sequence, Structure, and Function Analysis of the Bar-Headed Goose Hemoglobin

Hemoglobin is one of the most well-studied proteins in the last century. The sequence, structure, and function of several vertebrates have been investigated during the past 50 years. More than 200 hundreds of hemoglobin protein sequences have been deposited into the Swiss-Prot database. Three-dimensional structure wild type and mutants from dozens of species have been solved. This provides us a good opportunity to study the relationship among sequence, structure, and function of hemoglobins.

Bar-headed goose is a special species of migration birds. They live in the Qinghai Lake during summer time and fly to India all the way along over the Tibetan plateau in autumn and come back in spring. Interestingly, a close relative of bar-headed goose, the graylag goose, lives in the low land of India all year round and do not migrate. Sequence alignment of bar-headed goose hemoglobin with that of graylag goose shows that there are only 4 substitutions. The Pro 119 in the alpha subunit of graylag goose has been changed to Ala in bar-headed goose. This residue is located in the surface of the alpha/beta interface. In 1983, Perutz proposed that this substitution reduces the contact between the alpha and beta subunits and increases the oxygen affinity, due to the relation of the tension status in the deoxy form [1].

During the past decade, a research group at Peking University has solved the crystal structure of both deoxy and oxy forms of bar-headed goose as well as the oxy form of the graylag goose hemoglobin [2–4]. We will use this example to learn how to analyze the sequence and structure of hemoglobin molecules.

9.9.2   Exercises

- Find literature on bar-headed goose hemoglobin from PubMed.
- Find protein sequences of bar-headed goose and graylag goose hemoglobins from Swiss-Prot; make sequence alignment to compare the differences between these two molecules.
- Find protein sequences which share 90 % similarity with bar-headed hemoglobin alpha chain, make multiple sequence alignment, and construct a phylogenetic tree for the above protein sequences using the maximum parsimony method.
- Retrieve three-dimensional structure oxy (1A4F) and deoxy (1HV4) forms of bar-headed goose hemoglobin; compare the difference of the heme molecule with Swiss PDB Viewer.
- Make superimposition for the alpha and beta subunit of bar-headed goose (1A4F) and greylag goose (1FAW) hemoglobins, find out the differences of the substitution site (Pro119-Ala119), and measure the contact distance between the side chain of this residues and the side chain of Ile55 of the beta subunit.
- Make a summary about the sequence, structure, and function of bar-headed goose hemoglobin.

9.10   Literature

There are many online materials dedicated to bioinformatics education including courses, tutorials, and documents. You are most encouraged to get access to these self-educational websites during the course and for your further study.

9.10.1   Courses and Tutorials

- 2CAN – EBI bioinformatics support portal which provides short and concise introductions to basic concepts in molecular and cell biology and bioinformatics
- EMBER – an online practical course with multiple choice quiz designed and maintained by Manchester University, UK (free registration needed)

- Science Primer – NCBI science primer for various topics including bioinformatics, genome mapping, and molecular biology
- SADR – an online course for Sequence Analysis with Distributed Resources, Bielefeld University, Germany
- SWISS-MODEL – an online course for principles of protein structure, comparative protein modeling, and visualization maintained by Nicolas Guex and Manuel C. Peitsch at Glaxo Wellcome
- Swiss-PDB Viewer – an extensive tutorial for the molecular visualization and modeling program Swiss PDB Viewer, created and maintained by Gale Rhodes at the University of Southern Maine

9.10.2 Scientific Stories

- Molecular of the Month – A website which presents short accounts on selected molecules from the Protein Data Bank. It was created in January 2000 and is being maintained by David S. Goodsell at Scripps.
- Protein Spotlight – A website which tells short stories on protein molecules as well as the scientific research and the scientists behind these interesting stories. It was started in January 2000 and is being maintained by Vivienne B. Gerritsen at Swiss Institute of Bioinformatics.

9.10.3 Free Journals and Books

- NCBI Bookshelf – a growing collection of free online biomedical books that can be searched directly through the NCBI Entrez system.
- PubMed Central – the US National Institutes of Health (NIH) free digital archive of biomedical and life sciences journals.
- PubMed Central China mirror – the PubMed Central China mirror maintained by the Center for Bioinformatics, Peking University.
- BioMed Central – the website of more than 180 open access biomedical journals freely available, started by a UK publisher in London since 2001.
- PLOS – the website of several significant open access biological and medical journals freely available, started by the Public Library of Science, a nonprofit organization composed of many famous scientists.
- HighWire – a website maintained by the Stanford University Libraries. It gives a list of biomedical journals which provide either immediate or 6/12/18/24 months delay of free online full-text articles.
- Amedeo – a website maintained by Bernd Sebastian Kamps in Europe. It provides extensive links to various biomedical journals and books freely available.
- AnaTax – an online book chapter of the anatomy and taxonomy written in 1981 and being updated by Jane Richardson at Duke University.

9.11   Bioinformatics Databases

There are huge amount of online bioinformatics databases available on the Internet. The databases listed in this page are extensively accessed during this course.

9.11.1   List of Databases

- NAR databases – the most extensive list of biological databases being maintained by the international journal *Nucleic Acids Research* which publishes a special issue for molecular biology databases in the first issue of each year since 1996. All these database papers can be accessed freely. You may find links to the website of the databases described in the chapter.
- NCBI databases – the molecular databases maintained by NCBI. A Flash flowchart for 24 databases connected by lines shows the relationships and internal links among all these databases. These databases are divided into 6 major groups: nucleotide, protein, structure, taxonomy, genome, and expression. It also provides links to the individual database description page.
- EBI databases – the main portal to all EBI databases divided in several groups, such as literature, microarray, nucleotide, protein, structure, pathway, and ontology. Links to database query and retrieval systems can be found in this portal.

9.11.2   Database Query Systems

- NCBI Entrez – the unique interface to search all NCBI databases. Query results are displayed with entry numbers along the database names.
- EBI SRS – the database query system maintained by EBI. SRS stands for sequence retrieval system which was originally developed by Thure Etzold at the European Molecular Biology Laboratory in the early 1990s and moved to EBI. It was originally an open system and installed in a dozen of institutions with different databases. In the late 1990s, SRS became a commercial package but still free to academic use. In 2006, SRS was acquired by BioWisdom, a software company based in Cambridge, UK.
- EMBL SRS – the database query system (ver. 8.2) maintained by EMBL, at Germany.
- DKFZ SRS – the database query system (ver. 8.2) maintained by the German Cancer Research Center.
- Columbia SRS – the database query system (ver. 8.2) maintained by Columbia University, USA.

- CMBI MRS – the open-source database query system developed by Marteen Hekkelman at the Center for Molecular and Biomolecular Information (CMBI), the Netherlands.
- CBI MRS – the MRS installed at the Center for Bioinformatics, Peking University.

9.11.3   Genome Databases

Due to the rapid progress of DNA sequencing technology, hundreds of organisms have been sequenced at the genome scale. Genome databases and related analysis platforms can be accessed on the Internet. We select some of them for our teaching purpose of this course.

9.11.3.1   List of Genome Databases

- GOLD – Genomes Online Database, a comprehensive information resource for complete and ongoing genome sequencing projects with flowcharts and tables of statistical data. Created and maintained by
- Karyn's Genome – a collection of sequenced genomes with brief description and references for each genome. The direct links to the sequence data in EMBL or annotations in ENSEMBL make it very convenient for the user community.
- CropNet – the website of the UK Crop Plant Bioinformatics Network to the development, management, and distribution of information relating to comparative mapping and genome research in crop plants.

9.11.3.2   Genome Browsers and Analysis Platforms

- NCBI Genome – the entry portal to various NCBI genomic biology tools and resources, including the Map Viewer, the Genome Project Database, and the Plant Genomes Central
- GoldenPath – the genome browser website containing the reference sequence and working draft assemblies for a large collection of genomes at the University of California at Santa Cruz (UCSC)
- ENSEMBL – the web server of the European eukaryotic genome resource developed by EBI and the Sanger Institute [PDF]
- VISTA – a comprehensive suite of programs and databases for comparative analysis of genomic sequences
- TIGR Plant Genomics – TIGR plant genome databases and tools
- TIGR Gene Indices – TIGR gene indices maintained at Harvard

9.11.3.3   Genome Database of Model Organisms

- Gramene – a curated open-source data resource for comparative genome analysis in the grasses including rice, maize, wheat, barley, and sorghum, as well as other plants including arabidopsis, poplar, and grape. Cross-species homology relationships can be found using information derived from genomic and EST sequencing, protein structure and function analysis, genetic and physical mapping, interpretation of biochemical pathways, gene and QTL localization, and descriptions of phenotypic characters and mutations.
- TAIR – the Arabidopsis information resource maintained by Stanford University. It includes the complete genome sequence along with gene structure, gene product information, metabolism, gene expression, DNA and seed stocks, genome maps, genetic and physical markers, publications, and information about the Arabidopsis research community.
- AtENSEMBL – a genome browser for the commonly studied plant model organism Arabidopsis thaliana.
- Oryzabase – a comprehensive rice science database maintained by the National Institute of Genetics, Japan. It contains genetic resource stock information, gene dictionary, chromosome maps, mutant images, and fundamental knowledge of rice science.
- FlyBase – a comprehensive database of Drosophila genes and genomes maintained by Indiana University.
- CyanoBase – the genome database for cyanobacteria developed by Kazusa Institute, Japan.

9.11.4   Sequence Databases

DNA and protein sequence databases are the fundamental resources for bioinformatics research, development, and application.

9.11.4.1   DNA Sequence Databases

- GenBank – the web portal to the NIH genetic sequence database maintained by NCBI, also a part of the International Nucleotide Database Collaboration. Literature citation, release notes, and an example record can be found in this page.
- EMBL – the web portal to EMBL nucleotide sequence database maintained by EBI, also a part of the International Nucleotide Database Collaboration. Various documentations such as release notes, database statistics, user guide, feature table definition and sample entry, and FAQs are provided.

- RefSeq – the Reference Sequence collection constructed by NCBI to provide a comprehensive, integrated, non-redundant set of DNA and RNA sequences and protein products. It provides a stable reference for genome annotation, gene identification and characterization, mutation and polymorphism analysis, expression studies, and comparative analyses.
- UniGene – an Organized View of the Transcriptome created by NCBI. Each UniGene entry is a set of transcript sequences that appear to come from the same transcription locus, together with information on protein similarities, gene expression, cDNA clone reagents, and genomic location.
- dbSNP – the database of single nucleotide polymorphism maintained by NCBI.
- EMBLCDS – a database of nucleotide sequences of the coding sequence from EMBL.

9.11.4.2 Protein Sequence Databases

- Swiss-Prot – the entry site for the well-annotated protein sequence knowledge database maintained by SIB at Geneva, Switzerland. A list of references, a comprehensive user manual, and database statistics with tables and flowcharts are provided.
- UniProt – the main website for international protein sequence database which consists of the protein knowledgebase (UniProtKB), the sequence clusters (UniRef), and the sequence archive (UniParc).
- HPI – the Human Proteomics Initiative, an EBI project to annotate all known human sequences according to the quality standards of UniProtKB/Swiss-Prot. It provides for each known protein a wealth of information that includes the description of its function, its domain structure, subcellular location, posttranslational modifications, variants, and similarities to other proteins.
- IPI – the website of the International Protein Index database which provides a top level guide to the main databases that describe the proteomes of higher eukaryotic organisms.

9.11.5   Protein Domain, Family, and Function Databases

Protein molecules play functions in living organisms. They are usually classified into families based on the different functions they play. Proteins in the same family or subfamily often have conserved sequence motifs or fingerprints as well as ungapped blocks or functional domains. With the great amount of available sequence data, secondary databases of protein molecules have been constructed. We select some of them for this course.

9.11.5.1   Protein Domain Databases

- Prosite – a database of protein domains, families, and functional sites, created and maintained by the Swiss Institute of Bioinformatics
- PRINTS – a database of protein fingerprints consisting of conserved motifs within a protein family, created and maintained by Manchester University, UK
- BLOCKS – a database of multiple aligned ungapped segments corresponding to the most highly conserved regions of proteins, created and maintained by the Fred Hutchinson Cancer Research Center, USA
- CDD – a database of conserved protein domains created and maintained by the NCBI structure group
- ProDom – a database of comprehensive set of protein domain families automatically generated from the Swiss-Prot and TrEMBL sequence databases, developed and maintained by the University Claude Bernard, France

9.11.5.2   Protein Family Databases

- Pfam – a database of protein families represented by multiple sequence alignments and hidden Markov models, constructed and maintained by the Sanger Institute, UK

9.11.5.3   Protein Function Databases

- IMGT – the international immunogenetics information system, a high-quality integrated knowledge resource of the immune system of human and other vertebrate species, created and maintained by the University of Montpellier, France
- HPA – a website for the human protein atlas which shows expression and localization of proteins in a large variety of normal human tissues, cancer cells, and cell lines with the aid of immunohistochemistry images, developed and maintained by Proteome Resource Center, Sweden

9.11.6   Structure Databases

The central point of the protein structure database is Protein Data Bank (PDB) which was started in the late 1970s at the US Brookhaven National Laboratory. In 1999, the Research Collaboratory for Structural Bioinformatics (RSCB) was formed to manage the PDB. In 2003, an international collaboration among RSCB, MSD-EBI at Europe, and PDBj in Japan was initiated to form wwPDB, and the Magnetic

Resonance Data Bank (BMRB) joined wwPDN in 2006. Although RSCB is the main entry point to access macromolecular structures, we may also find protein structures through the NCBI Entrez system and the EBI MSDLite server.

9.11.6.1 Main Portals for the Protein Structures Database

- RSCB – the main repository of macromolecular structures maintained by the Research Collaboration for Structural Bioinformatics
- MMDB – the macromolecular database maintained by NCBI
- MSD – the entry point for the EBI macromolecular structure database
- MSDLite – the EBI web server providing simple search of protein structures
- PDBSUM – the EBI web server which provides overview, schematic diagrams, and interactions of structure
- BMRB – the biological magnetic resonance data bank maintained at University of Wisconsin-Madison
- ModBase – the database of comparative protein structure models developed and maintained at University of California, San Francisco

9.11.6.2 Classification of the Protein Structures

- SCOP – the database of Structure Classification of Proteins developed and maintained by Cambridge University
- CATH – the database of Calcification, Architecture, Topology, and Homologous superfamily developed and maintained by the University College London

9.11.6.3 Visualization of Protein Structures

- JenaLib – the Jena Library of Biological Macromolecules which provides information on macromolecular structures with an emphasis on visualization and analysis

9.12   Bioinformatics Tools

There are many bioinformatics tools over the Internet. Thanks to the great contribution of the scientific community of bioinformatics research and development. They make the bioinformatics programs and packages freely available to the end user biologists. The web-based tools can be accessed through the Internet using the web browsers such as Firefox and Internet Explorer. Users may also download and install some of the packages on a local machine to run the programs. We use both web-based platforms as well as command line tools integrated in a Linux-based bioinformatics environment Bioland developed locally at our center. The three major international bioinformatics centers, NCBI, EBI, and ExPASy, develop, collect, and maintain hundreds of bioinformatics tools. They also provide online service for most of these web-based tools.

9.12.1 List of Bioinformatics Tools at International Bioinformatics Centers

- ExPASy tools – a comprehensive list of online web-based bioinformatics tools provided by ExPASy and worldwide
- EBI tools – the entry page for the EBI bioinformatics tools
- NCBI tools – the entry page for the NCBI bioinformatics tools

9.12.2 Web-Based Bioinformatics Platforms

- WebLab – the comprehensive and user-friendly bioinformatics platform developed by the Center for Bioinformatics, Peking University. WebLab provides user spaces to store and manage input data and analysis results as well as literature references. The analysis protocols and macros allow users to process the job in a batch mode.
- EMBOSS explorer – the web interface for the EMBOSS package, maintained by Ryan Golhar at the University of Medicine and Dentistry of New Jersey.
- EMBOSS – the web interface for the EMBOSS package, maintained by the University of Singapore.
- SRS Tools – the EBI SRS database query server integrates several analysis packages such as EMBOSS and HMMER which can be launched directly with retrieved data from the SRS server or with external data provided by users.

9.12.3 Bioinformatics Packages to be Downloaded and Installed Locally

- EMBOSS – the main portal for the open-source bioinformatics project EMBOSS (European Molecular Biology Open Software Suite) headed by Peter Rice and Alan Bleasby at EBI. EMBOSS is a comprehensive package with some 200 individual programs for DNA and protein sequence analysis.
- PISE – the main page to learn and download the PISE software designed by the Pasteur Institute, France. It can generate the web interface for the programs of the EMBOSS package.

- wEMBOSS – the entry page for a simple description and download of the EMBOSS graphics user interface.
- wEMBOSS – the web page for the HMMER package which uses profile hidden Markov models to detect conserved domains in protein sequences, developed and maintained by Sean Eddy at Howard Hughes Medical Institute.

This page lists the most comprehensive packages such as EMBOSS. For other packages and programs, please find them in the individual pages list in the Tools menu.

9.13   Sequence Analysis

DNA and protein sequence analysis is the most fundamental approach in bioinformatics.

9.13.1   Dotplot

- Dotlet – the Java-based web server for the comparison of DNA and protein sequences using the dot plot approach, maintained by the Swiss Institute of Bioinformatics.

9.13.2   Pairwise Sequence Alignment

- Align – the web server for the pairwise sequence alignment, maintained by EBI. Either global (Needle) or local (Water) alignment can be performed.
- BLAST 2 – the web server for the alignment of two sequences using BLAST maintained at NCBI.

9.13.3   Multiple Sequence Alignment

- ClustalW – the web server for the global multiple sequence alignment program maintained at EBI. It uses a new version of ClustalW 2.0.
- Muscle – the web server for the multiple sequence comparison by log-expectation, maintained at EBI. It claimed to achieve better accuracy with higher speed than ClustalW depending on the chosen options.
- MAFAT – the web server of high speed multiple sequence alignment using fast Fourier transformation, maintained at EBI.
- KALIGN – the web server of fast and accurate multiple sequence alignment, maintained at EBI.

- T-Coffee – the web server of several tools for computing, evaluating, and manipulating multiple alignments of DNA and protein sequences and structures, developed and maintained by Cedric Notredame at the Center for Genomic Regulation, Spain.
- DIALIGN – the web server for the multiple sequence alignment, developed and maintained by Burkhard Morgenstern at Bielefeld University, Germany. It uses a local alignment approach to compare a whole segment of sequences without gap penalty.

9.13.4   Motif Finding

- SMART – the web server for motif discovery and search, developed, and maintained by the University of California at San Diego. It provides
- MEME – the web server for motif discovery and search, developed and maintained by the University of California at San Diego. It provides
- TMHMM – the web server for the prediction of transmembrane helices in proteins, developed and maintained by Denmark Technical University.

9.13.5   Gene Identification

- GENSCAN – the web server for the identification of complete gene structures in genomic DNA, maintained by MIT
- GenID – the web server for the prediction of genes in anonymous genomic sequences, developed and maintained by the University of Pompeu Fabra, Spain
- HMMGene – the web server for the prediction of vertebrate and C elegans genes, developed and maintained by the Denmark Technical University
- SoftBerry – the web server for gene prediction, limited free use for academic users

9.13.6   Sequence Logo

- WebLogo – the web server for the generation of sequence logos, developed and maintained by the University of California at Berkeley.

9.13.7   RNA Secondary Structure Prediction

- MFOLD – the web server for the prediction of RNA secondary structure, developed and maintained by Michael Zuker at Rensselaer Polytechnic Institute.

9.14   Database Search

Sequence similarity search against DNA or protein sequence database is one of the most extensively used approaches in the application of bioinformatics to molecular biology and genome research. Results of database search can deduce biological function for the newly identified sequence or to infer evolutionarily relationship among the query sequence and a group of matched subject sequences. The most famous program for database search is BLAST – the Basic Local Alignment Search Tool. Thanks to the BLAST team at NCBI who continuously develop this package and make it freely available for the community. Here, we introduce several BLAST web servers which are mostly accessed as well as the FASTA, BLAT, and MPSearch servers.

9.14.1   BLAST Search

- NCBI BLAST – the central point of the NCBI BLAST server which provides whole functionality of DNA and protein sequence database search programs including PSI-BLAST and PHI-BLAST and the whole list of databases in different scales and different genomes.
- EBI BLAST – the entry page of the EBI BLAST facility which provides both NCBI BLAST and WU-BLAST programs as well as a set of specialized BLAST programs to search the alternative splicing database, the cloning vector database, and the parasite database.
- WU-BLAST – the entry page of the WU-BLAST maintained by Warren Gish at Washington University. It provides several links to other BLAST servers worldwide.
- Sanger BLAST – the entry page of the sequence projects BLAST search services at the Sanger Institute, featured with the special genome databases generated by the completed or ongoing sequencing projects.

9.14.2   Other Database Search

- FASTA – the original website for the FASTA set programs maintained by William Pearson at the University of Virginia. It provides extensive documents and help materials as well.
- EBI FASTA – the entry page of EBI FASTA services. FASTA has a higher sensitivity in some cases comparing with BLAST with a slightly lower speed. It is specific to identify low similarity long regions for highly diverged sequences.

- MPSrch – the EBI MPSrch server which uses the Smith-Waterman algorithm to obtain the optimal sequence alignment results.
- GoldenPath BLAT – the web server for the genome database search, integrated in the GoldenPath platform.
- ENSEMBL BLAT – the web server for the genome database search, integrated in the ENSEMBL platform.

9.15   Molecular Modeling

Molecular modeling is one of the important disciplines of bioinformatics. The latest development of both hardware and software makes it possible for molecular visualization which is fundamental for molecular modeling. Currently, there are quite few plug-ins for the real-time display and manipulation of three-dimensional structures with Internet browsers such as Jmol and WebMol provided in the PDB web server. You may also install stand-alone tools on your PC such as Swiss PDB Viewer and PyMOL which have more functionalities. On the other hand, homology-based protein modeling web servers may help you to predict the three-dimensional structure of your protein based on sequence similarity between your protein and the templates with known 3D structures.

9.15.1   Visualization and Modeling Tools

- Swiss PDB Viewer – a comprehensive molecular visualization and modeling package developed by Nicolas Guex at GlaxoSmithKline. It provides a graphical interface and a user-friendly control panel for users to analyze several proteins at the same time. There are lots of useful functions such as superimposition, mutation, and energy minimization for simple molecular simulations.
- PyMOL – the home page of the molecular visualization system written in Python. It is a user-sponsored open-source software maintained by DeLano Scientific LLC.
- Cn3D – the web page of the molecular visualization program Cn3D developed and distributed by NCBI. It can be installed on your PC to display the three-dimensional structures obtained from the NCBI MMDB database.
- Kinemage – the home page of the protein visualization package Kinemage developed and maintained at the laboratory of Jane Richardson and David Richardson, Duke University.
- RasMol – the home page of molecular visualization freeware RasMol, maintained by the University of Massachusetts, Amherst.

9.15.2   Protein Modeling Web Servers

- Swiss Model – the home page of the automated comparative protein modeling server, hosted by the University of Basel and the Swiss Institute of Bioinformatics. Extensive documentation and help materials are provided.
- 3D Jigsaw – the web server to build three-dimensional models for protein molecules based on homologues of known structure, developed and maintained by the Cancer Research UK.
- CASP – the entry page for the international protein structure prediction center, hosted at University of California at Davies. It provides the means of objective testing of evaluation of different methods in protein structure prediction.

9.16   Phylogenetic Analysis and Tree Construction

There are more than 300 hundreds phylogeny programs available on the Internet. Most of them can be downloaded and installed freely on your own machine. Due to the great need of computing power, it is difficult to maintain online phylogenetic analysis web servers. The best way to do phylogenetic analysis is to use command line for the PHYLIP programs integrated in EMBOSS or install MEGA on your PC Windows.

9.16.1   List of Phylogeny Programs

- Phylogeny software – the whole list of phylogeny programs collected and classified in groups by Joe Felsenstein.

9.16.2   Online Phylogeny Servers

- WebLab Protocols – the WebLab platform we develop and maintain has integrated the PHYLIP package. The protocols and macros for both Neighbor Joining and maximum parsimony methods are extremely useful for biologists to construct phylogeny trees with well-defined data sets.
- NUS EMBOSS interface – the web interface of the PHYLIP programs integrated in the EMBOSS package, maintained by the University of Singapore.
- EBC interface – the web interface of some PHYLIP programs, maintained by Uppsala University, Sweden.

9.16.3   Phylogeny Programs

- PHYLIP – the website for the comprehensive package Phylogeny Inference Package (PHYLIP) created and maintained by Joe Felsenstein at the University of Washington. This package can be downloaded and installed on Linux, Windows, and Mac freely.
- TREE-PUZZLE – the web page for the phylogeny program which uses the maximum likelihood method to analogize nucleotide and amino acid sequences as well as other two-state data.
- PAML – the website for the Phylogenetic Analysis by Maximum Likelihood package developed and maintained by Ziheng Yang, at University College, London.
- MEGA – the website for the Molecular Evolutionary Genetics Analysis package developed and maintained by Masatoshi Nei and his colleagues. It was originally designed for the Windows platform with a graphics interface and uses the distance method to construct phylogenetic trees [PDF].

9.16.4   Display of Phylogenetic Trees

iTOL – the website of the Interactive Tree of Life for the display and manipulation of phylogenetic trees, developed and maintained by the European Molecular Biology Laboratory.

References

1. Perutz MF (1983) Species adaptation in a protein molecule. Mol Biol Evol 1(1):1–28
2. Zhang J, Hua Z, Tame JR, Lu G, Zhang R, Gu X (1996) The crystal structure of a high oxygen affinity species of haemoglobin. J Mol Biol 255(3):484–493
3. Liang YH, Liu XZ, Liu SH, Lu GY (2001) The structure of greylag goose oxy haemoglobin: the roles of four mutations compared with bar-headed goose haemoglobin. Acta Crystallogr D Biol Crystallogr 57(Pt 12):1850–1856
4. Liang Y, Hua Z, Liang X, Xu Q, Lu G (2001) The crystal structure of bar-headed goose hemoglobin in deoxy form: the allosteric mechanism of a hemoglobin species with high oxygen affinity. J Mol Biol 313(1):123–137

Chapter 10
Foundations for the Study of Structure and Function of Proteins

Zhirong Sun

10.1 Introduction

Proteins are the most abundant biological macromolecules, occurring in all cells and all parts of cells. Moreover, proteins exhibit enormous diversity of biological function and are the most final products of the information pathways. Protein is a major component of protoplasm, which is the basis of life. It is translated from RNA and composed of amino acid connected by peptide bonds. It participates in a series of complicated chemical reactions and finally leads to the phenomena of life. So we can say it is the workhorse molecule and a major player of life activity. Biologists focus on the diction of structure and function of proteins by the study of the primary, secondary, tertiary, and quaternary dimensional structures of proteins, posttranscriptional modifications, protein-protein interactions, the DNA-proteins interactions, and so on.

10.1.1 Importance of Protein

DNA, RNA, proteins, etc. are the basic components of life. DNA is the vector of genetic information and is transcribed into RNA which is in turn translated into protein. Protein is the expression of genetic information, the performer of kinds of biological functions, and the sustainer of metabolic activities in the organisms. Protein plays an important role in the whole processes of life, including the appearance of life to the growth of life to apoptosis.

There are two examples illustrating the importance of protein. The first one is about the SARS. One protein is found to increase the self-copy efficiency for 100

Z. Sun (✉)
School of Life Sciences, Tsinghua University, Beijing 100084, China
e-mail: sunzhr@mail.tsinghua.edu.cn

R. Jiang et al. (eds.), *Basics of Bioinformatics: Lecture Notes of the Graduate Summer School on Bioinformatics of China*, DOI 10.1007/978-3-642-38951-1_10,
© Tsinghua University Press, Beijing and Springer-Verlag Berlin Heidelberg 2013

times or so, which makes the viruses propagate at a high rate. Another example is a flu virus protein whose structure looks like a narrow-neck bag. This strange structure of the protein can help the virus resist drugs.

10.1.2 Amino Acids, Peptides, and Proteins

All proteins, whether from the most ancient lines of bacteria or from the most complex forms of life, are constructed from the same ubiquitous set of 20 amino acids, covalently linked in characteristic linear sequences. Proteins are the polymers of 20 amino acids. Different combinations of these 20 amino acids result in varied structures and functions of proteins. Protein structures are studied at primary, secondary, tertiary, and quaternary levels. Proteins have widely diverse forms and functions, including enzymes, hormones, antibodies, transporters, muscle, lens protein of eyes, spider webs, rhinoceros horn, antibiotics, and mushroom poisons.

10.1.2.1 Protein Research: What to Study and How to Study?

What should we study? What is the core problem? How can we study? What is most remarkable is that cells can produce proteins with strikingly different properties and activities by joining the same 20 amino acids in many different combinations and sequences. Nowadays, biologists study protein from these aspects: structure and function, the transfer of information.

10.1.2.2 Amino Acid

All 20 standard amino acids found in proteins are α-amino acids. Figure 10.1 shows the structure formula of α-amino acids. Each amino acid has a different side chain (or R group, R = "remainder of the molecule") and is given a three-letter abbreviation and a one-letter symbol. Biologists often use the first three letters or the first letter. The 20 amino acids of proteins are often referred to as the standard amino acids. All proteins in all species (from bacteria to human) are constructed from the same set of 20 amino acids. All proteins, no matter how different they are in structure and function, are made of the 20 standard amino acids. Figure 10.1 shows the structure formulae of all the 20 amino acids.

10.1.2.3 Protein Structure Hierarchy

Protein structures are studied at primary, secondary, tertiary, and quaternary levels. There are tight correlations among these levels.

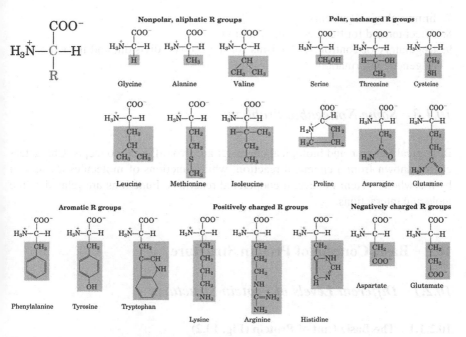

Fig. 10.1 Structure formulae of all the 20 amino acids

10.1.2.4   Different Classes of Proteins

From the aspect of chemical structures of proteins, proteins can be classified into two classes. If proteins are completely composed of amino acids, these proteins are called simple proteins, such as insulin; if there are other components, they are named conjugated proteins like hemoglobin.

According to the symmetry of proteins, proteins can be divided into globin and fibrin. Globins are more symmetric and similar to balls or ovals in shape. Globins dissolve easily and can crystallize. Most proteins are globins. Comparatively, fibrins are less symmetric and look like thin sticks or fibers. They can be divided into soluble fibrins and unsolvable fibrins.

Simple proteins can be subdivided into seven subclasses: albumin, globulin, glutelin, prolamine, histone, protamine, and scleroprotein. Conjugated proteins can also be subdivided into nucleoprotein, lipoprotein, glycoprotein and mucoprotein, phosphoprotein, hemoprotein, flavoprotein, and metalloprotein. Different classes of proteins have various functions. These include serving as:

1. Catalyzers of metabolism: enzyme
2. Structural component of organisms
3. Storage component of amino acid
4. Transporters
5. Movement proteins
6. Hormonal proteins

7. Immunological proteins
8. Acceptor and for transfer of information
9. Regulatory or control mechanisms for the growth, division, and the expression of genetic information

10.1.3   Some Noticeable Problems

Biological function and biological character are two different concepts. Characters can be shown from a chemical reaction, while functions of molecules are shown by the whole system in several cooperated reactions. Functions are related to the molecule interactions.

10.2   Basic Concept of Protein Structure

10.2.1   Different Levels of Protein Structures

10.2.1.1   The Basic Unit of Protein (Fig. 10.2)

10.2.1.2   Polypeptide Chain

Peptide and Peptide Bond

A peptide bond is made up by connecting an α-COOH of an amino acid and the α-NH$_3$ (Figs. 10.3 and 10.4). The simplest peptide composed of two amino acids is called dipeptide, containing one peptide bond. Those containing three, four, and five peptide bonds are called tripeptide, tetrapeptide, and pentapeptide, respectively. The peptide chain loses a molecule of H$_2$O when forming a peptide bond. In a polypeptide, an amino acid unit is called a residue.

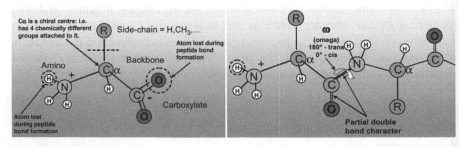

Fig. 10.2 Common structure of amino acid (*left*) and formation of polypeptide chain (*right*)

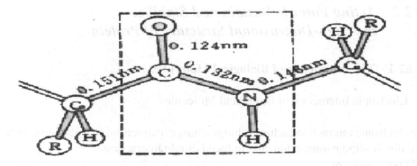

Fig. 10.3 The formation of a peptide bond

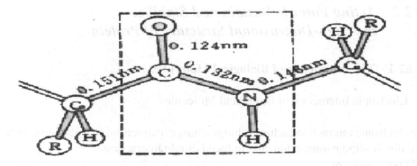

Fig. 10.4 The rigidity and coplanarity of peptide bond

The Fold of Polypeptide Chain and Dihedral Angle

The repeated structure on the backbone of polypeptide is called peptide unit or planar unit of peptide. Peptide bond cannot turn freely because of its double-bond character. The bonds beside peptide unit can wheel freely, which are described using dihedral angles ϕ and ψ.

10.2.1.3   The Imagination of a Polypeptide Chain (Fig. 10.5)

10.2.1.4   The Peptide Chain Is Directional

1. An amino acid unit in a peptide chain is called a residue.
2. The end having a free α-amino group is called amino-terminal or N-terminal.
3. The end having a free α carboxyl group is called carboxyl-terminal or C-terminal.
4. By convention, the N-terminal is taken as the beginning of the peptide chain and put at the left (C-terminal at the right). Biosynthesis starts from the N-terminal.

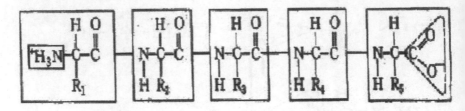

N-end residue --→ C-end residue

Fig. 10.5 The structure of a pentapeptide

10.2.2 Acting Force to Sustain and Stabilize the High-Dimensional Structure of Protein

10.2.2.1 The Interaction of Biological Molecules

The Electronic Interaction of Biological Molecules

The electronic interaction includes charge-charge interaction, charge-dipole interaction, dipole-dipole interaction, and induced dipole interaction.

Dipole moment

$$\mu = g \cdot l, \ u = -\mu \cdot E$$

Charge-charge interaction

Charge-dipole interaction (Fig. 10.6)

Dipole-Dipole Interaction

When the radius vector between two dipoles and the center is far bigger than the length of dipoles, namely, $r \gg l$, the interaction of these two dipoles is:

$$U = \frac{1}{\varepsilon r^3} \left[\mu_A \cdot \mu_B - \frac{3 \left(\mu_A \cdot r \right) \left(\mu_B \cdot r \right)}{r^2} \right] \quad l \ll r$$

Induced dipoles

Fig. 10.6 The interaction of
a positive charge and dipole μ

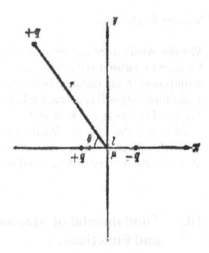

The neutral molecules or groups with overlapped positive and negative charges will be polarized by electric field and become induced dipoles. The dipole moment:

$$\mu_{ind} = -a_{ind}E$$

The Hydration of Polar Groups

Hydration is the process of the subject interacting or combining with water.

10.2.2.2   The Force to Sustain and Stabilize the High-Dimensional Structure of Proteins

The forces that sustain the structure of proteins are the so-called weak interaction, non-covalent bond, or inferiority bond, including hydrogen bond, hydrophobic interaction, electrostatic interaction, and van der Waals force. When these weak interactions present independently, they are weak bond, but when these bonds are added together, a strong force will form to sustain the protein structure space.

Electrostatic Force

Under the physiological condition, the side chain of acidic amino acid can be broken down into negative ions, while the side chain of basic amino acid can be disassociated into positive ions. Some atoms will form dipoles because of polarization. These interaction forces between charges or dipoles are called electrostatic force and it meets the Coulomb's law.

Van der Waals Force

Van der Waals force can also be called van der Waals bond. It includes attractive force and repulsion force. Van der Waals attractive force is in inverse ratio to the sixth power of the distance between atoms or groups. When they are too close to each other, they will repel each other. The van der Waals bond length is 0.3–0.5 nm. The bond energy is 1–3 kcal/mol.

Although the van der Waals force is weak, when the surfaces of two big molecules are close enough to each other, this force is very important. It contributes to sustain the tertiary structure and quaternary structure.

10.3   Fundamental of Macromolecules Structures and Functions

10.3.1   Different Levels of Protein Structure

Protein structures have conventionally been understood at four different levels (Fig. 10.7):

1. The primary structure is the amino acid sequence (including the locations of disulfide bonds).
2. The secondary structure refers to the regular, recurring arrangements of adjacent residues resulting mainly from hydrogen bonding between backbone groups, with α-helices and β-pleated sheets as the two most common ones.
3. The tertiary structure refers to the spatial relationship among all amino acid residues in a polypeptide chain, that is, the complete three-dimensional structure.
4. The quaternary structure refers to the spatial arrangements of each subunit in a multi-subunit protein, including nature of their contact.

10.3.1.1   The Formation of Protein Structure Level and the Folding of Peptide Chain

In protein solution, if the environment changes, for example, pH, ion strength, or temperature changes, the natural structure of protein may disintegrate and leads to the denaturation of proteins. This process is called protein denaturation. When the condition is normal, if the denatured protein can have their natural structure and character back, then the protein will renature.

The way to make bean curd by heating the solution of bean protein and adding a little salt is an example to make use of the protein denaturation to deposit protein.

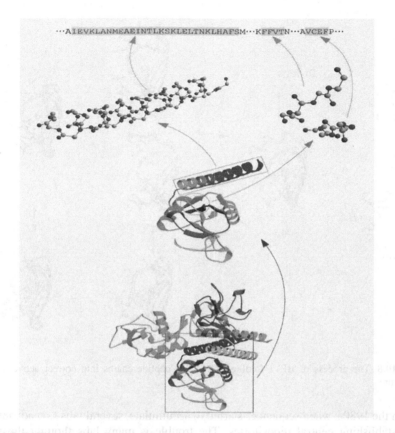

···AIEVKLANMEAEINTLKSKLELTNKLHAFSM···KFFVTN···AVCEFP···

Fig. 10.7 Different levels of protein structure

Table 10.1 The nucleic and protein databases

Nucleic database	Protein database
EMBL	SWISS-PROT
GenBank	PIR
DDBJ	MIPS
	TrEMBL
	NRL-3D

10.3.2   Primary Structure

According to the classical view, the primary structure of protein decides the high-level structure of proteins. So the high-level structure can be inferred from the primary structure. We can align multiple protein sequences (Table 10.1, Figs. 10.8 and 10.9).

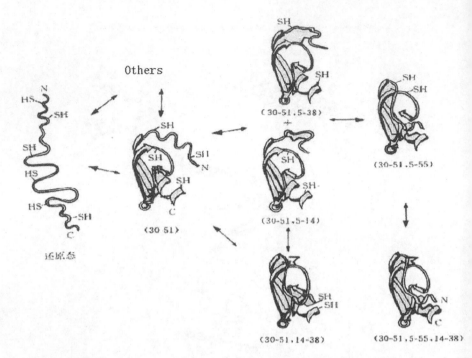

Others

(30-51,5-38)

(30-51,5-55)

(30-51,5-14)

(30 51)

还原态

(30-51,14-38)   (30-51,5-55,14-38)

Fig. 10.8 The process of BPT1 folding from loose peptide chains into correct active tertiary structure

In the 1980s, when sequences started to accumulate, several labs saw advantages to establishing central repositories. The trouble is many labs thought the same and made their own. The proliferation of databases causes problems. For example, do they have the same format? Which one is the most accurate, up-to-date, and comprehensive? Which one should we use?

10.3.3 Secondary Structure

10.3.3.1 Various Kinds of Protein Secondary Structure

Local organization of protein backbone is α-helix, β-strand (which assembles into β-sheet), turn, and interconnecting loop.

α-Helix is in a shape of stick. Tightly curled polypeptide backbone forms the inner side of the stick; the side chains expand outside in the form of helix. α-Helix tends to be stable because the hydrogen in NH and the oxygen in the fourth residue CO form hydrogen bond. Each helix contains 3.6 residues. The helix distance is 0.54 nm.

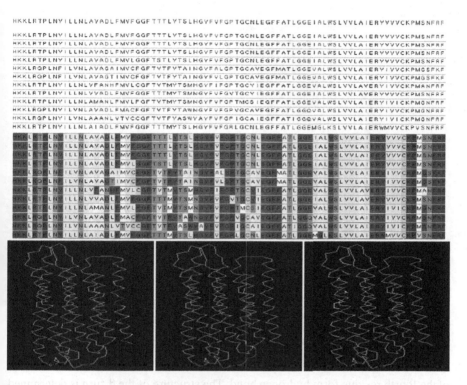

Fig. 10.9 An alignment of protein primary structure

β-Sheet is another frequently occurrence structure. Two or more fully expended polypeptides cluster together laterally. Hydrogen bond is formed by –NH and C=O on the neighboring peptide backbones. These polypeptide structures are β-sheet. In the β-sheets, all peptides join in the cross-linking between hydrogen bonds. The hydrogen bonds are almost vertical to the long axis of peptide chains. Along the long axis of peptide chain, there are repeated units.

β-Sheet includes two types. One is the parallel sheet. The arrangement polarization of its peptide chain (N–C) is unidirectional. The N-end of all the peptide chains is in the same direction. Another one is antiparallel. The polarization of the peptide chain is opposite for the neighboring chains.

In the backbones of polypeptide chain, the structures which are different from the α-helix and β-sheet are called random coil. Random coils mean the irregular peptide chain. For most globins, they often contain a great amount of random coils besides α-helix and β-sheet. In random coils, β-turn is a very important structure.

β-Turn can also be called reverse turn, β-bend, and hairpin structure. It is composed of four successive amino acids. In this structure, the backbone folds in a degree of 180°. The oxygen on C=O of the first residue and hydrogen on the N–H

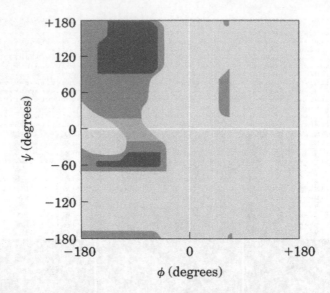

Fig. 10.10 Ramachandran plot for L-Ala residues. *Dark blue* area reflects conformations that involve no steric overlap and thus are fully allowed; *medium blue* indicates conformations allowed at the extreme limits for unfavorable atomic contacts; the *lightest blue* area reflects conformations that are permissible if a little flexibility is allowed in the bond angles (Color figure online)

of the fourth residue form hydrogen bond. The structure of the β-turn is determined by the dihedral angel (ϕ_2, ψ_2; ϕ_3, ψ_3) made of the second residue and third residue.

10.3.3.2 The Variability of Protein Secondary Structure (Figs. 10.10, 10.11, and 10.12)

10.3.4 Supersecondary Structure

Two or several secondary structure units connected by connecting peptides can form special space structures. They are called protein supersecondary structures.

10.3.4.1 High-Frequency Supersecondary Structure Motif

Protein databases (PDB)

1. Analysis of main-chain conformations in known protein structure.
2. 12,318 residues from 84 proteins, structure 5,712 fell outside the regions of regular structure.
3. Torsion angles (φ,ψ) for the 5,712 residues were calculated and allocated to seven classes in the Ramachandran plot: a,b,e,g,l,p,t $\Rightarrow$ a,b,e,l,t H, E.

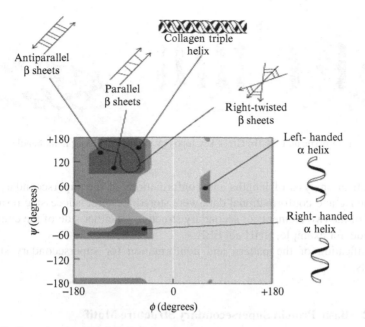

Fig. 10.11 Ramachandran plots for a variety of structures

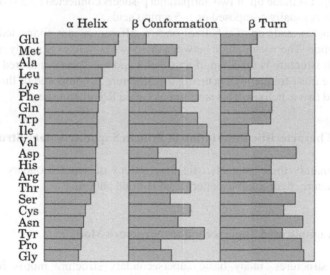

Fig. 10.12 Relative probabilities that a given amino acid will occur in the three common types of secondary structure

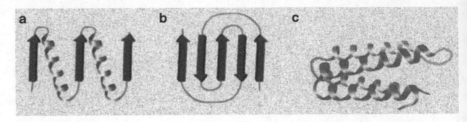

Fig. 10.13 (a) Rossmann fold; (b) Greek key topology structure; (c) four-helix bundle

4. Statistical analysis of lengths and conformations of the supersecondary motif. Sequence and conformational data were stored for three successive residues in each of the two elements of secondary structure on either side of the connecting peptide, for example, HHH abl EEE.
5. Classification of the pattern and conformation for supersecondary structure motifs.

10.3.4.2   Basic Protein Supersecondary Structure Motif

α-α-Hairpin is made up of two almost antiparallel α-helixes connected by a short peptide. This short peptide is usually composed of 1–5 amino acids.

β-β-Hairpin is made up of two antiparallel β-sheets connected by a short peptide. This peptide is usually composed of 1–5 amino acids.

α-α-Corner is made up of α-helixes on two different planes connected by a connecting peptide. The vector angle between these two α-helixes is nearly right angle.

α-β-Arch structure is made up of an α-helix and a β-sheet connected by a short peptide. The most frequently occurring α-β-structure is composed of three parallel β-sheets and two α-helixes. This structure is called Rossmann sheet.

10.3.4.3   Characteristic Description of Protein Supersecondary Structure

There are mainly three characteristic descriptions of supersecondary structures: sequence pattern, hydrophobic pattern, and H-bond pattern.

10.3.4.4   Complicated Supersecondary Structure Motif

In protein structures, many basic supersecondary structure motifs form some more complicated complexes motif, which are called complicated supersecondary structures.

The commonly occurring complicated supersecondary structures include Rossmann fold (Fig. 10.13a), Greek Key topology structure (Fig. 10.13b), and four-helix bundle (Fig. 10.13c), etc. (Figs. 10.14 and 10.15)

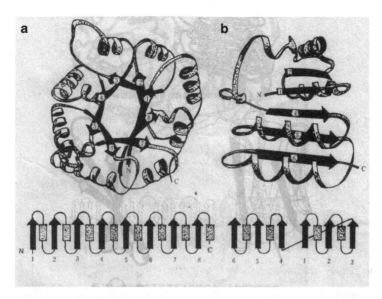

Fig. 10.14 Two different α/β-domains frequently observed in many proteins (**a**) closed β-barrel and (**b**) open curled β-sheet

Fig. 10.15 Up-down β-barrel structure and up-down open β-sheet structure

10.3.4.5   Protein Tertiary Structure

Polypeptide chains further fold by non-covalent bond interaction and curl into more complicated configuration, which is called tertiary structure.

For bigger protein molecules, polypeptide chains are always composed of two or more independent three-dimensional entity. These entities are called domains.

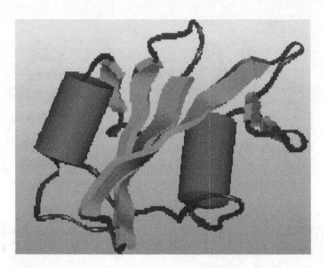

Fig. 10.16 Tertiary structure

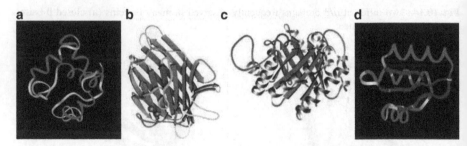

Fig. 10.17 (a) α-Protein, (b) β-protein, (c) α + β-protein, and (d) α/β-protein

According to the amount of α-helix and β-sheet, proteins can be divided into four types: α-protein, β-protein, α + β-protein, and α/β-protein (Fig. 10.16).

α-Protein contains more than 40 % of α-helix and less than 10 % of β-sheet (Fig. 10.17a). β-Protein contains more than 40 % of β-sheet and less than 10 % of α-protein (Fig. 10.17b). α + β-Protein contains more than 10 % of α-helix and β-sheet. α,β-Clusters in different regions. α/β-Protein (Fig. 10.17c) contains more than 10 % of α-helix and β-sheet. These two configurations appear in the peptide chain alternatively. The two configurations of different α/β-proteins (Fig. 10.17d) arrange face to face. The shape of the whole molecule varies a lot.

10.3.4.6 Protein Quaternary Structure

Spatial arrangement of subunits in a protein that contains two or more polypeptide chains is called quaternary structure. It often involves symmetry, but doesn't have to.

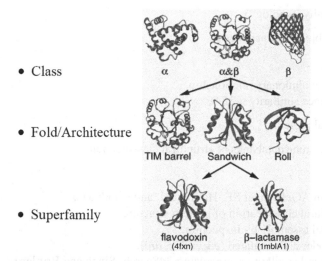

- Class
- Fold/Architecture
- Superfamily

Fig. 10.18   The hierarchy of structural classification

Subunits of proteins form quaternary structure by hydrophobic interaction, H-bond, and van der Waals. The number of most oligomeric proteins is even. There are always one or two types of subunits. The arrangement of most oligomeric protein molecules is symmetric.

Some globins contain two or more polypeptide chain. These polypeptide chains interact with each other, and each of them has their own tertiary structure. These polypeptide chains are subunits of proteins. From the view of structure, subunit is the smallest covalent unit of proteins. Proteins clustered by subunits are called oligomeric proteins. Subunit is the function unit of oligomeric proteins.

10.3.5   Folds

10.3.5.1   Structural Classification of Protein Structure

The hierarchy of structural classification (Fig. 10.18):

- Class
  - Similar secondary structure content
  - All α, all β, α + β, α/β, etc.
- Folds (architecture)
  - Core structure similarity
  - SSEs in similar arrangement

- Superfamily (topology)

  – Probable common ancestry

- Family

  – Clear evolutionary relationship
  – Sequence similarity >25 %

- Individual protein

  There are some databanks of structural classification:

- SCOP

  – Murzin AG, Brenner SE, Hubbard T, and Chothia C.
  – Structural classification of protein structures.
  – Manual assembly by inspection.
  – All nodes are annotated (e.g., all α, α/β).
  – Structural similarity search using 3dSearch (Singh and Brutlag).

- CATH

  – Dr. C.A. Orengo, Dr. A.D. Michie, etc.
  – Class-architecture-topology-homologous superfamily.
  – Manual classification at architecture level.
  – Automated topology classification using the SSAP algorithms.
  – No structural similarity search.

- FSSP

  – L.L. Holm and C. Sander.
  – Fully automated using the DALI algorithms (Holm and Sander).
  – No internal node annotations.
  – Structural similarity search using DALI.

- Pclass

  – A. Singh, X. Liu, J. Chang, and D. Brutlag.
  – Fully automated using the LOCK and 3dSearch algorithms.
  – All internal nodes automatically annotated with common terms.
  – JAVA-based classification browser.
  – Structural similarity search using 3dSearch.

10.3.5.2 Hierarchy of Structure

Homologous family: evolutionarily related with a significant sequence identity
Superfamily: different families whose structural and functional features suggest
    common evolutionary origin

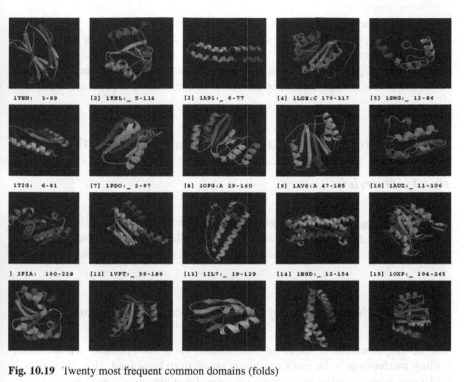

Fig. 10.19 Twenty most frequent common domains (folds)

Folds: different superfamilies having the same major secondary structures in the same arrangement and with the same topological connections (energetic favoring certain packing arrangements)

Class: secondary structure composition

10.3.5.3 Protein Molecule Movement and Function

Proteins have varieties of movements. Movement and structures are the basic elements of protein functions. Protein movement includes short-time and small-amplitude movement, median-time and median-amplitude movement, and long-time and big-amplitude movement (Fig. 10.19).

10.3.6 Summary

Five schemes of protein three-dimensional structures:

1. The three-dimensional structure of a protein is determined by its amino acid sequence.
2. The function of protein depends on its structure.

3. An isolated protein has a unique or nearly unique structure.
4. The most important forces stabilizing the specific structure of a protein are non-covalent interactions.
5. Amid the huge number of unique protein structures, we can recognize some common structural patterns to improve our understanding of protein architecture.

10.4 Basis of Protein Structure and Function Prediction

10.4.1 Overview

In the following part, we are going to talk about the comparative modeling, inverse folding, ab initio, secondary structure prediction, supersecondary structure prediction, structure-type prediction, and tertiary structure prediction.

10.4.2 The Significance of Protein Structure Prediction

The development and research of life science show that protein peptide chain-folding mechanism is the most important problem to be solved. How does protein fold from primary structure into active natural tertiary structure is waiting to be answered. The elucidation of the protein peptide chain-folding mechanisms is called decoding the second biological code.

As the human genome and other species genome sequencing plan start and finish, the capacity of databases (e.g., SWISS-PROT) collecting protein sequence increases exponentially. Meanwhile, the capacity of databases (e.g., PDB) collecting protein tertiary crystal structures increases slowly. The increasing rate of the protein sequence number is much greater than that of the known protein structure number. So we need the computational predictive tools to narrow the widening gap.

In the most genome era, one of the biggest challenges we face is to discover the structure and function of every protein in the genome plan. So, predicting protein structure theoretically becomes one way to decrease the disparity between protein structure and sequence.

Why should we predict secondary structure? Because it is an easier problem than 3D structure prediction (more than 40 years of history) and accurate secondary structure prediction can be important information for the tertiary structure prediction. Ever since the first work of prediction of secondary structure done by Chou-Fasman, it has been 30 years. The accuracy is around 60 %. Since 1990s, several machine learning algorithms have been successfully applied to the prediction of protein secondary structure and the accuracy reaches 70 %. From this, we can see a good method can help improve the prediction result significantly.

PHD — one of the
most accurate and reliable
prediction methods

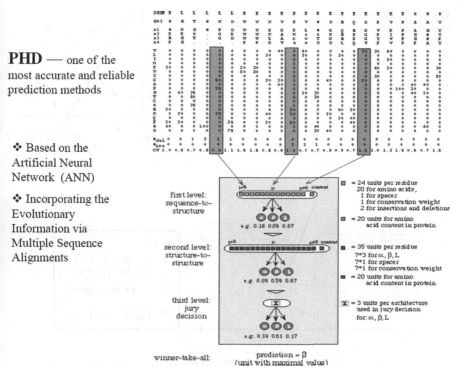

❖ Based on the
Artificial Neural
Network (ANN)

❖ Incorporating the
Evolutionary
Information via
Multiple Sequence
Alignments

Fig. 10.20 PHD method

There are a few prediction methods including statistical method (Chou-Fasman method, GOR I-IV), nearest neighbors (NNSSP, SSPAL, Fuzzy-logic-based method), neural network (PHD (Fig. 10.20), Psi-Pred, J-Pred), support vector machine (SVM), and HMM.

10.4.3   The Field of Machine Learning

10.4.3.1   Support Vector Machine

There are many researches in this field. V. Vapnik [1] developed a promising learning theory (Statistical Learning Theory (SLT)) based on the analysis of the nature of machine. Support vector machine (SVM) is an efficient implementation of SLT. SVM has been successfully applied to a wide range of pattern recognition problems, including isolated handwritten digit recognition, object recognition, speaker identification, and text categorization.

Fig. 10.21 The linearly
separable case

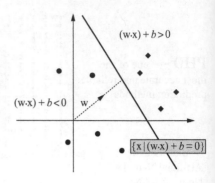

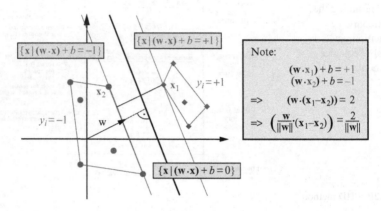

Fig. 10.22 Optimal separating hyperplane (OSH)

For the linearly separable case (Fig. 10.21), the SVM tries to look for one unique
separating hyperplane, which is maximal in the margin between the vectors of
the two classes. This hyperplane is called Optimal Separating Hyperplane (OSH)
(Fig. 10.22).

Introducing Lagrange multipliers and using the Karush-Kuhn-Tucker (KKT)
conditions and the Wolfe dual theorem of optimization theory, the SVM training
procedure amounts to solving a convex quadratic programming problem:

$$\text{Maximize} \quad \sum_{i=1}^{N} \alpha_i - \frac{1}{2} \sum_{i=1}^{N} \sum_{j=1}^{N} \alpha_i \alpha_j \cdot y_i y_j \cdot \vec{x}_i \cdot \vec{x}_j$$

$$\text{subject to} \quad \alpha_i \geq 0$$

$$\sum_{i=1}^{N} \alpha_i y_i = 0 \quad i = 1, 2, \ldots, N$$

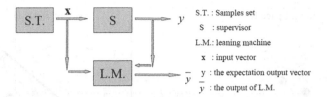

Fig. 10.23 Setting of the machine learning problem. A model of learning from samples. During the learning process, the learning machine observes the pairs (**x**,**y**) (the training set). After training, the machine must on any given x return a value. The goal is to return a value which is close to the supervisor's response y

The solution is a unique globally optimized result which can be shown to have an expansion (Fig. 10.23):

$$\vec{\mathbf{w}} = \sum_{i=1}^{N} y_i \alpha_i \cdot \vec{\mathbf{x}}_i$$

When an SVM is trained, the decision function can be written as:

$$f\left(\vec{\mathbf{x}}\right) = \mathrm{sgn}\left(\sum_{i=1}^{N} y_i \alpha_i \cdot \vec{\mathbf{x}} \cdot \vec{\mathbf{x}}_i + b\right)$$

For the linearly non-separable case, the SVM performs a nonlinear mapping of the input vectors from the input space R^d into a high-dimensional feature space H and the mapping is determined by a kernel function. Then like the linearly separable case, it finds the OSH in the higher-dimensional feature space H.

The convex quadratic programming problem:

$$\text{Maximize} \quad \sum_{i=1}^{N} \alpha_i - \frac{1}{2} \sum_{i=1}^{N} \sum_{j=1}^{N} \alpha_i \alpha_j \cdot y_i y_j \cdot K\left(\vec{\mathbf{x}}_i, \vec{\mathbf{x}}_j\right)$$

$$\text{subject to} \quad 0 \le \alpha_i \le C$$

$$\sum_{i=1}^{N} \alpha_i y_i = 0 \quad i = 1, 2, \ldots, N$$

The decision function:

$$f(\vec{\mathbf{x}}) = \mathrm{sgn}\left(\sum_{i=1}^{N} y_i \alpha_i \cdot K\left(\vec{\mathbf{x}}, \vec{\mathbf{x}}_i\right) + b\right)$$

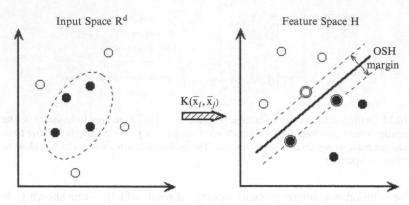

Fig. 10.24 Kernel function technology

The problem of risk minimization:

Given a set of functions

$$\{f\left(\vec{x},\alpha\right) : \alpha \in \Lambda\}, f\left(\vec{x},\alpha\right) : \vec{x} \rightarrow \{-1,+1\}, \vec{x} \in R^d$$

and a set of examples

$$\left(\vec{x}_i, y_i\right), \vec{x}_i \in R^d, y_i \in \{-1,+1\}, i = 1,2,\ldots,N$$

each one independently drawn from an unknown identical distribution.

The goal is to find an optimal function $f\left(\vec{x},\alpha^*\right)$ which minimizes the expected risk (or the actual risk) (Fig. 10.24).

$$R(\alpha) = \int L\left(f\left(\vec{x},\alpha\right), y\right) dP\left(\vec{x}, y\right)$$

$$\text{i.e. } R\left(\alpha^*\right) = \inf_{\alpha \in \Lambda} R(\alpha)$$

Here $L\left(f(\vec{x},\alpha^*), y\right)$ is the loss function. For this case one simple form is

$$L\left(f\left(\vec{x},\alpha\right), y\right) = \frac{1}{2}\left|y - f\left(\vec{x},\alpha\right)\right|, \vec{x} \in R^d, y \in \{-1,+1\}$$

10.4.3.2 The Empirical Risk Minimization (ERM)

The risk functional $R(\alpha)$ is replaced by the so-called empirical risk function constructed on the basis of the training set:

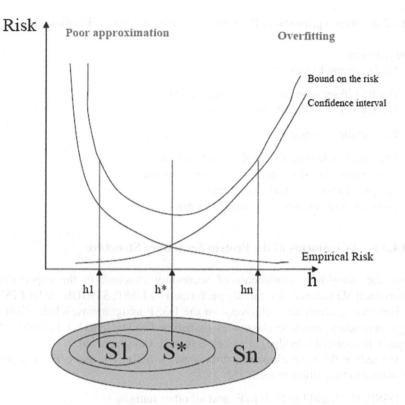

Fig. 10.25 SRM

$$R_{\text{emp}}(\alpha) = \frac{1}{N} \sum_{i=1}^{N} L\left(y_i, f\left(\vec{x}_i, \alpha\right)\right)$$

$$R_{\text{emp}}(\alpha_e{}^*) = \inf_{\alpha \in \Lambda} \left\{ R_{\text{emp}}(\alpha) \right\}$$

Note is $\alpha_e^* = \alpha^*$? No! The answer is not simple!

10.4.3.3 The Structural Risk Minimization (SRM)

The bound of generalization ability of learning machine (Vapnik & Chervonenkis):

$$R(\alpha) \leq R_{\text{omp}}(\alpha) + \Phi\left(\frac{N}{h}\right)$$

Here, N is the size of the training set; h, VC dimension, the measure of the capacity of the learning machine; and $\Phi(N/h)$, the confidence interval. When the N/h is larger, the confidence interval is smaller (Fig. 10.25).

10.4.3.4   New Approach to Protein Secondary Structure Prediction

The data sets
  Two nonhomologous data sets:

1. The RS126 set – percentage identity – 25 %
2. The CB513 set – the SD (or Z) score – 5

We exclude entries if:

1. They are not determined by X-ray diffraction.
2. The program DSSP could not produce an output.
3. The protein had physical chain breaks.
4. They had a resolution worse than 0.19 nm.

10.4.3.5   Assignments of the Protein Secondary Structure

Now the automatic assignments of secondary structure to the experimentally determined 3D structure are usually performed by DSSP, STRIDE, or DEFINE.
  Here we concentrate exclusively on the DSSP assignments, which distinguish eight secondary structure classes: H (α-helix), G (310-helix), I (π-helix), E (β-strand), B (isolated β-bridge), T (turn), S (bend), and (the rests).
  We reduce the eight classes to three states – helix (H), sheet (E), and coil (C) according to two different methods:

1. DSSP: H, G, and I to H; E to E; and all other states to C
2. DSSP: H and G to H, E and B to E, and all other states to C

10.4.3.6   Assessment of Prediction Accuracy

Cross-validation trials are necessary to minimize variation in results caused by a particular choice of training or test sets.
  A full jackknife test is not feasible, especially on the CB513 set for the limited computation power. We take the sevenfold cross-validation on both sets.

1. Q-index (Q_3, QH, QE,QC)
2. Matthews' Correlation Coefficient (CH, CE, CC)
3. Segment Overlap Measure (SOV)

Q-index gives percentage of residues predicted correctly as helix (H), strand (E), coil (C), or all three conformational states. The definition of Q-index is as follows:

1. For a single conformational state:

$$Q_I = \frac{\text{Number of residues correctly predicted in state } i}{\text{Number of residues observed in state } i} * 100$$

where I is either H, E, or C.

2. For all three states:

$$Q_3 = \frac{\text{Number of residues correctly predicted}}{\text{Number of all residues}} * 100$$

10.4.3.7   The Coding Scheme

For the case of the single sequence, each residue is coded by the orthogonal binary vector $(1,0,\ldots,0)$ or $(0,1,\ldots,0)$. The vector is 21-dimensional. If the window length is l, the dimensionality of the feature vector (or the sample space) is $21*l$. When we include the evolutionary information, for each residue the frequency of occurrence of each of the 20 amino acids at one position in the alignment is computed.

10.4.3.8   The Design of Binary Classifiers

We design six binary classifiers (SVMs) as follows:

1. Helix/non-helix – H/$\sim$ H
2. Sheet/non-sheet – E/$\sim$ E
3. Coil/non-coil – C/$\sim$ C
4. Helix/sheet – H/E
5. Sheet/coil – E/C
6. Coil/helix – C/H

10.4.3.9   The Design of Tertiary Classifiers

Assembly of the binary classifiers:

1. SVM_MAX_D
   We combined the three one-versus-rest classifiers (H/$\sim$H, E/$\sim$E, and C/$\sim$C) to handle the multiclass case. The class (H, E, or C) for a testing sample was assigned as that corresponding to the largest positive distance to the OSH.
2. SVM_TREE (Fig. 10.26)
3. SVM_NN (Fig. 10.27)

The tertiary classifiers we designed:

> 1. SVM_NN
> > 2. SVM_TREE1
> > > 3. SVM_TREE3
> > > > 4. SVM_TREE3
> > > > > 5. SVM_VOTE
> > > > > > 6. SVM_MAX_D
> > > > > > > 7. SVM_JURY

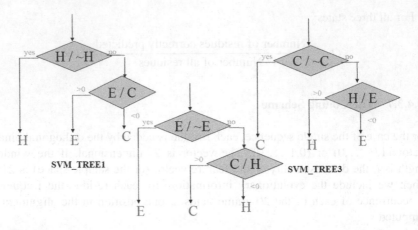

Fig. 10.26 SVM tree

Fig. 10.27 SVM-NN

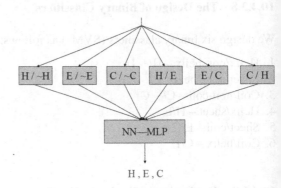

10.4.3.10 Results and Analysis

The selection of the optimal kernel function and the parameters:

$$\text{RBF} \quad k(\vec{x}, \vec{y}) = \exp(-r|\vec{x} - \vec{y}|^2)$$

We set the optimal $\gamma = 0.10$.

Accuracy measure

Three-state prediction accuracy: Q_3

$$Q_3 = \frac{\text{Correctly predicted residues}}{\text{Number of residues}}$$

A prediction of all loop: $Q_3 \sim 40\%$

Fig. 10.28 Neuron

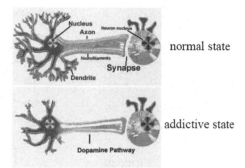

normal state

addictive state

Improvement of accuracy:

Chou and Fasman (1974)	~50–53 %
Garnier (1978)	63 %
Zvelebil (1987)	66 %
Qian and Sejnowski (1988)	64.3 %
Rost and Sander (1993)	70.8–72.0 %
Frishman and Argos (1997)	<75 %
Cuff and Barton (1999)	72.9 %
Jones (1999)	76.5 %
Petersen et al. (2000)	77.9 %
Hua and Sun (2001)	76.2 %
Guo and Sun (2003)	80 %

10.4.3.11 Neural Network (Figs. 10.28, 10.29, 10.30, 10.31, 10.32, 10.33, and 10.34)

10.4.4 Homological Protein Structure Prediction Method

Homology modeling is a knowledge-based protein structure prediction. These kinds of methods are based on the evolutional conservation of protein structure and sequence. They use the structure of known proteins to build the structure of the unknown homological proteins. They are the most mature protein structure prediction methods so far. When the homology is high, we will get reliable prediction results. In the whole genome, only about 20–30 % sequences can be predicted using these methods.

One difficult point in the homology modeling method is the prediction of the circle region on the protein surface. That is because the circle region on the surface is very flexible. But because the circle region is usually the active part of the protein, the prediction of the structure of circle region is quite important to the protein structure modeling.

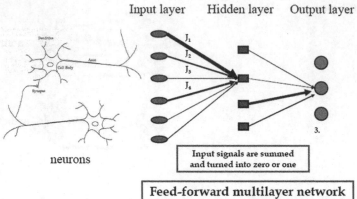

Feed-forward multilayer network

Fig. 10.29 Neural network

Fig. 10.30 Neural network
training

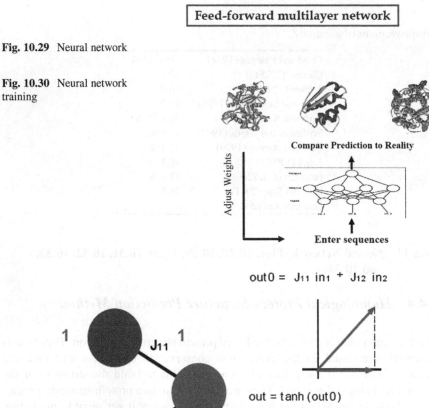

$$out0 = J_{11}\ in_1 + J_{12}\ in_2$$

$$out = \tanh(out0)$$

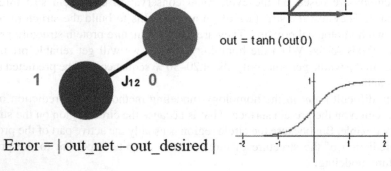

$$Error = |\ out\_net - out\_desired\ |$$

Fig. 10.31 Simple neural network

Fig. 10.32 Train a neural network

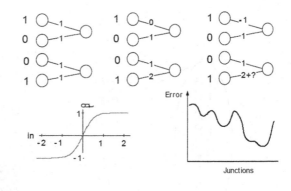

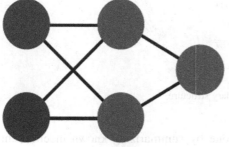

$$\text{out}_i = f\left(\sum_j J_{ij}^2 \cdot f\left(\sum_k J_{jk}^1 \cdot \text{in}_k\right)\right)$$

Fig. 10.33 Simple neural network with hidden layer

The protein homology modeling includes:

1. Matching of object protein sequence and model sequence
2. Modeling object protein structure model according to the model structure
3. Modeling the conserved region in the object protein
4. Modeling the SCRs backbone
5. Predicting the side chain structure
6. Optimizing and estimating the modeling structure

10.4.4.1   Threading Method

Threading (or inverse folding) method can be used to predict structure without homology information. The basic assumption is that the folding type of natural protein is limited. So we can align the sequence of proteins whose structures are unknown and those proteins whose structures are known. And then predict on the best alignment. This method cannot predict new types of proteins correctly.

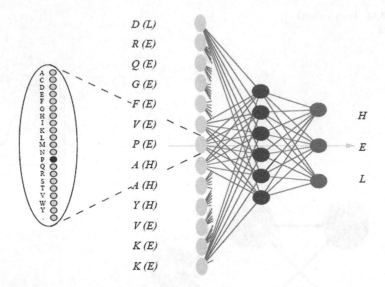

Fig. 10.34 Neural network for secondary structure

Threading method can be done by summarizing known independent protein structure patterns as the model of unknown structure and then by learning known database to summarize average potential function which can distinguish correction and error. In this way we can get the best alignment way.

Protein sequence incrustation:

1. Basing on the experience method. Build various potential functions by analyzing protein of known structure, and see if it can align with known structure by using the standard of lowest folding configuration to guide the object protein sequence incrustation.
2. Basing on the 3D profile. Predict sequence space structure by building a 3D profile, using dynamic programming, comparing new sequence with those in profile databases, and seeking optimal alignment.

10.4.5 Ab Initio Prediction Method

10.4.5.1 Protein Secondary Structure Prediction

Protein secondary structure prediction research has developed for more than three decades. From the progression of research method, there are three different periods. The first period is statistic prediction basing on single residue; the second period, statistic prediction basing on sequence segments; and the third period, statistic prediction combining evolutionary information.

Rost and Sander (1993) promoted prediction basing on neural network – PHD (Profile fed neural network systems from HeiDelberg). It is the first method with the prediction accuracy over 70 %, first efficient method bringing in evolutionary method, and one of the most accurate methods so far.

PHD is a complicated method basing on neural network. It includes poly-sequence information. Recently, Cuff and Barton synthesize many good secondary prediction methods, such as DSC, NNSSP, PREDATOR, and PHD. Up to now, there are still some other artificial intelligence methods to predict secondary structure, such as expert system and nearest neighbor method.

Recently, it is a good opportunity to predict protein secondary structure. For one thing, structural genomic plan is carried out throughout the world to increase the speed of measuring the number of protein structure and fold type. For another, the field of machine learning develops fast. For example, in recent 2 years, the building and perfecting of famous statistic learning theory of V. Vapnik make it possible for us to use the latest machine learning method to improve the prediction accuracy of secondary structure.

Our paper published on JMB (*J. Mol. Biol.*) applied SVM to predict protein secondary structure and got an accuracy of 76.2 %.

10.4.5.2 Chou-Fasman Method

1. Residue propensity factor

$$P_{ij} = \frac{f_{ij}}{f_j}$$

j: configuration
i: one of the twenty amino acids
f_j: fraction of the jth configuration
f_{ij}: jth configuration fraction of the ith amino acid residue. $f_{ij} = n_{ij}/N_i$
n_{ij}: the total appearance of a residue in a certain configuration
N_i: the total number of a residue in the statistical samples. $f_j = N_j/N_t$
N_t: the total number of residues in the statistical samples

2. The tendentiousness of folding-type related secondary structure

   (a) Protein folding type: all α, all β, $\alpha + \beta$, and α/β
   (b) Analysis of secondary structure tendentiousness: α-helix propensity factor P_α, β-sheet propensity factor P_β, and irregular curl propensity factor P_C

3. Chou-Fasman method

   (a) α-Helix rule
       In a protein sequence, there are at least four residues in the neighboring six residues tending to form α-helix kernel. The kernel extends laterally until the

average value of α-helix tendentiousness factor in the polypeptide segment $P_\beta < 1.0$. Lastly, drop three residues at each end of α-helix. If the rest part is longer than six residues, $P_\alpha > 1.03$, it will be predicted as helix.

(b) β-Sheet folding rule

If three residues in five tend to form β-sheet, we think it is the folding kernel. The kernel extends laterally until the average of the tendentiousness of the polypeptide segment $P_\beta < 1.0$. Lastly, discard two residues from each end; if the rest part is longer than the four residues and $P_\alpha > 1.05$, then it is predicted as β-sheet.

Reference

1. Vapnik VN (1995) The Nature of Statistical Learning Theory, Springer-Verlag

Chapter 11
Computational Systems Biology Approaches for Deciphering Traditional Chinese Medicine

Shao Li and Le Lu

11.1 Introduction

Traditional Chinese medicine (TCM) is a system with its own rich tradition over 3,000 years. Compared to Western medicine (WM), TCM is holistic with emphasis on regulating the integrity of the human body. However, understanding TCM in the context of "system" and TCM modernization both remain to be problematic. Along with the "Omics" revolution, it comes the era of system biology (SB). After years of studies in the cross field of TCM and SB, we find that bioinformatics and computational systems biology (CSB) approaches may help for deciphering the scientific basis of TCM. And the previous difficulty in the direct combination of WM and TCM, two distinct medical systems, may also be overcome through the development of systems biology, which tends toward preventive, predictive, and personalized medicine [1].

This course is designed to offer an introduction of the concepts and developments in the CSB approaches for deciphering TCM. Firstly, we placed "disease" under the backdrop of systems biology so as to seek some agreements to TCM in concept and in methodology (Sect. 11.2). Then, some pilot studies for CSB and TCM are reviewed, such as a CSB-based case study for TCM "ZHENG" (Sect. 11.3) and a network-based case study for TCM *Fu Fang* (Sect. 11.4).

Section 11.2, *Disease-related network*, introduces the biological basis and some CSB methods in the disease-related network construction and modularity analysis. LMMA (literature mining and microarray analysis) approach and CIPHER (Correlating Interactome and PHEnome to pRedict disease genes) approach, which are both developed by our lab, are given as examples here. LMMA constructs biological

S. Li (✉) • L. Lu
MOE Key Laboratory of Bioinformatics and Bioinformatics Division, TNLIST/Department of Automation, Tsinghua University, Beijing 100084, China
e-mail: shaoli@tsinghua.edu.cn

R. Jiang et al. (eds.), *Basics of Bioinformatics: Lecture Notes of the Graduate Summer School on Bioinformatics of China*, DOI 10.1007/978-3-642-38951-1_11,
© Tsinghua University Press, Beijing and Springer-Verlag Berlin Heidelberg 2013

network for a specific disease (e.g., "angiogenesis") by using the literature-based co-occurrence method and refines the proposed network through the integration of microarray data. CIPHER interprets biological network modularity and relation to phenotype network and performed a global inference of human disease genes.

Section 11.3, *TCM ZHENG-related network*, first introduces the TCM ZHENG and the possible relation between systems biology-based medicine and TCM. Then, we mainly focus on a CSB-based case study for TCM ZHENG – a systems biology approach with the combination of computational analysis and animal experiment to investigate Cold ZHENG and Hot ZHENG in the context of the neuroendocrine–immune (NEI) system.

Section 11.4, *Network-based study for TCM "Fu Fang,"* advocates the application of system biology in drug discovery, especially *Fu Fang* in TCM. We further illustrate a network-based case study for a Cold *Fu Fang*, QLY, and a combination of herbal compounds extracted from it, which has anti-angiogenesis synergic effects and can modulate both ZHENG- and angiogenesis-related networks.

11.2 Disease-Related Network

Systems biology mainly consists of two key elements, namely, large-scale molecular measurements and computational modeling [2]. The molecular constituents of a system and their variations across a series of dynamic phenotypic changes can be measured, and the measurements are collectively referred to as "Omics," such as genomics, transcriptomics, proteomics, metabolomics, pharmacogenomics, physiomics, and phenomics. They are quantitative study of sequences, expression, metabolites, and so on. "Omics" – data integration helps to address interesting biological questions on the systems level. A systems biology experiment consists of large-scale molecular measurements as well as computational modeling, so not every "Omics" experiment falls into systems biology. Omics experiment itself sometimes could only be regarded as large-scale reductionism [2].

The computational systems biology (CSB) uses computational approaches to understand biological systems in system level, which integrates not only various types of data in multiple levels and phases but also the computational model and experimental investigation (Fig. 11.1). Integration across the two dimensions, i.e., levels of structure as well as scale and phases of processes, allows the systems approach to analyze the hierarchical structure and temporal dynamics of the underlying networks.

11.2.1 *From a Gene List to Pathway and Network*

High-throughput approaches such as microarray can measure the expression of thousands of genes simultaneously and then genes could be ordered, according to their differential expression, in a ranked list [3]. The genes showing the largest

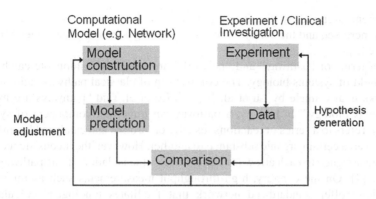

Fig. 11.1 Combination of computational and experimental approaches

difference will appear at the top or bottom of the list. Subramanian et al. suggest that common approaches focusing on a handful of genes appeared at the top and bottom of the list have several major limitations [3]. For certain statistical significance threshold, there may result in no individual genes after filtration of multiple hypotheses testing, if the biological differences are relatively modest given the noise nature of the microarray technology; on the other hand, a long list of genes without any unifying biological theme may be left, the interpretation of which has to depend on a biologist's area of expertise. And analysis based on individual genes may miss important effects on pathways, since cellular processes are always affected by sets of genes. Moreover, for studies from different research groups, there may not be overlap between the lists of statistically significant genes [3].

To overcome these analytical challenges, Subramanian et al. proposed a Gene Set Enrichment Analysis (GSEA) approach to interpret gene expression data at the level of gene sets [3]. Given an a priori defined gene set, the goal of GSEA is to determine whether the members of the gene set are randomly distributed throughout the ranked list or mainly found at the extreme (top or bottom). The sets related to the phenotypic distinction are expected to show the latter distribution. Subramanian et al. created a catalog of 1,325 gene sets, including 319 cytogenetic sets, 522 functional sets, 57 regulatory-motif sets, and 427 neighborhood sets. The Estimation Score (ES) is the maximum deviation from 0 met in the random walk; and it is equivalent to a weighted KS (Kolmogorov–Smirnov)-like statistic. ES is evaluated in the ranked list through a top-down approach, and it reflects the degree to which a gene set is overrepresented at top or bottom of the ranked list. And the statistical significance of the ES is estimated by an empirical phenotype-based permutation test procedure that preserves the complex correlation structure of the gene expression data. The estimated significance level is adjusted to account for multiple hypothesis testing. The ability of GSEA is explored with six examples with biological background information, and it could reveal many biological pathways in common whereas single-gene analysis could only find little similarity between two independent studies. Through GSEA approach, the explanation of a large-scale experiment is erased by identifying pathways and processes. The signal-to-noise ratio is increased and the detection of modest changes in individual genes becomes possible [3].

Both terms of "pathway" and "network" are currently fashionable catchwords in the field of systems biology. The comparison of classical pathways and modern networks is also made by Lu et al. [4]. As Lu et al. said "pathways are integral to systems biology." In biology, a pathway, an important component of systems biology, refers to a series of reactions, usually controlled and catalyzed by enzymes that convert a certain organic substance to another. However, the inconsistence in the meaning assigned to each arrow (or edge) implies the isolation of one pathway from another [4]. On the contrary, high-throughput measurements such as microarray technology offer standardized network that facilitates topological calculations. Compared to network studies, pathway studies have several limitations, such as the inconsistence between the same pathways documented in different databases. Moreover, the pathways are isolated in classical representations which use symbols that lack a precise definition [4]. One approach to overcome these limitations, as is reported by Lu et al. [4], is to embed the classical pathways within large-scale networks and demonstrate the cross talk between them [4]. Thus, in the post-genomic era, a novel concept is expected to regard a living cell as a complex network which consists of various biomolecules.

11.2.2 Construction of Disease-Related Network

Networks are found in biological systems of varying scales as diverse as molecular biology and animal behavior. In a certain sense, systems biology is "network biology" in viewing of the integration across the topological structure and spatiotemporal dynamics of various interactions. Biological networks are believed to be abstract representations of biological systems, which capture many of their essential characteristics [5].

In general, there are four parameters for a network, i.e., node, edge, directed edge, and degree (or connectivity). Node represents a gene, protein, metabolite, or any subsystem. Edge represents an association, connection, co-expression, or any interaction. Directed edge stands for the modulation (regulation) of one node by another, e.g., arrow from gene X to gene Y means that gene X affects expression of gene Y. Degree of a node is the number of links (edges) it has.

Network reconstruction of biological entities such as genes, transcription factors, proteins, compounds, and other regulatory molecules is very important for understanding the biological processes and the organizational principles of the biological systems [5]. Biological networks related to complex disease can be built based on literature and the "Omics" data, respectively.

Rapid progress in the biomedical domain has resulted in enormous amount of biomedical literatures. Along with the booming growth of biomedical researches, literature mining (LM) has become a promising direction for knowledge discovery.

Various techniques have been developed which make it possible to reveal putative networks hidden in the huge collection of individual literatures [6]. Literature-mining tools enable researchers to identify relevant papers and two fundamentally different approaches to extracting relationships from biological texts are currently being used, namely, co-occurrence and natural-language processing (NLP). For example, the co-occurrence-based biological network construction is based on the assumption that if two genes are co-mentioned in a MEDLINE record, there is an underlying biological relationship between them. Jenssen et al. carried out automated extraction of explicit and implicit biomedical knowledge from publicly available gene and text databases to create a gene-to-gene co-citation network for 13,712 named human genes by automated analysis of titles and abstracts in over 10 million MEDLINE records [7]. Genes have been annotated by terms from the medical subject heading (MeSH) index and terms from the gene ontology (GO) database [7]. Jenssen et al. evaluated the quality of the network by manual examination of 1,000 randomly chosen pairs of genes as well as comparison with the Database of Interacting Proteins (DIP) and the Online Mendelian Inheritance in Man (OMIM) database. They further analyzed publicly available microarray data and suggested that their approach could be a complement to conventional clustering analysis. Jenssen et al. then linked the signature gene list to disease terms in MeSH to detect diseases associated with the signature genes. The results have shown that the top-ranked terms were those related to Fragile X and Angelman syndromes, lymphoma, leukemia, and tuberculosis [7].

Despite the successes, there exists a principal question in respect to the literature-mining approaches – whether correctly determined co-occurrences in the title and abstract precisely reflect meaningful relationships between genes. Jenssen et al. suggested that based on their studies, a substantial number of meaningful biological relationships could be detected by their approach [7]. Jenssen et al. further suggested that tools for mining the literature will probably have a pivotal role in systems biology in the near future [8].

Recently, Omics data-based disease networks containing physical, genetic, and functional interactions have arrested an increasing attention and various methods have been addressed. The data resource for these networks mainly includes genetic interaction, gene expression profiles (such as from microarray), protein–protein interaction (PPI) data, and protein–DNA interaction data. Herein we take microarray-based gene expression network construction for instance. A common assumption for such networks is that if two genes have similar expression profiles, they are co-regulated and functionally linked. There are many methods which could be used to reconstruct the gene expression networks, and each has its own advantage and disadvantage, respectively (Table 11.1).

There are in essence some problems if the biological networks are reconstructed solely on "Omics" data. For example, it is well known that microarray data is of high noise and the chosen model as long as the final network structure may not properly reflect the complex biological interactions. Therefore, the combination of literature mining and "Omics" data is an advanced method to improve the biological network reconstruction. Two examples of this integrative method are given below.

Table 11.1 Comparison among some commonly used methods for the reconstruction of gene expression networks

Methods	Advantage	Disadvantage
Clustering	Straightforward	Inconsistent
Linear modeling	Simplifying power	Rough network models
Boolean networks	Logic and general	Rough and determinism
Differential equations	Exactitude	Less training data, time delay, and high computational cost
Bayesian networks	Local	Limitation in network structure (e.g., self-feedback)
Reverse engineering	Powerful	Depended on parameters

In 2005, Calvano et al. presented a structured network knowledge-base approach to analyze genome-wide transcriptional responses in the context of known functional interrelationships among proteins, small molecules, and phenotypes and applied it to the study on changes in blood leukocyte gene expression patterns in human subjects receiving an inflammatory stimulus (bacterial endotoxin) [9]. The response to endotoxin administration in blood leukocytes can be viewed as an integrated cell-wide response, namely, propagating and resolving over time. The sample included gene expression data in whole blood leukocytes measured before and at 2, 4, 6, 9, and 24 h after the intravenous administration of bacterial endotoxin to four healthy human subjects. Four additional subjects under the same condition but without endotoxin administration were used as control. A total of 3,714 unique genes were found to express significantly. Calvano et al. used 200,000 full-text scientific articles, and a knowledge base of more than 9,800 human, 7,900 mouse, and 5,000 rat genes was manually curated and supplemented with curated relationships parsed from MEDLINE abstracts. As a result, Calvano et al. proposed a biological network of direct physical, transcriptional, and enzymatic interactions observed between mammalian which was computed from this knowledge base, and every gene interaction in the network is supported by published information [9].

Another example is from our lab. As has been elucidated in the foregoing, the co-occurrence (co-citation) literature-mining approach [7, 10] is the simplest and most comprehensive in implementation. It can also be easily adopted to find the association between biological entities, such as the genes relations [7] and the chemical compound–gene relations [11]. In our early study [12], we combined prior knowledge from biologists and developed a subject-oriented mining technique. This technique could be used to extract subject-specific knowledge, and despite the different individual experimental conditions, it is capable of retrieving information from lots of literatures. However, the reservoir of literatures is in essence a diverse investigation collection, such that realistic analysis of various types of relations is always unavailable. If a network is merely constructed from literatures, it is rarely specific in respect to certain biological process and commonly contains redundant relations. The resulting networks are usually large, densely connected without significant biological meaning. Therefore, the crudeness and redundancy are two main disadvantages of the literature-derived network.

Network reconstruction from the high-throughput microarray data is another active area in the past decade as mentioned above [13, 14]. Microarray technology that records large-scale gene expression profiles allows characterizing the states of a specific biological system, providing a powerful platform for assessing global gene regulation and gene function. So far, a number of methods are available on reconstructing gene networks using microarray such as deterministic Boolean networks [15] and ordinary differential equations. However, it is difficult to build a reliable network from a small number of array samples owing to the nonuniform distribution of gene expression levels among thousands of genes. Such a technique is also insufficient for detailed biological investigation when prior knowledge is absent [16].

The literature mining and microarray analysis approaches both develop to identify the underlying networks consisting of biological entities. The combination of the information from experiment and literature sources may lead to an effective method for modeling biological networks [16]. Many studies have been focused on topics related to gene clustering, literature mining [17], and modeling biological processes such as neuroendocrine–immune interactions [18]. We proposed a novel approach to reconstruct gene networks through combining literature mining and microarray analysis (LMMA), where a global network is first derived using the literature-based co-occurrence method, and then refined using microarray data. The LMMA approach is applied to build an angiogenesis network. The network and its corresponding biological meaning are evaluated in multiple levels of KEGG Gene, KEGG Orthology, and pathway. The results show that the LMMA-based network is more reliable and manageable with more significant biological content than the LM-based network [19].

The first step in the LMMA approach is to derive co-occurrence dataset by using literature mining. To find co-citations, a pool of articles and a dictionary contained gene symbols and their synonyms are required. In LMMA approach, the literature information is mainly obtained from the PubMed. We performed LM by sharing assumption with many existing LM systems that there should be a potential biological relationship between the two genes co-cited in the same text unit [7, 10]. In practice, we regard two Human Genome Organization (HUGO) gene symbols as correlated if they are co-cited in the same sentence. Microarray datasets related to a biological process are collected from experiments or public repositories such as SMD (Stanford Microarray Database, http://www.bmicc.org/web/english/search/dbnei) that stores a large volume of raw and normalized data from public microarray information [20]. The LMMA approach employs a statistical multivariate selection for gene interaction analysis. This is based on the hypothesis that if a co-cited gene pair is positively or negatively co-expressed, they will indeed interact with each other [21, 22]. Taking the values of n genes as variables $x_1, x_2, \ldots x_n$, the dataset with m observations (i.e., m microarray experiments) and the n variables are denoted by

$$[x_1, x_2, \ldots, x_n] = \begin{bmatrix} x_{11} & x_{12} & \cdots & x_{1n} \\ x_{21} & x_{22} & \cdots & x_{2n} \\ \vdots & \vdots & \ddots & \vdots \\ x_{m1} & x_{m2} & \cdots & x_{mn} \end{bmatrix}.$$

Assuming the relations of variables in gene expression data following a linear model [23], an LM-based network can be refined through multiple variable selection, resulting in a network called LMMA network. We defined each node coupling with its neighboring nodes as a "unit network." The linear approximation of a sub-model is expressed as

$$x_k = \beta_{k0} + \beta_{k1} x_{k1} + \beta_{k2} x_{k2} + \cdots + \beta_{kl} x_{kl} + e,$$

where the variables $x_{k1}, x_{k2} \cdots x_{kl}$ denote the neighboring node of x_k in the LM-based network and e is the random error. Subsequently, stepwise multiple variables selection is used to add new variables and eliminate insignificant variables. The significance of a variable is measured by a P-value which is determined from an F-test [19].

Pathway information is essential for successful quantitative modeling of biological systems [24]. A well-known pathway database that provides the information of metabolic, regulatory, and disease pathways is deposited in KEGG (Kyoto Encyclopedia of Genes and Genomes, http://www.genome.ad.jp/kegg) [25]. The relationship recorded in KEGG database is known to be special on the conception KEGG Orthology (KO, http://www.genome.jp/dbget-bin/get_htext? KO+-s+F+-f+F), a classification of orthologous genes that links directly to known pathways defined by KEGG [26]. In order to take further insights on the underlying biological meanings of networks, we map the LM- and LMMA-based networks to KEGG pathway database. First, the KO hierarchy and the known associations between genes and their corresponding KO functional terms from the KO dataset were extracted. Second, all the annotated genes from the KEGG Genes (KG) dataset are also extracted. Both the KO hierarchical and the KG hierarchical relations are employed as benchmarks to validate the interactions in the networks [26].

Angiogenesis is the process of generating new capillary blood vessels and a key issue for various disorders especially for a variety of solid tumors and vascular and rheumatoid disease [27]. Angiogenesis is considered to have much more significant impact on people worldwide than most other pathological processes [28]. So far, the molecular basis of angiogenesis remains unclear and it is a sticking point to uncover its underlying molecular mechanism and biological pathways [28]. To address such problem from a viewpoint of network, we reconstructed angiogenesis-oriented networks using both LM and LMMA approaches. First, we collected all the angiogenesis-related PubMed abstracts (till July 24, 2005) using "angiogenesis" as a keyword. A total of 23,497 "angiogenesis"-related PubMed abstracts were indexed automatically. By putting HUGO glossary into this abstract pool, 1,929 angiogenesis-related genes were obtained. A total of 9,514 co-citations among these

Table 11.2 LM-based and LMMA-based angiogenesis network structures [19]

	LM-EC	LMMA-EC	LM-ST	LMMA-ST
Common nodes[a]	1,257	1,031	1,258	1,162
Connections[a]	6,761	2,848	6,884	3,935
Average path length[a]	2.9810	3.6101	2.9741	3.3487
Average degree[b]	5.3738	2.2777	5.4722	3.1375
SSE[c]	522.3206	380.1941	520.2295	479.0745
SS$_{mse}$[c]	0.0669	0.057	0.0614	0.0589
Microarray size	1,257*53	1,257*53	1,258*119	1,258*119

[a]In the largest connected sub-network
[b]In the whole network
[c]All nodes except for the isolated ones

genes were extracted to construct the co-occurrence-based angiogenesis network. We constructed an LM-based network with a co-occurrence number of at least one. This resulted in the network with the maximum gene interactions. Next, we selected the gene expression profiles of endothelial cells (EC) and solid tumors (ST) from SMD. It is believed that EC is responsible for the generation of blood vessels and ST is the majority of angiogenesis-dependent diseases [27, 29]. The EC microarray dataset contained 44,639 genes and 53 experiments, while the ST microarray dataset contained 39,726 genes and 119 experiments. The largest connected gene network in LM with its genes identified in the EC microarray dataset was called LM-EC network (1,257 genes and 6,761 connections). Similarly, LM-ST network is the largest connected gene network in LM and its genes were from ST microarray dataset (1,258 genes and 6,884 connections). Accordingly, two LMMA-based angiogenesis networks, LMMA-EC and LMMA-ST, were built [19].

Table 11.2 lists the network parameters for LM- and LMMA-based angiogenesis networks. It shows that redundant connections are eliminated after multivariate selection. The connections for LMMA-EC and LMMA-ST networks are much smaller than that of the predominant sub-networks of LM-EC and LM-ST, respectively.

If the connections are eliminated, the average degree of genes will be dramatically decreased and node number and path length will be slightly reduced. Moreover, as shown in Fig. 11.2a and b, when comparing with the LM-random filtering networks derived from the permutation test, the LMMA network results in not only significantly larger cluster size ($P < 0.0001$ by Kolmogorov–Smirnov test) but also smaller path length of the largest cluster ($P < 0.001$ by t-test). The results demonstrate that LMMA is more stable and integrative than that of the LM-random filtering. Thus, LMMA seems to maintain the backbone of the LM-based angiogenesis network [19].

Angiogenesis networks constructed by LM and LMMA are tested for accuracy on confident sets of interactions. Both precision and recall rates are calculated against KEGG, one commonly used benchmark.

LMMA improves the precision rate when comparing with LM alone. On the other hand, as Bork et al. [31] suggested, the choice of benchmark set is still a knotty

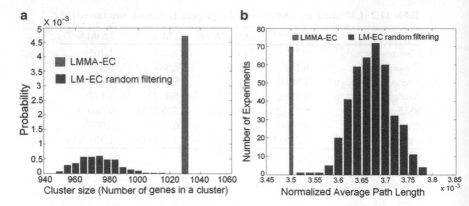

Fig. 11.2 (**a**) Comparison of the cluster sizes between the LMMA-EC network and the LM-EC random filtering networks ($P < 0.0001$ by Kolmogorov–Smirnov test). Other clusters are with <10 nodes (data not shown). (**b**) Comparison of the normalized average path length in the largest cluster between the LMMA-EC and the LM-EC random filtering networks ($P < 0.001$ by t-test) [30]

problem because the agreement among different benchmark sets is surprisingly poor [31]. Moreover, it is commonly known that co-occurrence in literature often describes or reflects more general relationships between genes. Some of these may be implicit and/or so novel that they have not yet reached the status of common knowledge or accepted fact that are often required for inclusion in databases such as KEGG. When calculated against KEGG, these two aspects may be the reason why both the LM and LMMA approaches led to a low recall rate (Fig. 11.3). Regardless, the integration with microarray data can significantly increase the reliability of gene co-occurrence networks extracted from the literature [19].

An EGF (epidermal growth factor) unit network derived respectively from the co-occurrence literature mining and the LMMA approaches is illustrated in Fig. 11.4. Moreover, 11 most statistically significant KEGG pathways such as MAPK and cytokine–cytokine receptor interaction pathways are extracted from the LMMA-based angiogenesis networks by Fisher Exact Test. And a multifocal signal modulation therapy for angiogenesis-related disorders could be proposed.

11.2.3 Biological Network Modularity and Phenotype Network

Many biological networks are complex and multilayered system that can be roughly disassembled into "network motif" in structure and "module" in function. Motifs represent recurrent topological patterns and modules are bigger building units that exhibit a certain functional autonomy. Network motif indicates the basic units organized into recurrent patterns of interconnections, which appear frequently throughout the network [32–34]. And the motifs cluster into semi-independent functional units called modules. In a variety of biological contexts, there exists

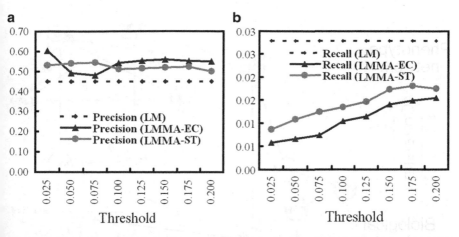

Fig. 11.3 Comparison of (**a**) precision and (**b**) recall in LM, LMMA-EC, and LMMA-ST angiogenesis networks at different thresholds. Here LM represents LM-EC and LM-ST since genes in LM-EC and LM-ST are identical when mapping to KEGG. The X axis denotes the P-value thresholds calculated from F-test in the step of statistical multivariate selection. Both the precision and the recall rates are calculated against KEGG [30]

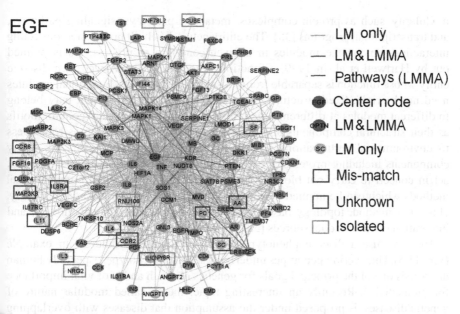

Fig. 11.4 An angiogenesis unit network. Totally 21 genes co-cited with EGF (epidermal growth factor) in LM are removed by LMMA. By manually revisiting the PubMed records, these 21 genes are found in false relations with EGF resulted from homonymic mismatches and confused lexical orders (in *blue* pane), unknown relations (in *purple* pane), and isolated relations (in *yellow* pane) [30] (Color figure online)

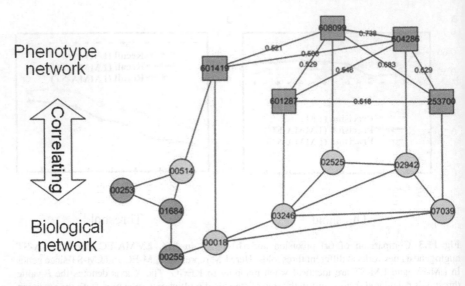

Fig. 11.5 Correlating phenotype network and biological network

modularity, such as protein complexes, metabolic pathways, signaling pathways, and transcriptional programs [35]. The entire network is built up by interconnecting interactions among the modules in a given condition or process. As is pointed out by Hartwell et al. in 1999, a functional module can be viewed as a discrete entity whose function is separable from those of other modules. Functional modules need not be rigid, fixed structures, for example, a given component may belong to different modules at different times [36]. Moreover, models may contain motifs as their structural components and maintain certain properties such as robustness to environmental perturbations and evolutionary conservations [35, 36]. Many components including proteins, DNA, and other molecules in a cellular network act in concert to carry out biological processes. There are several computational methods which can be applied in network modeling using "Omics" data, namely, clustering method, topology-based analysis, probabilistic graphical models, and integration of various data sources [35].

Here we use a disease phenotype-related biological module as an example (Fig. 11.5). Uncovering genotypes underlying specific phenotype, especially human disease, is one of the principal goals for genetics research and is of vital importance for biomedicine. Recently an interesting topic, the so-called modular nature of genetic diseases, is proposed under the assumption that diseases with overlapping clinical manifestations are caused by different gene mutations shared by the same functional module [37]. Lage et al. created a phenome–interactome network based on the data from both human protein complexes containing proteins involved in human diseases and quality-controlled protein interactions [38].

To further understand the disease-related module, we established CIPHER (Correlating Interactome and PHEnome to pRedict disease genes), a systems biology approach for phenotype–genotype association inference and disease gene discovery

(Fig. 11.5). CIPHER explains phenotype similarity using genotype distance in molecular network. By integrating protein interaction network, phenotype similarity network, and human disease network [39], CIPHER systematically quantifies the relevance between genes and diseases to prioritize candidate genes. We use CIPHER to infer genome-wide molecular basis for 1,126 disease phenotypes, systematically examining the genetic background of a wide spectrum of phenotypes. The resulting disease relevance profile is a rich resource for disease gene discovery, which can be used to guide the selection of disease candidates for gene mapping. It also provides comprehensive information on genetic overlaps of various complex phenotypes, which can be immediately exploited in the design of genetic mapping approaches that involve joint linkage or association analysis of multiple seemingly disparate phenotypes. Finally, clustering of the profile presents a global picture of the modularity of genetic diseases, suggesting the existence of disease module that comprises a set of functionally related genes and a set of genetically overlapped diseases, in which the gene set is highly relevant to the disease set.

Our disease relevance profile establishes a "multi-diseases multi-genes" view of genetic diseases and will facilitate the study of systems biomedicine [40]. We also believe the global concordance analysis may provide ways to better understand the association of different diseases, a holistic rule held in TCM as well, which places more emphasis on the association of phenotype and treats the same disease with different ZHENG while treating different diseases with the same ZHENG [30]. We shall explain this next.

11.3 TCM ZHENG-Related Network

The term "systems biology" is new but the idea of "system" in medicine is long-standing. TCM is a whole medical system, one of the "complete systems of theory and practice that have evolved independently from or parallel to allopathic (conventional) medicine" [41]. TCM is holistic with emphasis on regulating the integrity of the human body and treats the person as a whole, in essence in a systematic viewpoint. After the "Omics" revolution in modern life science, systems biology attempts to integrate various types of data in multiple levels and phases and has risen to be the key driver of the medical and biological sciences in the twenty-first century. Therefore, it is a golden opportunity now to uncover the essential principles of biological systems and applications backed up by in-depth understanding of system behaviors. The computational systems biology uses computational approaches to understand biological systems in system level. The previous difficulty in the direct combination of WM and TCM approaches may be overcome through the development of systems biology-based medicine approach which focuses on system, preventive, predictive, and personalized methods. Furthermore, the combination of modern WM and TCM can educe new concept and integrative medicine. It would be a great challenge for both TCM and modern systems biology (Fig. 11.6).

Fig. 11.6 Some agreements between systems biology-based medicine and TCM

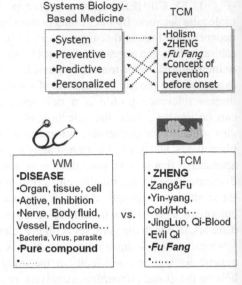

Fig. 11.7 TCM versus WM

11.3.1 "ZHENG" in TCM

Traditional Chinese medicine (TCM) is a system with its own rich tradition and over 3,000 years of continuous practice and refinement through observation, testing, and critical thinking [30]. There are many differences between TCM and Western medicine (WM) (Fig. 11.7). TCM can be characterized as holistic with emphasis on regulating the integrity of the human body and the interaction between human individuals and their environment. The fundamental belief in TCM is treating the person as a whole and searching for patterns of disharmony to put back into "Yin–Yang" balance. And TCM applies multiple natural therapeutic methods for patient management, with *Fu Fang* (herbal formula) as typical treatments [30].

Most importantly, "DISEASE" is the key pathological principle of WM, whereas TCM uses "ZHENG" as the key pathological principle to understand the human homeostasis and guide application of Chinese herbs. ZHENG, roughly described as syndrome of traditional Chinese medicine, is the basic unit and the key concept in TCM theory. All diagnostic and therapeutic methods in TCM are based on the differentiation of ZHENG, and this concept has been used for thousands of years in China [42]. Also, ZHENG is used as a guideline in disease classification in TCM [30].

ZHENG is characterized by the comprehensive analysis of "TCM phenotype" gained by four diagnostic methods, observation, listening, questioning, and pulse feeling, on the basis of governing exterior to infer interior. Thus, a same disease in WM may be diagnoses as different ZHENGs, while a same ZHENG may be present in different diseases. For example, Liu-Wei-Di-Huang pill could nourish Yin and restore balance; hence, it is always used to treat the Kidney-Yin deficiency ZHENG, which may be present in dozens of diseases.

ZHENG is also a leading exponent of Yin and Yang, the main principles of TCM holism. Yin is characterized by the property of Cold, weak, restricted, dark, heavy, turbid, downward, inward, and so forth. Yang is characterized by the property of Hot, excited, moving, strong, bright, light, clear, upward, outward, and so on. Yin and Yang represent the transfer between opposition and interdependence. Accordingly, the eight guiding principles for differentiating ZHENG are Yin and Yang, Exterior and Interior, Cold and Hot, and Deficiency and Excess.

Internationally, the practice of TCM, the main item of Complementary and Alternative Medicine (CAM), has risen dramatically in recent decades in the United States [43]. In 1992, US Congress established the Office of Unconventional Therapies, which later became the Office of Alternative Medicine (OAM), to explore "unconventional medicine practices." In 1998, OAM became the National Center for Complementary and Alternative Medicine (NCCAM). In 2006, an international cooperation in TCM is sponsored by China and EU. However, despite the risen practice, TCM still faces a lot of challenges. What is most important – the scientific basis of TCM is unclear and there still lack appropriate methods which could combine the WM and the TCM methods.

There are some clues that both TCM concept and practice may have potential biological basis, such as the well-known Yin–Yang principle. In 1995, Hunter indicated that processes reversibly controlled by protein phosphorylation require not only a protein kinase (PK) but also a protein phosphatase (PP) [44], which paralleled the ancient Yin–Yang principle. In 2003, Ou et al. attempted to define the physical meaning of Yin–Yang in TCM by correlating it with biochemical processes. They proposed that Yin–Yang balance is antioxidation–oxidation balance with Yin representing antioxidation and Yang as oxidation. Their proposal is partially supported by the fact that the antioxidant activity and polyphenolic contents of Yin tonic TCM herbs are much more than that of Yang tonic TCM herbs, say about six times [45]. In 2005, Lu et al. suggested that new insights into the "Yin and Yang" of neurotrophin activity could provide profound implications for the understanding of the role of neurotrophins in a wide range of cellular processes [46]. In 2007, Zhang applied Yin–Yang principle to illustrate the paradoxical roles and interplay of IFN-γ in inflammation and autoimmune disease by the familiar diagram of Yin and Yang, the "TaiJi" map [47]. Here Yang denotes an inflammatory process, while the Yin denotes the reduction of inflammation and activates a regulatory process [47].

Despite the new explanation of the ancient TCM concepts from biological basis, traditional Chinese medicine and Western medicine still face almost irreconcilable differences. Researches during the past decades suggested that there are more than single factors accounting for the molecular basis of TCM features such as Yin and Yang, ZHENG, as well as *Fu Fang* (herbal formula). Moreover, it is difficult to establish ZHENG-related animal models and hard to explain the "TCM phenotype." The biological mechanism for the relationship between ZHENG and ZHENG-oriented *Fu Fang* is also difficult to dig out, as Xue and Roy pointed out: "it follows that the reductionist approach of isolation of a single bioactive compound is not always appropriate for TCM" [48]. However, after the "Omics" revolution, it is a golden opportunity to uncover the essential principles of biological systems

Fig. 11.8 Cold–Hot ZHENG
and *Fu Fang*

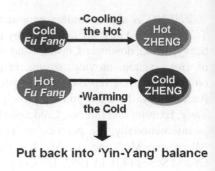

Put back into 'Yin-Yang' balance

and applications backed up by in-depth understanding of system behaviors, which also convert reductionism to system [49]. Accordingly, the culture and the research practice are both under the influence caused by the changes from reductionist paradigm to systems paradigm [50].

In the new era of science and technology, whether systems biology can bring TCM and WM together has become an issue of increasing concern. Someone believed that "the wind is now right for the development of traditional Chinese medicine" [51]. Leroy Hood, who is regarded as the systems biology's founding father, agreed that "It's conceivable that systems biology could find applications in trying to sort out components in Chinese herbal medicine, but it's very early days," and "It would be an enormous challenge at this point and time." Hood also suggested that there are barriers in contemplating the way to treat with TCM due to its complexity. For instance, the analysis of large and complex datasets and the accurate detection of metabolites, especially proteins, in blood require novel computational and detection approaches, respectively [52].

Although it is difficult to directly combine WM and TCM approaches, we believed that there may exist a way to combine them through the development of systems biology-based medicine approaches which toward the preventive, predictive, and personalized medicine in common (Figs. 11.7 and 11.8).

11.3.2   A CSB-Based Case Study for TCM ZHENG

Traditional Chinese medicine uses ZHENG as the key pathological principle to understand the human homeostasis and guide the applications of Chinese herbs [30]. If the way of treatment of WM can be described as "treatment according to DISEASE differentiation," then the way of treatment of TCM is the "treatment determination based on ZHENG differentiation." During the past decades, researchers have tried to explain the context of ZHENG on the biological basis and find there is more than a single factor involved in a ZHENG. Yet, it is difficult to establish ZHENG-related animal models and hard to explain the ZHENG-associated phenotype patterns. Thus, ZHENG is absent in most of the international TCM publications. Here we conducted a CSB-based case study for understanding TCM ZHENG [30].

Table 11.3 The major symptom profile terms of Cold ZHENG and Hot ZHENG

Subjects	Terms
Cold ZHENG-related symptom profile terms	Cold (chill, coldness), cold pain, tastelessness, clear abundant urine (clear urine in large amounts), loose stool, pale tongue, white fur (white moss), tight pulse, stringy pulse
Hot ZHENG-related symptom profile terms	Fever, heat (hot), diaphoresis, flushed face, burning pain, deep-colored urine, red eyes, thirst, desire for drinking, constipation, red tongue, dry tongue, thin fur (thin moss), yellow fur (yellow moss), rapid pulse

Firstly, we selected a pair of typical ZHENGs: "Cold" ZHENG and "Hot" ZHENG. The Cold ZHENG ("HAN ZHENG" in Mandarin) and Hot ZHENG ("RE ZHENG" in Mandarin) are two key statuses of ZHENG, which therapeutically direct the use of Chinese herbs in TCM. Cold ZHENG and Hot ZHENG are widely applied in the diagnosis and the treatment of patients suffering from inflammation, infection, stress, and immune disorders. As listed in Table 11.3, each ZHENG can be expressed as a certain symptom profile according to the authoritative and standard TCM terminology issued by State Administration Bureau of TCM [53] and State Bureau of Technical Supervision [54]. And the typical symptoms of Cold ZHENG are chill without fever, whereas the typical symptoms of Hot ZHENG are fever without chill.

According to the Yin–Yang theory in TCM, Cold ZHENG is also described as the preponderance of Yin or the deficiency of Yang, while Hot ZHENG is resulted from the preponderance of Yang or the deficiency of Yin. Both Cold and Hot are signs of the Yin–Yang imbalance. For restoring the Yin–Yang balance, many Chinese herbs are correspondingly categorized as either Hot-Cooling type or Cold-Warming type: Hot-Cooling herbs (i.e., Cold *Fu Fang*) are used to remedy Hot ZHENG, and Cold-Warming herbs (i.e., Hot *Fu Fang*) are used to remedy Cold ZHENG (Fig. 11.9).

Next, we attempt to explore the molecular basis of Cold–Hot ZHENG in the context of neuroendocrine–immune (NEI) system. In modern Western medicine (WM), NEI system is conceived as a pivot in modulating host homeostasis and naturally optimizing health through complex communications among chemical messengers (CMs), including hormones, cytokines, and neuron transmitters [30, 55, 56]. NEI system also acts as the host homeostasis mediator during the course of various body disorders, such as rheumatoid arthritis.

If CMs are considered as the biochemical ingredients of the NEI system, then genes that (directly or indirectly) encode these CMs can be considered as the genic ingredients of the NEI system. In 2006, we established a dbNEI database, which serves as a web-based knowledge resource special for the NEI systems. dbNEI collects 1,058 NEI-related signal molecules, 940 interactions between them, and 72 affiliated tissues from the Cell Signaling Networks database. We manually select 982 NEI papers from PubMed and give links to 27,848 NEI-related genes from UniGene database. NEI-related information, such as signal transductions, regulations, and control subunits, is all integrated (Fig. 11.9) [57].

Fig. 11.9 dbNEI (http://www.bmicc.org/web/english/search/dbnei)

dbNEI (http://www.bmicc.org/web/english/search/dbnei) provides a knowledge environment for understanding the main regulatory systems of NEI in molecular level. With a standardized inquiry platform, dbNEI smoothes the disunion of selections, structures, and information of NEI-related data among different databases. Through the unified inquiry interface offered by searching system, users can make a uniformed inquiry that facilitates the information mining. dbNEI deposits NEI-related productions, proteins, genes, their interactions, and the related information from public databases including CSNDB, UniGene, and OMIM. The references are also restricted to the citation in those cases from PubMed. Especially, dbNEI automatically visualizes the corresponding network according to the needs of users.

which will be helpful for the integration of the main physiological regulatory systems of nervous, endocrine, and immune, as well as their relationship with dysregulatory diseases [57].

Three characteristics make NEI a passable bridge for connecting WM and TCM and a background for understanding TCM ZHENG. First, the NEI system acts as the host homeostasis mediator during the course of various body disorders. Second, there are some clues that patients with Cold ZHENG and Hot ZHENG, two representative and mutually controlled ZHENGs, present abnormal NEI functions [58–60]. Third, NEI can be viewed as an existed experiential system for the future systematic biomedicine since systems biology currently is only able to handle relatively simple networks [61]. Therefore, we used NEI as an ideal context to explore the underlying molecular basis of ZHENG.

To perform the network analysis for ZHENG, we collected the Cold ZHENG-related symptom profile and Hot ZHENG-related symptom profile (Table 11.3). By using both ZHENG-related symptom profiles as keywords and NEI literature as background, we built two types of literature-derived networks for Cold–Hot ZHENG, respectively: a gene network using the HUGO genes and a CM network using NEI-related CMs. Each connection present in each network was manually validated. Through network analysis and topological comparison, we find interesting features of these two ZHENGs in the NEI system, which show a putative correspondence between ZHENGs and different aspects of the NEI system (Fig. 11.9). Hormones and immune factors are predominant in the Cold and Hot ZHENG networks, respectively, and these two networks are connected by neuron transmitters.

To evaluate the above findings through computational approach, we selected two groups of diseases with their typical symptom profiles corresponding to Cold ZHENG and Hot ZHENG, respectively, in the view of TCM [30]. We performed a pathway analysis for both groups of diseases and found that genes related to Hot ZHENG-related disease are mainly present in the cytokine–cytokine receptor interaction pathway whereas genes related to both the Cold ZHENG-related and Hot ZHENG-related diseases are linked to the neuroactive ligand–receptor interaction pathway [30]. Thus, gene investigation about ZHENG-related diseases further validates patterns derived from the literature-mining approach.

Furthermore, an experimental follow-up was conducted to investigate the effects of the ZHENG-oriented herbal treatments, Cold *Fu Fang* and Hot *Fu Fang*, on the hub nodes in Cold ZHENG and Hot ZHENG networks. Then, we found that the Cold ZHENG and Hot ZHENG network are scale-free. The number of edges a node has in a network is called the degree of that node [62–64], which indicates how many genes/CMs one gene/CM is related to. If the degree of a node is more than twofold of the median degree of all nodes in a network, this gene or CM is believed to play a critical role in the network structure, and it is regarded as a hub gene or a hub CM. Since the direct biological measurement of ZHENG is hardly available, we carried out an animal experiment to evaluate the effects of the Cold-Warming and Hot-Cooling TCM herbal formulas (CWHF and HCHF, respectively) on the key CMs of both the Cold ZHENG and Hot ZHENG networks. It is known that rheumatoid arthritis (RA), a long-term autoimmune disease that causes chronic inflammation of the joints and surrounding tissues, has a close relationship with the dysfunction of NEI system and can be divided into Cold ZHENG and Hot ZHENG in TCM [58, 65, 66]. Cold ZHENG-related RA and Hot ZHENG-related RA will be treated by Hot *Fu Fang* and Cold *Fu Fang*, respectively. Consequently, we choose RA as a disease model to explore ZHENG within the context of the NEI system and conducted experiments on the rat model of collagen-induced arthritis (CIA, the most widely used model for RA) [67, 68].

Experimental results show that the Cold ZHENG-oriented herbs (Hot *Fu Fang*) tend to affect the hub nodes in the Cold ZHENG network and the Hot ZHENG-oriented herbs (Cold *Fu Fang*) tend to affect the hub nodes in the Hot ZHENG network. The results are harmonious with the traditional ZHENG management philosophy of "Warming the Cold and Cooling the Hot" (Fig. 11.10) in TCM. Thus, the computational findings of Cold–Hot ZHENG-related network were subsequently verified [30]. This work suggests that the thousand-year-old concept of ZHENG may have a network system as its molecular basis. The preliminary study is only the first step to reveal the biological foundation of ZHENG and may help to design a tailored diagnosis and treatment for patients in the future. To sum up, CSB approaches can reach the underlying complex system for TCM ZHENG by integration of the computational prediction and experimental validation and thus lead to step-by-step improvements of the modernization of TCM.

Whole-network analyses are necessary to elucidate the global properties (Fig. 11.10). Moreover, studying Cold–Hot-related network dynamics and the corresponding functional emergence may contain four stages, i.e., molecular interaction, ZHENG network, ZHENG patients, and health and environment conditions. Recent genomic technologies have resulted in more detailed descriptions of signaling mechanisms, and reconstructions of ever-larger signaling networks have become possible. Such reconstructions will enable a systemic understanding of signaling network function, which is crucial for studying various diseases [69]. A systematic approach for the reconstruction and mathematical analysis of large-scale signaling networks requires the integration of events which happen at diverse spatiotemporal scales [70, 71]. The integrated genomic and proteomic analyses have many components, high interconnectivity, different spatiotemporal scales, and signaling networks under complex controls [72].

On this account, we provided a preliminary entropy-based, holistic model for evaluating and unveiling the rule of TCM life system. The change of entropy of open systems has two parts: entropy flow and entropy production. Based on the nonequilibrium thermodynamics, it has

$$dS = d_e S + d_i S,$$

where $d_e S$ is the transfer of entropy across the boundaries of the system and the $d_i S$ is the entropy produced within the systems [73].

Last but not least, it is hopeful to understand TCM, an ancient systematic medicine, from the viewpoints of informatics and systems.

a Network structure of CM-based ZHENG network

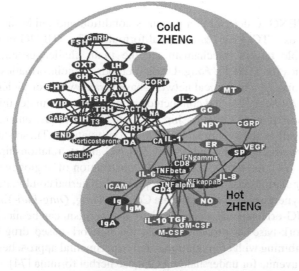

b CM-based ZHENG network

c Gene-based ZHENG network

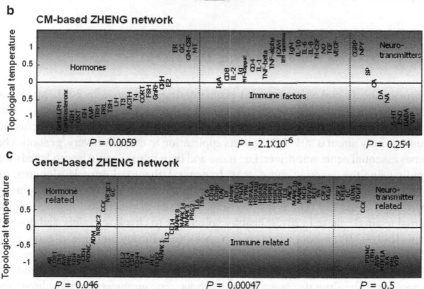

Fig. 11.10 Network patterns of Cold–Hot ZHENG. (**a**) CM-based ZHENG network: Color of a node signifies its ZHENG state (*blue*, Cold ZHENG; *red*, Hot ZHENG; *purple*, both Cold and Hot ZHENGs. The same to (**b**). Network can be divided into two major network clusters: Cold ZHENG network (composed of all Cold ZHENG-related nodes) and Hot ZHENG network (composed of all Hot ZHENG-related nodes). The Yin–Yang background here is to indicate the possible relationship between Cold–Hot ZHENG networks and Yin–Yang theory of TCM; a mutual transformation of both may occur in given quantitative changes. (**b**) "Topological temperature" (a transform of the topology distance between Cold ZHENG and Hot ZHENG networks) of each node in the CM-based Cold ZHENG (−1 to 0) and Hot ZHENG (0–1) networks. (**c**) Gene-based Cold ZHENG (−1 to 0) and Hot ZHENG (0–1) networks. Center of the CM name or gene name corresponds to its topological temperature [30]

11.4   Network-Based Study for TCM "Fu Fang"

TCM uses ZHENG to determine a patient's condition and guide the applications of Chinese herbs – TCM *Fu Fang* (herbal formula). The ZHENG-oriented effects and the multiple targets' mechanism are the main challenges encountered by recent researches for TCM *Fu Fang*. Using methods of bioinformatics and systems biology, we proposed a biological network-based framework for understanding the mechanism of *Fu Fang* [74] and conducted some case studies under this framework which aimed to explore the relationship between TCM *Fu Fang* and corresponding ZHENGs, as well as the synergism of herbal combinations. These studies include the network construction for Cold or Hot ZHENG and its relationship with Hot or Cold *Fu Fang* [74], the biological network construction of angiogenesis [19], and the network regulation-based emergent property of an herbal combination with anti-angiogenesis synergism extracting from a Cold *Fu Fang*, *Qing-Luo-Yin*. It is shown that the ZHENG-oriented effects and the herbal synergism can be nicely explicated by such network-based approaches. Thus, the network-based drug combination discovery, combining with computational and experimental approaches, is expected to open a new avenue for understanding Chinese herbal formula [74].

11.4.1   Systems Biology in Drug Discovery

Pharmaceuticals ranked as the most profitable industry again in 2001. More than 23,000,000 molecule and compound had been discovered till December 31, 1999 according to the Chemical Abstracts (CA, USA), while the proportion of effective drug was only about 0.001 %. Due to its application in drug discovery, genomics becomes essential economic driver, i.e., more and better drugs could be produced more rapidly using the genomic approaches. In general, the usual drug development life cycle takes about 7–15 years; say 2–10 years in the discovery where bioinformatics approaches could aid. The following stages consist of preclinical testing (lab and animal testing), phase I (20–30 healthy volunteers to check for safety and dosage), phase II (100–300 patient volunteers to check for efficacy and side effects), phase III (1,000–5,000 patient volunteers to monitor reactions to long-term drug use), FDA review and approval, and post-marketing testing. Many approaches could help to accelerate the drug discovery stage, such as gene engineering, high-throughput screening, computer-aided drug design, combinatorial chemistry, chemical biology, bioinformatics, and systems biology.

Systems biology looks to bridge the molecular and the physiological biology and has the potential to affect the entire drug discovery and development process [75–77]. It is useful in developing biomarkers for efficacy and toxicology which could lead to more efficient preclinical developments, screening for combination of drug targets and elucidating side effects in drugs, and so on. In drug discovery, the existing drug may have poor specificity and high side effects, and the detailed cell regulatory network model elucidated by systems biology could help to further understand the drug mechanism. Butcher et al. proposed that when applied to known, validated targets, the current "win-by-number" approach is very powerful. Disappointingly, when applied to less well biologically understood targets, for example, genome-derived targets, such an approach has led to few new drugs. In fact, the cost of the discovery of new drug keeps rising whereas the approval rates decline, given the great progresses in genomics and high-throughput screening technologies during the last two decades [75].

Hopefully, drug discovery through systems biology may significantly bring down the cost and time of new drug development. Various types of data, such as those from literatures, gene expression, protein interactions, potential functional network organization, and cell response information from human cell-based assays, could all be integrated to generate the model of disease. Models are iteratively evaluated and improved by comparison between the predictions and systems-level responses, including cell, tissue, and organism-level responses, that are measured by traditional experimental assays as well as profiles from human cell mixtures under distinct experimental conditions. Component level of "Omics" data could provide a scaffold, which limits the range of possible models at the molecular level [75]. Furthermore, assessing the effects of drugs or genes simultaneously is helpful for the uncovering of the complex biological responses to diverse diseases and the screening for new therapeutic clews. Once applied to biomedical research, systems biology could provide a novel way for resolving some of the most difficult problems which block the development of modern drug discovery: (a) rational target identification and recognition, (b) avoiding side effects of therapeutics before commitment to clinical trials, and (c) the development of the alternative markers to monitor the clinical efficacy and personal disease treatment.

11.4.2   Network-Based Drug Design

Many complex diseases are resulted from the deregulation of biological network, as mentioned in Sect. 1.2.2. Take cancer as an example. Cancer can be regarded as a systems biology disease [78]. Robustness is the ability to maintain stable functions in response to various perturbations. Robustness is highlighted by recent systems-level studies, such as the bacterial chemotaxis, the cell cycle, the circadian rhythms, the tolerance of stochastic fluctuations in fundamental biochemical processes – transcription and protein interactions – and large-scale biochemical networks [79]. However, robustness not only could keep homeostasis in organisms but also may maintain dysfunction under the same mechanism, as occurs in tumor resistance to anticancer drugs. Therefore, robustness could not be regulated without deep understanding and comprehensive study of system dynamics. For example, the cell cycle will become fragile and is easily disrupted by minor perturbations if certain feedback loops are removed; nevertheless, similar observation could be got by carefully modulating certain feedback loops without completely eliminating them.

The robustness of cancer comes from its heterogeneity, redundancy, and feed-back. The complexity of cancer results from the large number of interacting molecules, cross talk between pathways, feedback circuitry, and the nonlinear relations between the interacting molecules and the spatiotemporal resolution of signaling, which make it difficult if not impossible to predict the altered outcome of signaling on the basis of the changes in an interaction map alone [78]. Many cancer genes are known, as well as the composition of many pathways in which their gene products functioning and interacting between pathways [80]. The interconnections between the pathways give rise to large complex signaling networks. Mutations which cause cancer frequently occur directly in components of these pathways or in genes that indirectly affect the functioning of these pathways. Cancer studies have primarily taken the form of intensive investigation of individual genes and a small number of interactions between genes.

The aim of systems biology is to describe all the elements of the system, identify networks that correlate the biological elements of a system, and characterize the information flow that links these elements and networks into an emergent biological process [76]. Treating disease as the operation of perturbed networks, the network-based multicomponent drugs can be developed accordingly. Through a comparison between normal and disease network, critical nodal points (proteins) can be identified which are likely to reconfigure the perturbed network structure back toward its normal state or specifically kill the diseases cell. These nodal proteins represent potential drug targets [76].

As Keith et al. indicated "It is now time to revisit past experiences to identify multicomponent therapeutics for the treatment of complex diseases," the reduc-tionist approach, one-target, one-drug paradigm, is hard to exploit the network complexity and pathway redundancy that is engineered into living systems through evolutionary processes [81]. In fact, the network-based and molecule-based drug design approaches complement each other, i.e., network-based approach locates best targets and molecule-based approach makes the drug.

11.4.3 Progresses in Herbal Medicine

Analyses of the chemical components in TCM herbs have made great progresses. By individual or combined use of spectrum, chromatography, mass spectrometry, NMR, X-ray diffraction, and so on, Chinese herbs could be identified and qualified [82]. In the meanwhile, new drugs can be discovered through the testing of bioactive compounds of TCM herbs, with two outstanding examples as artemisinin and arsenic trioxide (As_2O_3). Li et al. reported that the addition of Chinese herbs artemisinin could greatly increase the rate of parasite clearance with no additional side effects [83]. Shen et al. reported that arsenic compounds, mainly arsenic trioxide (As_2O_3), an ancient drug used in the traditional medicine, yield a high-quality remission and survival in newly diagnosed acute promyelocytic leukemia [84]. Besides, Hoessel et al. identified indirubin, the active ingredient of Danggui

Longhui Wan which is a mixture of plants used in traditional Chinese medicine to treat chronic diseases, and its analogues as potential inhibitors of cyclin-dependent kinases (CDKs) [85]. Kong et al. identified berberine (BBR), a compound isolated from a Chinese herb, as a new cholesterol-lowering drug [86]. However, many other herbal extracts lose their activities when the extracts are fractionated into individual chemical components [75, 81]. Thus, whether such a reductionist approach can be widely applied to all TCM, especially help the modernization of *Fu Fang*, needs more consideration.

11.4.4 TCM Fu Fang (Herbal Formula)

TCM is characterized by the use of *Fu Fang* (herbal formula), mixtures of natural herbs and herbal extracts. *Fu Fang* is the integral treatment based on ZHENG differentiation and there are more than 100,000 *Fu Fang* in TCM. The elementary theory of herbs consists of (1) nature and flavor of Chinese herbs, (2) compatible application of herbs, and (3) principle–method–recipe–medicine. In TCM, herb has four natures, i.e., Cold, Hot, Warm, and Cool, and those without obvious nature are called Calm or Plain; it has five flavors, i.e., sour, bitter, sweet, pungent, salty, and tasteless. *Fu Fang* could be classified into Sovereign herb, Minister herb, Assistant herb, and Envoy herb. The seven relations in the compatible application of *Fu Fang* are herb used singly, mutual used singly, mutual promotion, incompatibility, counteract toxicity of another herb, mutual inhibition, and antagonism.

The systems biology approach is considered to have the potential to revolutionize natural product research and to advance the development of modern science-based herbal medicine [87]. For instance, Wang et al. [88] used metabolomics, the ultimate phenotyping, to study *Fu Fang* and proposed that it may open up the possibility of studying the effect of complex mixtures, such as those used in TCM, in complex biological systems.

11.4.5 A Network-Based Case Study for TCM Fu Fang

Angiogenesis, the aberrant growth of blood vessels, is a major pathological process of various diseases such as cancer and rheumatoid arthritis (Fig. 11.11). There are few other processes having such important impact as angiogenesis on the well-being of so many people worldwide [29]. Angiogenesis as well as its related disorders is also one of our principal objects of study.

Keeping angiogenesis under tight regulation is crucial to maintain normal physiological function and health. The local disequilibrium between positive and negative regulators of the growth of microvessels may result in the angiogenic phenotype switch. Angiogenesis could be found in physiological exceptions such as in the female reproductive system as well as during wound healing. The challenge

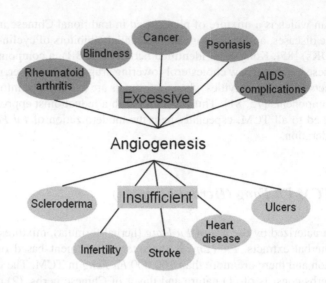

Fig. 11.11 Some of the angiogenesis-related diseases

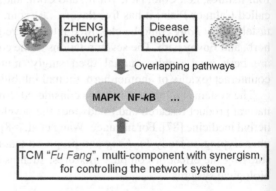

Fig. 11.12 Disease and ZHENG can be combined in the network level. Network-based disease or ZHENG system is suitable for multiple-target inference such as TCM *Fu Fang*

for understanding the molecular, genetic, and cellular mechanisms of angiogenesis is to define the molecular basis and pathways of angiogenic disorders in greater detail and in a more integrated manner [29]. Thus, the excitement of the science can be converted into the development of efficient, safe therapies. As described previously, we constructed an "angiogenesis" network with pathway annotations by using LMMA approach, which is helpful for analyzing the interactions of multiple pathways and providing the multiple targets for such complex biological processes [19]. By comparison between angiogenesis-related network [19] and TCM ZHENG-related network [30], it is shown that disease (e.g., angiogenesis) and ZHENG (e.g., Cold–Hot ZHENG) can be combined in the network level and some common pathways are identified (Fig. 11.12).

The following question is "how to control the network-based disease or ZHENG?" We believed that such complex systems are more suitable to be subjected to multiple-target inference. There is widespread evidence that combinations of compounds can be more effective than the sum of the effectiveness of the individual agents [81]. Csermely et al. suggested that partial inhibition of a surprisingly small number of targets can be more efficient than the complete inhibition of a single target based on network models [88]. In order to use the network efficiency as a measure for drug efficiency, two assumptions are necessarily made, namely, (1) a mechanism that is targeted by a drug could be represented as a network and (2) all elements of this network need to interact for the function of the targeted mechanisms [88]. The future success of the network-based drug design paradigm will depend on new models to identify the correct multiple targets and their multi-fitting, low-affinity drug candidates as well as on more efficient in vivo testing [88].

TCM has thousands of years experienced in *Fu Fang* and the multiple-target inference. Thus, TCM provides fertile ground for modern systems-based drug development, as long as it could pass along a pathway of discovery, isolation, and mechanistic studies before eventual deployment in the clinic [89]. In TCM, angiogenesis can be attributed to the dysfunction of "Luo" (venation). In general, the "Luo" in human body is just like veins in a leaf blade. The "Luo" suffering from pathogenic Cold or Hot is responsible for many long-time diseases. We identified the suppressive effects of a Cold *Fu Fang*, Chinese herbal medicine *Qing-Luo-Yin* (QLY) extracts on the angiogenesis of collagen-induced arthritis in rats, which indicates the *Qing-Luo-Yin*'s anti-angiogenesis effects [90]. This has been evaluated as a representative TCM *Fu Fang* with anti-angiogenesis effect.

An anti-angiogenesis medicine is an important way of treatment by blocking pathological changes of rheumatoid arthritis or "starving-to-death" tumor. For treatment of RA, major medicines include hormones, gold salts, nonsteroidal anti-inflammatory drugs (NSAIDs), and Chinese patent medicines. The first three kinds of medicines cannot block development of RA, and long-term administration of them will produce substantive side effect. QLY is a traditional Chinese medicine recipe for the treatment of rheumatoid arthritis. It is mainly composed of four herbs and has the effect of anti-inflammatory, reducing synovial capillary vessel angiogenesis, preventing formation of pannus, alleviating cartilage damage, and reducing secretory function and metabolic activity of synovial cell hyperfunction. It also shows remarkable effect in suppressing angiogenesis in terms of molecular, organism, and pathology viewpoints, and it is well recognized as one of the representative research results of Chinese medicine recipes for treating angiogenesis [90]. The traditional Chinese medicine QLY, having prominent characteristics and addressing both the symptoms and root causes for treating angiogenesis-related disease (such as rheumatoid arthritis), will obtain stronger momentum and wider prospect.

However, the ingredients of QLY are very complex and there are complex interactions among the effect of individual ingredient; hence, the pharmacological effect of QLY is not produced by the simple addition of its component. We identified a combination of two substances from the decoction of QLY, which has synergic actions of efficacy enhancing, toxicity reducing, and clear substance basis and maintains the characteristic of traditional Chinese medicines of recipe association.

Such a traditional Chinese herb-derived composition can be used as effective constituent of a medicine with anti-angiogenesis synergy effect and applied for the invention patent both in China (2006) and the USA (2007). Furthermore, we found that such a combination of two substances from QLY can regulate the TNF-alpha. Although this study only reaches a few aspects of QLY in regard to corresponding network, it arrived at two meaningful conclusions related to the biological basis of TCM. First, network-based system can help to explain the effect of TCM *Fu Fang*. Second, the synergy effect of herbal combination may come from the network interactions [90].

In practice, unlike allopathic medicine and its usual one-drug–one-disease approach, TCM makes use of complex herbal mixtures for the tailored and holistic treatment of patient given his ZHENG conditions. To find out how TCM works and facilitates a better understanding of TCM, many new plans are sponsored, such as "Outline of TCM Innovation and Development Plan (2006–2020)" (issued by MOST joined other 15 departments including MOH, SATCM, and SFDA, China) and "Outline for TCM International Science & Technology Cooperation (2006–2020)" (issued by MOST, MOH, and SATCM, China). Many international cooperation and research such as the EU–China work program named "Traditional Chinese Medicine in the post-genomic era" are also launched. The goal of these plans focuses on the development of new methodologies, which can be used to test the traditional features and principles of TCM, and places particular emphasis on the relation between ZHENG and *Fu Fang*. Our preliminary investigations described in this chapter indicate that bioinformatics and computational systems biology may help the modernization of TCM. And the ongoing purposes in this cross field, namely, "traditional Chinese medicine bioinformatics (TCMB)" and "TCM computational systems biology," include the integration of TCM and WM, future personalized medicine, and multicomponent drug development.

References

1. Hood L, Heath JR, Phelps ME, Lin B (2004) Systems biology and new technologies enable predictive and preventative medicine. Science 306(5696):640–643
2. Hiesinger PR, Hassan BA (2005) Genetics in the age of systems biology. Cell 123(7): 1173–1174
3. Subramanian A, Tamayo P, Mootha VK, Mukherjee S, Ebert BL et al (2005) Gene set enrichment analysis: a knowledge-based approach for interpreting genome-wide expression profiles. Proc Natl Acad Sci 102(43):15545–15550
4. Lu LJ, Sboner A, Huang YJ, Lu HX, Gianoulis TA et al (2007) Comparing classical pathways and modern networks: towards the development of an edge ontology. Trends Biochem Sci 32(7):320–331
5. Barabasi A-L, Oltvai ZN (2004) Network biology: understanding the cell's functional organization. Nat Rev Genet 5(2):101–113
6. Shatkay H, Feldman R (2003) Mining the biomedical literature in the genomic era: an overview. J Comput Biol 10(6):821–855

7. Jenssen T-K, Laegreid A, Komorowski J, Hovig E (2001) A literature network of human genes for high-throughput analysis of gene expression. Nat Genet 28(1):21–28
8. Jensen LJ, Saric J, Bork P (2006) Literature mining for the biologist: from information retrieval to biological discovery. Nat Rev Genet 7(2):119–129
9. Calvano SE, Xiao W, Richards DR, Felciano RM, Baker HV et al (2005) A network-based analysis of systemic inflammation in humans. Nature 437(7061):1032–1037
10. Stapley BG, Benoit G (2000) Biobibliometrics: information retrieval and visualization from co-occurrences of gene names in Medline abstracts. Pac Symp Biocomput 2000:529–540
11. Zhu S, Okuno Y, Tsujimoto G, Mamitsuka H (2005) A probabilistic model for mining implicit 'chemical compound-gene' relations from literature. Bioinformatics 21(Suppl 2):ii245–ii251
12. Zhang C, Li S (2004) Modeling of neuro-endocrine-immune network via subject oriented literature mining. Proc BGRS 2:167–170
13. Someren EPV, Wessels LFA, Backer E, Reinders MJT (2002) Genetic network modeling. Pharmacogenomics 3(4):507–525
14. de Jong H (2002) Modeling and simulation of genetic regulatory systems: a literature review. J Comput Biol 9(1):67–103
15. Liang S, Fuhrman S, Somogyi R (1998) Reveal, a general reverse engineering algorithm for inference of genetic network architectures. Pac Symp Biocomput 1998:18–29
16. Le Phillip P, Bahl A, Ungar LH (2004) Using prior knowledge to improve genetic network reconstruction from microarray data. In Silico Biol 4(3):335–353
17. Kuffner R, Fundel K, Zimmer R (2005) Expert knowledge without the expert: integrated analysis of gene expression and literature to derive active functional contexts. Bioinformatics 21(Suppl 2):ii259–ii267
18. Wu L, Li S (2005) Combined literature mining and gene expression analysis for modeling neuro-endocrine-immune interactions. In: Advances in intelligent computing 2005. Springer, Heidelberg/Berlin, pp 31–40
19. Li S, Wu L, Zhang Z (2006) Constructing biological networks through combined literature mining and microarray analysis: a LMMA approach. Bioinformatics 22(17):2143–2150
20. Sherlock G, Hernandez-Boussard T, Kasarskis A, Binkley G, Matese JC et al (2001) The Stanford microarray database. Nucleic Acids Res 29(1):152–155
21. D'Haeseleer P, Liang S, Somogyi R (2000) Genetic network inference: from co-expression clustering to reverse engineering. Bioinformatics 16(8):707–726
22. Ge H, Liu Z, Church GM, Vidal M (2001) Correlation between transcriptome and interactome mapping data from *Saccharomyces cerevisiae*. Nat Genet 29(4):482–486
23. D'Haeseleer P, Wen X, Fuhrman S, Somogyi R (1999) Linear modeling of mRNA expression levels during CNS development and injury. Pac Symp Biocomput 1999:41–52
24. Cary MP, Bader GD, Sander C (2005) Pathway information for systems biology. FEBS Lett 579(8):1815–1820
25. Goto M, Kanehisa S (2000) KEGG: Kyoto encyclopedia of genes and genomes. Nucleic Acids Res 28(1):27–30
26. Mao X, Cai T, Olyarchuk JG, Wei L (2005) Automated genome annotation and pathway identification using the KEGG Orthology (KO) as a controlled vocabulary. Bioinformatics 21(19):3787–3793
27. Folkman J (1995) Angiogenesis in cancer, vascular, rheumatoid and other disease. Nat Med 1(1):27–30
28. Carmeliet P, Jain RK (2000) Angiogenesis in cancer and other diseases. Nature 407(6801): 249–257
29. Carmeliet P (2003) Angiogenesis in health and disease. Nat Med 9(6):653–660
30. Li S, Zhang ZQ, Wu LJ, Zhang XG, Li YD et al (2007) Understanding ZHENG in traditional Chinese medicine in the context of neuro-endocrine-immune network. IET Syst Biol 1(1):51–60
31. Bork P, Jensen LJ, von Mering C, Ramani AK, Lee I et al (2004) Protein interaction networks from yeast to human. Curr Opin Struct Biol 14(3):292–299

32. Milo R, Shen-Orr S, Itzkovitz S, Kashtan N, Chklovskii D et al (2002) Network motifs: simple building blocks of complex networks. Science 298(5594):824–827
33. Shen-Orr SS, Milo R, Mangan S, Alon U (2002) Network motifs in the transcriptional regulation network of *Escherichia coli*. Nat Genet 31(1):64–68
34. Lee TI, Rinaldi NJ, Robert F, Odom DT, Bar-Joseph Z et al (2002) Transcriptional regulatory networks in *Saccharomyces cerevisiae*. Science 298(5594):799–804
35. Qi YHG (2006) Modularity and dynamics of cellular networks. PLoS Comput Biol 2(12):e174
36. Hartwell LH, Hopfield JJ, Leibler S, Murray AW (1999) From molecular to modular cell biology. Nature 402:47–52
37. Oti M, Brunner HG (2007) The modular nature of genetic diseases. Clin Genet 71:1–11
38. Lage K, Karlberg EO, Storling ZM, Olason PI, Pedersen AG et al (2007) A human phenome-interactome network of protein complexes implicated in genetic disorders. Nat Biotechnol 25(3):309–316
39. Goh K-I, Cusick ME, Valle D, Childs B, Vidal M et al (2007) The human disease network. Proc Natl Acad Sci 104(21):8685–8690
40. Wu X, Zhang M, Li S (2007) Global inference of human phenotype-genotype associations by integrating interactome, phenome and diseasome. In: The 8th international conference on systems biology, Long Beach, 1–6 Oct 2007
41. FDA (2006) Draft guidance for industry: complementary and alternative medicine products and their regulation by the Food and Drug Administration. FDA, Rockville
42. Gu CD (1956) The inner classic of the yellow emperor, essential questions (Huangdi Neijing, Suwen). People's Medical Publishing House, Beijing, pp 186–191
43. Eisenberg DM, Davis RB, Ettner SL, Appel S, Wilkey S et al (1998) Trends in alternative medicine use in the United States, 1990–1997: results of a follow-up national survey. JAMA 280:1569–1575
44. Hunter T (1995) Protein kinases and phosphatases: The Yin and Yang of protein phosphorylation and signaling. Cell 80(2):225–236
45. Ou B, Huang D, Hampsch-Woodill M, Flanagan JA (2003) When east meets west: the relationship between yin-yang and antioxidation-oxidation. FASEB J 17(2):127–129
46. Lu B, Pang PT, Woo NH (2005) The yin and yang of neurotrophin action. Nat Rev Neurosci 6(8):603–614
47. Zhang J (2007) Yin and yang interplay of IFN-γ in inflammation and autoimmune disease. J Clin Invest 117(4):871–873
48. Xue TRR (2003) Studying traditional Chinese medicine. Science 300(5620):740–741
49. Kitano H (ed) (2001) Foundations of systems biology. The MIT Press, Cambridge, MA
50. Palsson B (2000) The challenges of in silico biology. Nat Biotechnol 18(11):1147–1150
51. Qiu J (2007) China plans to modernize traditional medicine. Nature 446(7136):590–591
52. Qiu J (2007) Traditional medicine: a culture in the balance. Nature 448(7150):126–128
53. State Administration of Traditional Chinese Medicine (SATCM) of The People's Republic of China (ed) (1994) The criteria of diagnosis and therapeutic effect of diseases and ZHENGs in TCM. Nanjing University Press, Nanjing, p 29
54. China State Bureau of Technical Supervision (ed) (1997) The clinic terminology of traditional Chinese medical diagnosis and treatment – ZHENG. Standard Press of China, Beijing, p 52
55. Besedovsky H, Sorkin OE (1977) Network of immune-neuroendocrine interactions. Clin Exp Immunol 27:1–12
56. Roth J, LeRoith D, Shiloach J, Rosenzweig JL, Lesniak MA et al (1982) The evolutionary origins of hormones, neuro-transmitters, and other extracellular chemical messengers: implications for mammalian biology. N Engl J Med 306:523–527
57. Zhuang Y, Li S, Li Y (2006) dbNEI: a specific database for neuro-endocrine-immune interactions. Neuroendocrinol Lett 27(1/2):53–59
58. Li S (2002) Advanced in TCM symptomatology of rheumatoid arthritis. J Tradit Chin Med 22:137–142
59. Hsu CH, Yu MC, Lee TC, Yang YS (2003) High eosinophil cationic protein level in asthmatic patients with "Heat", Zheng. Am J Chin Med 31:277–283

60. Zhang X, Ji B, Chen B, Xie Z (1999) Relationship of cytokines and cold-heat syndrome differentiation in patients of duodenal ulcer. Zhongguo Zhong Xi Yi Jie He Za Zhi 19: 267–269
61. Pennisi E (2005) How will big pictures emerge from a sea of biological data? Science 309:94
62. Song C, Havlin S, Makse HA (2005) Self-similarity of complex networks. Nature 433:392–395
63. Albert R, Jeong H, Barabasi AL (2000) Error and attack tolerance of complex networks. Nature 406:378–382
64. Jeong H, Tombor B, Albert R, Oltvai ZN, Barabasi AL (2000) The large-scale organization of metabolic networks. Nature 407:651–654
65. Wilder RL (2002) Neuroimmunoendocrinology of the rheumatic diseases, past, present, and future. Ann N Y Acad Sci 966:13–19
66. Choy EH, Panayi SGS (2001) Cytokine pathways and joint inflammation in rheumatoid arthritis. N Engl J Med 344:907–916
67. Trentham DE, Townes AS, Kang AH (1977) Autoimmunity to type II collagen an experimental model of arthritis. J Exp Med 146(3):857–868
68. Staines NA, Wooley PH (1994) Collagen arthritis – what can it teach us? Br J Rheumatol 33:798–807
69. Papin JA, Hunter T, Palsson BO, Subramaniam S (2005) Reconstruction of cellular signalling networks and analysis of their properties. Nat Rev Mol Cell Biol 6(2):99–111
70. Weng G, Bhalla US, Iyengar R (1999) Complexity in biological signaling systems. Science 284(5411):92–96
71. Sivakumaran S, Hariharaputran S, Mishra J, Bhalla US (2003) The database of quantitative cellular signaling: management and analysis of chemical kinetic models of signaling networks. Bioinformatics 19(3):408–415
72. Ge H, Walhout AJM, Vidal M (2003) Integrating 'omic' information: a bridge between genomics and systems biology. Trends Genet 19(10):551–560
73. Kang G-L, Li S, Zhang J-F (2007) Entropy-based model for interpreting life systems in traditional Chinese medicine. eCAM 5(3):273–279
74. Li S (2007) Framework and practice of network-based studies for Chinese herbal formula. J Chin Integr Med 5(5):489–493
75. Butcher EC, Berg EL, Kunkel EJ (2004) Systems biology in drug discovery. Nat Biotechnol 22(10):1253–1259
76. Hood L, Perlmutter RM (2004) The impact of systems approaches on biological problems in drug discovery. Nat Biotechnol 22(10):1215–1217
77. Butcher EC (2005) Can cell systems biology rescue drug discovery? Nat Rev Drug Discov 4:461–467
78. Hornberg JJ, Bruggeman FJ, Westerhoff HV, Lankelma J (2006) Cancer: a systems biology disease. Biosystems 83(2–3):81–90
79. Kitano H (2004) Cancer as a robust system: implications for anticancer therapy. Nat Rev Cancer 4(3):227–235
80. Hanahan D, Weinberg RA (2000) The hallmarks of cancer. Cell 100(1):57–70
81. Keith CT, Borisy AA, Stockwell BR (2005) Multicomponent therapeutics for networked systems. Nat Rev Drug Discov 4(1):71–78
82. Sinha G (2005) Medicine of the masses. Nat Med 11:9–10
83. Li GQ, Arnold K, Guo XB, Jian HX, Fu LC (1984) Randomised comparative study of mefloquine, qinghaosu, and pyrimethamine-sulfadoxine in patients with falciparum malaria. Lancet 2(8416):1360–1361
84. Shen Z-X, Shi Z-Z, Fang J, Gu B-W, Li J-M et al (2004) Inaugural article: all-trans retinoic acid/As$_2$O$_3$ combination yields a high quality remission and survival in newly diagnosed acute promyelocytic leukemia. Proc Natl Acad Sci 101(15):5328–5335
85. Hoessel R, Leclerc S, Endicott JA, Nobel MEM, Lawrie A et al (1999) Indirubin, the active constituent of a Chinese antileukaemia medicine, inhibits cyclin-dependent kinases. Nat Cell Biol 1(1):60–67
86. Kong W, Wei J, Abidi P, Lin M, Inaba S et al (2004) Berberine is a novel cholesterol-lowering drug working through a unique mechanism distinct from statins. Nat Med 10(12):1344–1351

87. Wang M, Lamers R-JAN, Korthout HAAJ, Nesselrooij JHJV, Witkamp RF et al (2005) Metabolomics in the context of systems biology: bridging traditional Chinese medicine and molecular pharmacology. Phytother Res 19(3):173–182
88. Csermely P, Agoston V, Pongor S (2005) The efficiency of multi-target drugs: the network approach might help drug design. Trends Pharmacol Sci 26(4):178–182
89. Corson TW, Crews CM (2007) Molecular understanding and modern application of traditional medicines: triumphs and trials. Cell 130(5):769–774
90. Li S, Lu A-P, Wang Y-Y, Li Y-D (2003) Suppressive effects of a Chinese herbal medicine qing-luo-yin extract on the angiogenesis of collagen-induced arthritis in rats. Am J Chin Med 31:713–720

Chapter 12
Advanced Topics in Bioinformatics and Computational Biology

Bailin Hao, Chunting Zhang, Yixue Li, Hao Li, Liping Wei, Minoru Kanehisa, Luhua Lai, Runsheng Chen, Nikolaus Rajewsky, Michael Q. Zhang, Jingdong Han, Rui Jiang, Xuegong Zhang, and Yanda Li

12.1 Prokaryote Phylogeny Meets Taxonomy

Bailin Hao (✉)
Fudan University, Shanghai, China
The Santa Fe Institute, Santa Fe, NM, USA

Phylogeny defined as the context of evolutionary biology is the connections between all groups of organisms as understood by ancestor/descendant relationships. Since many groups of organisms are now extinct, we can't have as clear a picture

B. Hao (✉)
T - Life Research Center, Fudan University, Shanghai, China

The Santa Fe Institute, Santa Fe, NM, USA
e-mail: hao@mail.itp.ac.cn

C. Zhang
Department of Physics, Tianjin University, Tianjin, China
e-mail: ctzhang@tju.edu.cn

Y. Li
Shanghai Center for Bioinformatics Technology, Shanghai, China
e-mail: yxli@sibs.ac.cn

H. Li
Department of Biochemistry and Biophysics, UCSF, University of California,
San Francisco, CA, USA
e-mail: haoli@genome.ucsf.edu

L. Wei
Center for Bioinformatics, Peking University, Beijing, China
e-mail: weilp@mail.cbi.pku.edu.cn

M. Kanehisa
Institute for Chemical Research, Kyoto University, Kyoto, Japan
e-mail: kanehisa@kuicr.kyoto-u.ac.jp

R. Jiang et al. (eds.), *Basics of Bioinformatics: Lecture Notes of the Graduate Summer School on Bioinformatics of China*, DOI 10.1007/978-3-642-38951-1_12,
© Tsinghua University Press, Beijing and Springer-Verlag Berlin Heidelberg 2013

ofhow modern life is interrelated without their fossils. Phylogenetics, the science of phylogeny, is a useful tool severed as one part of the larger field of systematic including taxonomy which is a practice and science of naming and classifying the diversity of organisms.

All living organisms on the earth are divided into prokaryotes and eukaryotes. Prokaryotes are unicellular organisms that do not have a nucleus in the cell, and DNA molecules encoding the genetic information just float in the cells. Prokaryotes are the most abundant organisms on earth and have been thriving for more than 3.7 billion years. They shaped most of the ecological and even geochemical environments for all living organisms. Yet our understanding of prokaryotes, particularly their taxonomy and phylogeny, is quite limited. It was the Swedish naturalist Carolus Linnaeus (1707–1778) who firstly introduced the taxonomic hierarchy made of kingdom, phylum, class, order, family, genus, and species. In 1965, Zukerkandl and Pauling suggested that evolutionary information may be extracted from comparison of homologous protein sequences in related species, thus opening the field of molecular phylogeny. A breakthrough in molecular phylogeny of prokaryotes was made by Carl Woese and coworkers in the mid-1970s. They compared the much conserved RNA molecule in the tiny cellular machines that make proteins, the so-called small-subunit ribosomal RNAs (SSU rRNAs), to infer the distance between species. This method has led to a reasonable phylogeny among many prokaryote species by the alignment of the symbolic sequences of about 1,500 letters long. The modern prokaryotic taxonomy as reflected in the new edition of *Bergey's Manual of Systematic Bacteriology* is now largely based on 16S rRNA analysis.

Recently proposed CVTree approach, using entirely different input data and methodology, supports most of the 16S rRNA results and may put the prokaryotic

_____

L. Lai
College of Chemistry and Molecular Engineering, Peking University, Beijing, China
email: lhlai@pku.edu.cn

R. Chen • J. Han
Chinese Academy of Sciences, Beijing, China
e-mail: chenrs@sun5.ibp.ac.cn; jdhan@picb.ac.cn

N. Rajewsky
Max-Delbrück-Center for Molecular Medicine, Berlin, Germany
e-mail: rajewsky@mdc-berlin.de

M.Q. Zhang
Department of Molecular and Cell Biology, The University of Texas at Dallas,
800 West Campbell Rd, RL11, Richardson, TX, USA 75080

Tsinghua National Laboratory for Information Science and Technology,
Tsinghua University, Beijing 100084, China
e-mail: mzhang@cshl.edu

R. Jiang • X. Zhang • Y. Li
Department of Automation, Tsinghua University, Beijing, China
e-mail: ruijiang@tsinghua.edu.cn; zhangxg@tsinghua.edu.cn; daulyd@mail.tsinghua.edu.cn

branch of the tree of life on a secure footing. The CVTree approach was first announced in 2002 and has been described in several follow-up works. A web server has been installed for public access. In brief, the input to CVTree is a collection of all translated amino acid sequences from the genome of an organism downloaded from NCBI database. Then the number of K-peptides is counted by using a sliding window, shifting one letter at a time along all protein sequences. These counts are kept in a fixed lexicographic order of amino acid letters to form a vector with 20^k components. A key procedure leading to the final composition vector is the subtraction of a background caused mainly by neutral mutations in order to highlight the shaping role of natural selection. As mutations occur randomly at molecular level, this is done by using a (K-2)th-order Markovian prediction based on the number of (K-2)- and (K-1)-peptides from the same genome. A distance matrix is calculated from these composition vectors, and the standard neighbor-joining program from the PHYLIP package is used to generate the final CVTrees.

CVTree method has many new characteristics compared with more traditional methods. It is an alignment-free method as each organism is represented by a composition vector with 20^k components determined by the number of distinct K-peptides in the collection of all translated protein sequences. This strategy overcomes the huge computational complexity caused by the difference in genome size and gene number of prokaryotes in sequence alignment method.

Moreover, the CVTree provides a parameter-free method that takes the collection of all proteins of the organisms under study as input and generates a distance matrix as output. In fact, the CVTree method has shown rather high resolution to elucidate the evolutionary relationship among different strains of one and the same species. It does not require the selection of RNA or protein-coding gene(s) as all translated protein products in a genome are used. Since there may be large number of gene transfers in species, this phenomenon may show more affection on original methods than CVTree.

The CVTree results are verified by direct comparison with systematic bacteriology. The CVTrees constructed from many organisms bear a stable topology in major branching patterns from phyla down to species and strains. As compared to some traditional phylogenetic tree construction methods, the CVTree approach gets a nice feature of "the more genomes the better agreement" with Bergey's taxonomy. The high-resolution power of the CVTrees provides a means to elucidate evolutionary relationships among different strains of one and the same species when the 16S rRNA analysis may not be strong enough to resolve too closely related strains.

While the 16S rRNA analysis cannot be applied to the phylogeny of viruses as the latter do not possess a ribosome, the CVTree method has been successfully used to construct phylogeny of coronaviruses including human SARS virus and double-strand DNA viruses and chloroplasts as well.

Many experimental results have supported CVTree method can output very high precision of phylogenetic tree construction agreeing well with the taxonomy which is based more and more on the 16S rRNA analysis. The CVTree approach and the 16S rRNA analysis use orthogonal data from the genome and utilize different methodology to infer phylogenetic information. Yet they support each other in an overwhelming majority of branchings and clusterings of taxa, thus providing a reliable framework to demarcate the natural boundaries among prokaryote species. It would be useful to see the CVTree method applied to such eukaryotes as fungi. This method also gives new calculation of phylogenetic distance between different species and can be further used in many other bioinformatics problems.

12.2 Z-Curve Method and Its Applications in Analyzing Eukaryotic and Prokaryotic Genomes

Chunting Zhang (✉)
Tianjin University, Tianjin, China

At the stage of post-genomics, ever-increasing genomic sequences across various prokaryotic and eukaryotic species are available for the mechanism exploration of biological systems. The computer-aided visualization of the long sequences with complex structures which could provide us the intuitive clues at understanding biological mechanisms is urgently needed for both biologists and scientists from other disciplines. One representative tool of visualizing methods is the Z-curve, which makes us capable of analyzing the genomic sequences in terms of geometric approach as the complement of algebraic analysis. The Z-curve is a three-dimensional curve which is a unique representation for a given DNA sequence in the sense that each can be uniquely reconstructed given the other. The Z-curve has been proven that it is the generalized formula of several proposed visualization methods, for example, the H-curve, the game representation, the W-curve, and the two-dimensional DNA walk. The Z-curve has been used in several biological applications, including gene recognition and Isochore prediction, furthermore, succeeded at helping identifying novel biological mechanisms. The Z-curve database for the available sequences of archaea, bacteria, eukaryote, organelles, phages, plasmids, viroids, and viruses is established and open for biological researches. This is a review of the Z-curve and the successful applications for both prokaryotic and eukaryotic species.

The Z-curve is a unique three-dimensional curve representation for a given DNA sequence in the sense that each can be uniquely reconstructed given the other. The Z-curve is composed of a series of nodes $P_1, P_2, P_3, \ldots, P_N$, whose coordinates x_n, y_n and z_n ($n = 1, 2, \ldots, N$, where N is the length of the DNA sequence being studied) are uniquely determined by the Z-transform of DNA sequence:

$$\begin{cases} x_n = (A_n + G_n) - (C_n + T_n) \\ y_n = (A_n + C_n) - (G_n + T_n) \\ z_n = (A_n + T_n) - (C_n + G_n) \end{cases}$$

$$x_n, y_n, z_n \in [-N, N], \quad n = 0, 1, 2, \ldots, N$$

where A_n, C_n, G_n, and T_n are the cumulative occurrence numbers of A, C, G, and T, respectively, in the subsequence from 1st base to the nth base in the sequence. The Z-curve is defined as the connection of the nodes $P_1, P_2, P_3, \ldots, P_N$ one by one sequentially with straight lines starting from the origin of the three-dimensional coordinate system. Once the coordinates x_n, y_n, and z_n ($n = 1, 2, \ldots, N$) of a Z-curve are given, the corresponding DNA sequence can be reconstructed from the inverse Z-transform. In terms of biology, the three components of the Z-curve make sense that x_n, y_n, and z_n represent the distributions of purine/pyrimidine (R/Y), amino/keto (M/K), and strong H bond/weak H bond (S/W) bases along the sequence, respectively. The three components of the Z-curve uniquely describe the DNA sequence being studied and contain all the information in the original sequences.

The perceivable form of the Z-curve provides an intuitive insight to the researches of genomic sequences. The database of the Z-curves for archaea, bacteria, eukaryote, organelles, phages, plasmids, viroids, and viruses is established and contains pre-calculated coordinates of more than 1,000 genomes. As the complement of GenBank/EMBL/DDBJ, the Z-curve database provides a variant resolution geometric insight into some features of the nucleotide composition of genomes, ranging from the local scale to the global scale. It has been shown that the visualization and the complete information of the Z-curve offer benefits to Bioinformatics community. The joint effect of three components of the Z-curve has been identified at recognizing genes in available genomes across various prokaryotic species, including *S. cerevisiae*, bacteria, archaea, coronavirus, and phages. The respective software services are established in the Z-curve database. The linear combinations of the x_n and y_n components of the Z-curve are defined as a family of disparity curves, which could be used to analyze the local deviations from Chargaff Parity Rule 2 showing that globally both $\%A \approx \%T$ and $\%G \approx \%C$ are valid for each of the two DNA strands. The AT- and GC-disparity curves calculated by the Z-curves have been applied to predict the replication origins and terminations of some bacterial and archaeal genomes. The individual z_n component has a significant advantage at the calculation of the G + C content. With the Z-curve, the calculation of the G + C content of genomes is windowless in contrast to the previous methods and could be performed at any resolution. The z'_n-curve, a transform of the Z-curve, termed GC profile, is defined to describe the distribution of the G + C content along the DNA sequences. Intuitively, a jump in the z'_n-curve indicates an A + T-rich region, whereas a drop means a G + C-rich region. A sudden change in the z'_n-curve might imply a transfer of foreign DNA sequence from other species. A typical example is the z'_n-curve for the smaller chromosome of *Vibrio cholera*, where the position of the integron island is precisely identified by observing a sudden jump in the z'_n-curve. Recent studies have shown that the windowless calculation and analysis of the G + C content of the eukaryotic genomes obtain several impressive achievements at the Isochore edge determination, the Isochore structure exploration, and the Isochore predictions.

12.3 Insights into the Coupling of Duplication Events and Macroevolution from an Age Profile of Transmembrane Gene Families

Yixue Li (✉)
Shanghai Center for Bioinformatics Technology, Shanghai, China

This study stemmed from another project focusing on the evolution of transmembrane proteins. We noticed the noncontinuous distribution of duplication events over evolutionary time, and the pattern to some extent overlapped with fossil evidences of macroevolution, which led us to think about the relationship between molecular evolution and macroevolution. The neutral evolution theory for molecular level and Darwin's macroevolution theory conflict with each other, but both are supported by evidences at different levels. What's the connection?

We tried to answer this question by studying the duplication of transmembrane proteins. The evolution of new gene families subsequent to gene duplication may be coupled to the fluctuation and dramatically alternations of population and environment variables. By using the transmembrane gene family, which is a key component for information exchange between cells and the environment, it is possible to find the cycles and patterns in the gene duplication event records and the relationship between the evolutionary patterns on the molecular level and the species level.

We started by building transmembrane gene family. First, we predicted transmembrane proteins from 12 eukaryotes by including proteins with at least one predicted transmembrane helix. Then we developed a pipeline to build gene families by integrating strategy of COG, HOBACGEN, and other additional steps. After a manual check we had 863 homology families of eukaryote transmembrane proteins. We then constructed the phylogenetic tree for these families by neighbor-joining method. The molecular time scale of the inferred tree was calibrated and adjusted by both fossil data and molecular data with known molecular age. Finally we were able to detect 1,651 duplication events in the final dataset with 786 gene families. All of the identified duplication events were recorded with the corresponding ages.

The overall age distribution was determined on the basis of 1,620 transmembrane gene duplication events. Similar to previous report, this distribution clearly shows three peaks (0.13 billion years [Gyr], 0.46 Gyr, and 0.75 Gyr ago approximately).

We next examined the relationship of the apparent disturbances of the age distribution with oxidation event records reconstructed from geochemical and fossil research. Interestingly enough, the time point at which the density of the duplicates increases distinctly under the baseline distribution is completely consistent with the reliable minimum age for the advent of oxygenic photosynthesis (2.75 Gyr ago) and the age of the domain Eucarya concluded from molecular fossils (about 2.7 Gyr ago). Our findings imply the linkage between the oxygen level and the transmembrane gene duplicates.

We performed decomposition in Fourier series of the detrended density trace of the duplicates in the Phanerozoic phase. We identified three potential peaks, which are 60.92-Myr, 27.29-Myr, and 10.32-Myr cycles. The 60.92-Myr cycle has the strongest cyclicity in the density trace of the transmembrane gene duplicates. Consistent with the opinion that the macroevolutionary time series have characteristics of a random walk, because these cycles are not statistically significant, they cannot reject the null hypothesis of a random walk. The existence of the most potential cycle of 60.92 Myr in the age distribution of transmembrane gene families is a very interesting discovery, because it is not indistinguishable from the 62 ± 3-Myr cycle that is the most statistically significant cycle detected in biodiversity recently reported.

We had clearly shown that the duplication events of transmembrane genes are coupled with the macroevolution measurement and asynchronous with the animal biodiversity. The evolution history is a coevolution process of the environment and life. The overall shape of the age distribution is driven by the oxygen level in the atmosphere, while the waves of the distribution might be driven by some rhythmic external force. Furthermore, we proposed a plausible evolutionary scenario to explain these findings based on the factors finally determining the fate of the duplicates, which implies that the environment alternation would induce the redundancy of the existent genome system that is beneficial for survival in a rigorous condition. In addition, we presented a methodology to provide a unique, temporally detailed understanding of the interaction of the transmembrane gene duplication events and the environment variables. Since the sequence data are thoroughly independent from the fossil record and more readily attainable, this methodology may give us a new strategy to validate patterns such as the 62-Myr cycle, which was detected from fossil or other geophysical records. Further studies using this method may offer important insights into the interplay of the microevolution and macroevolution factors.

12.4   Evolution of Combinatorial Transcriptional Circuits in the Fungal Lineage

Hao Li (✉)
University of California, San Francisco, CA, USA

Now it's a good time to study basic mechanism of evolution, because the whole-genome sequences of an increasing large number of species are available. Bioinformatics provides important tools for research on evolution. We study the evolution of combinatorial control in yeast transcriptional regulation network.

Transcription network can response to external environmental changes, and the regulation often involves combinatorial control of multiple transcriptional factors (TFs). One can view the transcriptional network as a black box, with the activity of TFs as input and the transcript levels of all genes as the output.

We are interested in two questions: first, how to construct the transcriptional network? And second, why should it be like this? What are the functional and evolution constraints? Here we focus on the later, that is, how transcriptional networks evolve? What are the basic patterns or steps? Here we will first show a study on the transcriptional circuits controlling yeast mating types.

Mating type in the yeasts *Saccharomyces cerevisiae* and *Candida albicans* is controlled by the MAT locus, which has two versions: MATa and MATα. Cells that express only the MATa- or MATα-encoded proteins are a-cells and α-cells, respectively. The a-cells express a-specific genes (asgs), which are required for a-cells to mate with α-cells. On the other hand, α-cells express the α-specific genes (αsgs). In *S. cerevisiae*, the asgs are on by default and are repressed in other cells by protein α2 encoded by MATα. In *C. albicans*, however, the asgs are off by default and are activated in a-cells by protein a2 encoded by MATa. Both molecular mechanisms give the same logical output: asgs are expressed only in a-cells. By comparative genomics analysis, we show that a2-activation most likely represents the ancestral state and the a2 gene was recently lost in the *S. cerevisiae*, now using the α2-repressing mode of asg regulation. In the promoters of several asgs, we found a regulatory element with several distinctive features. First, the sequence contains a region that closely resembles the binding site of Mcm1, a MADS box sequence-specific DNA-binding protein that is expressed equally in all three mating types, and is required for the regulation of both asgs and αsgs in *S. cerevisiae*. We also show that the Mcm1 residues that contact DNA are fully conserved between *C. albicans* and *S. cerevisiae*, strongly implicating this region of the element as a binding site for Mcm1 in *C. albicans*. Second, the putative Mcm1 site in *C. albicans* asg promoters lies next to a motif of the consensus sequence CATTGTC. The spacing between this motif and the Mcm1 site is always 4 bp. This motif is similar to demonstrated binding sites for a2 orthologues in *Schizosaccharomyces pombe* and *Neurospora crassa* and to the α2 monomer site of *S. cerevisiae*.

These evidences suggest the following changes in *cis*- and *trans*-elements can lead to a profound evolutionary change in the wiring of a combinatorial circuit: (1) "tuning up" of a binding site for a ubiquitous activator, making gene expression independent of a cell-type-specific activator; (2) a small change in an existing DNA-binding site, converting its recognition from one protein to that of an unrelated protein; and (3) a small change in the amino acid sequence of a sequence-specific DNA-binding protein, allowing it to bind DNA cooperatively with a second protein.

In a second study, we center on Mcm1. Besides the regulatory role in mating, it also involves in many other processes such as cell cycle. By comparing data from *Saccharomyces cerevisiae*, *Kluyveromyces lactis*, and *Candida albicans*, we find that the Mcm1 combinatorial circuits undergone substantial changes. This massive rewiring of the Mcm1 circuitry has involved both substantial gain and loss of targets in ancient combinatorial circuits as well as the formation of new combinatorial interactions. We have dissected the gains and losses on the global level into subsets of functionally and temporally related changes. One particularly dramatic change is the acquisition of Mcm1 binding sites in close proximity to Rap1 binding sites at 70 ribosomal protein genes in the *K. lactis* lineage. Another intriguing and very recent gain occurs in the *C. albicans* lineage, where Mcm1 is found to bind in combination with the regulator Wor1 at many genes that function in processes associated with adaptation to the human host, including the white-opaque epigenetic switch.

12.5   Can a Non-synonymous Single-Nucleotide Polymorphism (nsSNP) Affect Protein Function? Analysis from Sequence, Structure, and Enzymatic Assay

Liping Wei (✉)
Peking University, Beijing, China

After the completion of the human genome project, increasing attention has focused on the identification of human genomic variations, especially single-nucleotide polymorphisms (SNPs). It is estimated that the world population contains a total of ten million SNP sites, resulting in an average density of one variant per 300 bases. SNPs in coding and regulatory regions may play a direct role in diseases or differing phenotypes. Among them, the single amino acid polymorphisms (SAPs, conventionally known as non-synonymous SNPs or nsSNPs), which cause amino acid substitutions in the protein product, are of major interest because they account for about 50 % of the gene lesions known to be related to genetic diseases. Through large-scale efforts such as the HapMap project (http://www.hapmap.org), The Cancer Genome Atlas (TCGA, http://cancergenome.nih.gov), and whole-genome association studies, available SAP data is accumulating rapidly in databases such as dbSNP, HGVbase, Swiss-Prot variant page, and many allele-specific databases. However, because of the high-throughput nature of these efforts, many SAPs could not be experimentally characterized in terms of their possible disease association. Furthermore, the underlying mechanisms that explain why a SAP may be associated with disease and have deleterious functional effect are not yet fully understood.

In the past 5 years, several bioinformatics methods have been developed to use sequence and structural attributes to predict possible disease association or functional effect of a given SAP. A popular sequence-based method is SIFT, which predicts whether an amino acid substitution is deleterious or tolerated based on the evolutionary conservation of the SAP site from multiple sequence alignment. More recent methods incorporate both sequence and structural attributes and use a range of classifiers such as rule-based, decision trees, support vector machines (SVMs), neural networks, random forests, and Bayesian networks to annotate SAPs. Zhi-Qiang Ye et al. recently employed machine learning method, named SAPRED (http://sapred.cbi.pku.edu.cn/), and obtained better performance (82.6 %) than early methods. SAPRED first constructed a relatively balanced dataset from the Swiss-Prot variant pages, then investigated the most complete set of structural and sequence attributes to date, and identified a number of biologically informative new attributes that could explain why a SAP may be associated with disease. Finally, the method incorporated these attributes into an SVM-based machine learning classifier.

SAPRED investigated a large set of structural and sequence attributes including both commonly used ones such as residue frequency and solvent accessibility and new ones that are novel to this study. These attributes include residue frequency and conservation, solvent accessibilities, structural neighbor profiles, nearby functional sites, structure model energy, hydrogen bond, disulfide bond, disordered region, aggregation properties, and HLA family. They accessed these attributes and get many new findings. They confirmed that residue frequencies provided the best discrimination reported as early researches. Previous studies found solvent accessibilities to be the type of attributes with the second most predictive power. However, their study identified two new types of attributes that showed higher predictive power than solvent accessibilities.

The new attributes studied in this work may further the understanding of the biological mechanism underlying the functional effect and disease association of SAPs. At the same time, SAPRED also contribute to the increase in accuracies of predicting the disease association of SAPs. In particular, the predictive power of structural neighbor profile is almost as high as that of residue frequencies, highlighting the importance of the microenvironment around a SAP. In addition, the predictive power of nearby functional sites is higher than solvent accessibilities, the second most powerful type of attributes in previous studies. By considering residues both at and near functional sites in terms of both sequence and structure, SAPRED significantly enlarged the coverage and overcame the limitations in previous work that used only the functional site residues themselves. The other new attributes, such as disordered regions and aggregation properties, also provided direct biological insights into the study of SAPs and contributed to the overall accuracy of prediction.

After prediction, further biological experiments can be used to describe the association between SAP and disease. An excellent example in SAP study is R41Q associated with certain severe adverse reactions to oseltamivir. The use of oseltamivir, widely stockpiled as one of the drugs for use in a possible avian influenza pandemic, has been reported to be associated with neuropsychiatric disorders and severe skin reactions, primarily in Japan. R41Q, near the enzymatic active site of human cytosolic sialidase, a homologue of virus neuraminidase is the target of oseltamivir. This SNP occurred in 9.29 % of Asian population and none of European and African American population. Structural analyses by SAPRED and Ki measurements using in vitro sialidase assays indicated that this SNP could increase the unintended binding affinity of human sialidase to oseltamivir carboxylate, the active form of oseltamivir, thus reducing sialidase activity. In addition, this SNP itself results in an enzyme with an intrinsically lower sialidase activity, as shown by its increased Km and decreased Vmax values. Theoretically administration of oseltamivir to people with this SNP might further reduce their sialidase activity. The reported neuropsychiatric side effects of oseltamivir and the known symptoms of human sialidase-related disorders are correlated. This Asian-enriched sialidase variation caused by the SNP, likely in homozygous form, may be associated with certain severe adverse reactions to oseltamivir.

The "sequence → structure → function" model for describing protein activity states that the amino acid sequence determines the higher structures of a protein molecule, including its secondary and tertiary conformations, as well as quaternary complexes and further states that the formation of a definite ordered structure represents the foundation for the function of the protein. Different attributes including sequence and structure can give more contribution to the final function prediction and show the association between disease and SAP. However, to better predict the possible disease association of SAPs, existing methods still need to be improved in several aspects. First, more biologically informative structural and sequence attributes need to be investigated to further understand the underlying mechanism of how a SAP may be associated with a disease. Second, several studies used imbalanced datasets which impeded the performance of their classifiers. Third, by using more biologically informative attributes and a better dataset, the overall accuracy of the prediction can be improved.

12.6 Bioinformatics Methods to Integrate Genomic and Chemical Information

Minoru Kanehisa (✉)
Kyoto University, Kyoto, Japan

Although the comprehensive genome sequence has only recently been revealed, biologists have been characterizing the roles played by specific proteins in specific processes for nearly a century. This information spans a considerable breadth of knowledge and is sometimes exquisitely detailed and stored as primary literature, review articles, and human memories. Recent efforts have established databases of published kinetic models of biologic processes ranging in complexity from glycolysis to regulation of the cell cycle. These chemical information and genomic information can be found in many databases such as NCBI, KEGG, PID, and Reactome, which allow researchers to browse and visualize pathway models and, in some cases, to run simulations for comparison with experimental data.

KEGG (Kyoto Encyclopedia of Genes and Genomes) is a knowledge base for systematic analysis of gene functions, linking genomic information with higher-order function information. KEGG mainly consists of the PATHWAY database for the computerized knowledge on molecular interaction networks such as pathways and complexes, the GENES database for the information about genes and proteins generated by genome sequencing projects, and the LIGAN database for the information about chemical compounds and chemical relations that are relevant to cellular processes. KEGG BRITE is a collection of hierarchies and binary relations with two interrelated objectives corresponding to the two types of graphs: to automate functional interpretations associated with the KEGG pathway reconstruction and to assist discovery of empirical rules involving genome-environment interactions. In addition to these main databases, there are several other databases including EX-

PRESSION, SSDB, DRUG, and KO. KEGG can be considered as a complementary resource to the existing database on sequences and three-dimensional structures, focusing on higher-level information about interaction and relations of genes and proteins.

The KEGG databases are highly integrated. In fact, KEGG should be viewed as a computer representation of the biological system, where biological objects and their relationships at the molecular, cellular, and organism levels are computerized as separate database entries. Cellular functions result from intricate networks of molecular interactions, which involve not only proteins and nucleic acids but also small chemical compounds. The genomic and chemical information in database can be integrated by bioinformatics methods which are summarized as two procedures: decomposition into building blocks and reconstruction of interaction networks. First, chemical structure can be decomposed into small building blocks by comparison of bit-represented vectors (fingerprints) or comparison of graph objects. Conserved substructures can be viewed as building blocks of compounds and variable substructures as building blocks of reactions. Here small building blocks are divided into two categories. One is metabolic compounds that are subject to enzyme-catalyzed reactions that maintain the biological system. The other category is regulatory compounds that interact with proteins, DNA, RNA, and other endogenous molecules to regulate or perturb the biological system. In the same way, the proteins can be decomposed into domains or conserved motifs considered as small building blocks. Second, using the existed pathway information (metabolic pathways) and genomic information (such as operon structure in bacterial), interaction networks (or network modules) can be reconstructed.

Extracting information from the chemical structures of these small molecules by considering the interactions and reactions involving proteins and other biological macromolecules, a knowledge-based approach for understanding reactivity and metabolic fate in enzyme-catalyzed reactions in a given organism or group was presented by Mina Oh et al. They first constructed the KEGG RPAIR database containing chemical structure alignments and structure transformation patterns, called RDM patterns, for 7,091 reactant pairs (substrate-product pairs) in 5,734 known enzyme-catalyzed reactions. A total of 2,205 RDM patterns were then categorized based on the KEGG PATHWAY database. The majority of RDM patterns were uniquely or preferentially found in specific classes of pathways, although some RDM patterns, such as those involving phosphorylation, were ubiquitous. The xenobiotics biodegradation pathways contained the most distinct RDM patterns, and a scheme was developed to predict bacterial biodegradation pathways given chemical structures of, for example, environmental compounds.

If the chemical structure is treated as a graph consisting of atoms as nodes and covalent bonds as edges, two chemical structures of compounds can be compared by using related graph algorithms. On the basis of the concept of functional groups, 68 atom types (node types) are defined for carbon, nitrogen, oxygen, and other atomic species with different environments, which has enabled detection of biochemically meaningful features. Maximal common subgraphs of two graphs can be found by searching for maximal cliques (simply connected common subgraphs) in the association graph. The procedure was applied to the comparison and clustering of 9,383 compounds, mostly metabolic compounds, in the KEGG/LIGAND database. The largest clusters of similar compounds were related to carbohydrates, and the clusters corresponded well to the categorization of pathways as represented by the KEGG pathway map numbers. When each pathway map was examined in more detail, finer clusters could be identified corresponding to subpathways or pathway modules containing continuous sets of reaction steps. Furthermore, it was found that the pathway modules identified by similar compound structures sometimes overlap with the pathway modules identified by genomic contexts, namely, by operon structures of enzyme genes.

With increasing activity of metabolomics, chemical genomics, and chemical biology in which the biological functions of a large number of small molecules are uncovered at the molecular, cellular, and organism levels, new bioinformatics methods have to be developed to extract the information encoded in the small molecular structures and to understand the information in the context of molecular interactions and reactions involving proteins and other biomolecules. As above shown results, the integrated analysis of genomic (protein and nucleic acid) and chemical (small molecule) information can give more effective results in inferring gene functions or cellular processes. Furthermore, combined with more other database information such as microarray data, more results and higher performance can be got.

12.7 From Structure-Based to System-Based Drug Design

Luhua Lai (✉)
Peking University, Beijing, China

For the purpose of better understanding the structure-based and the later system-based drug design, we first introduce the general procedure of drug discovery. This procedure, cost 10–12 years averagely, is composed of lead compound discovery, lead compound optimization, activity test, preclinical research, clinical research, and medicaments listing. Though the drug design usually refers to the first two steps, currently there is a very important trend that the subsequent steps should be considered as from the beginning. That is, from the early stage of the drug research and design, we may consider the absorption, metabolism, and toxicity of a drug, which can reduce the waste as much as possible.

Targeted drug design needs to study the targets. Most of our known drug targets are proteins, and the combination of drugs and targets follows a "lock and key" model, which is also a basic hypothesis of the structure-based drug design, meaning that the drug design is actually finding targets interacted with specific proteins. This structure-based drug design method is facing two main difficulties now: conformational change in the interaction and the accurate and fast calculation of binding free energies. Other design may also include the mechanism-based drug design, but it is less successfully applied compared with that by the structure-based drug design.

When doing molecular recognition, the following issues need to be taken into account, including how to identify the interaction targets among so many molecules, faraway attraction, conformational changes and induced fit, energy interactions, desolvation interactions, dipole-dipole interactions, and so on. As is known that drug design can be regarded as a molecular recognition process, the main effects that influence the molecular recognition include the van der Waals interaction, the Coulomb law, the hydrogen bond interaction, as well as the hydrophobic interaction.

The drug design process is classified into two cases, based on whether the three-dimensional structure of the receptor is known. If the structure is known, the relevant researches include finding new types of lead compounds according to the structures of the receptors, studying the interactions between receptors and known compounds as well as optimizing the lead compounds. Otherwise if the structure is unknown, the relevant researches include studying the QSAR (quantitative structure-activity relationship) in series of compounds and analyzing the pharmacophore model. Specifically, the new lead compounds can be found through database algorithms, segment connecting method, and de novo design approach. Even though it is universally regarded that the structure-based drug design has achieved great success on the aspects of lead compound finding by means of database screening, there also exist big limitations; for example, if a newly discovered protein cannot be found in the existing database, it will fail to find a molecule to combine. Under this circumstance, though there does not exist any complete molecule which is suitable for being combined with the protein, there may exist several already synthesized molecular segments. In some manner, we can connect these molecular segments with each other and obtain a complete molecule, and this method is called segment connecting method. The disadvantage of this method is that in reality the segment combination is not always the conformation of the lowest energy, rendering the combination hard to be connected in reality. For people who are studying the drug design approaches, the most challengeable approach may be the de novo design approach, which is expected to be independent of existing compounds and databases. By knowing only the protein's shape at the combination site, we are able to grow a new molecular, which can fit the shape very well.

In the computation of binding free energy, we need a scoring function, and it is usually given in the forms of empirical or semiempirical formula, for example, the force field-based scoring functions (D-Score, G-Score, GOLD, Autodock, and DOCK), the empirical scoring function (LUDI, F-Score, ChemScore, Fresno, SCORE, and X-SCORE), and the knowledge-based scoring function (PMF, DrugScore, and SMoG).

The network-based drug design approach is distinguished from the "one target, one molecule" approach in the structure-based drug design. In the network-based drug design, the target lies in a complex network of certain cell of certain organ of certain human bodies, which is much more complicated than a single molecule. However, with our gradually deep understanding toward the complex biological systems, some analysis can be done at the molecular network level, including the mathematical modeling and the dynamical property simulation of the disease-related networks, finding the key nodes in use of the multiparameter analysis, implementing regulations using multiple nodes in network control, as well as simulating of the effect of multi-node control. Last but not least, the network-based drug design does not mean that we do not need the structure-based drug design; instead, the latter is an indispensable step in no matter what kinds of drug design. It is appropriate to say that the network-based drug design provides a fresh idea for the structure-based drug design.

12.8   Progress in the Study of Noncoding RNAs in *C. elegans*

Runsheng Chen (✉)
Chinese Academy of Sciences, Beijing, China

With the completion of human genome project, more and more genomes are sequenced and more secret of life is discovered. Important properties of human genome are that it has lower protein number ($\sim$25,000) than expected and more regions in human genome are noncoding sequences. Noncoding sequences are segments of DNA that does not comprise a gene and thus does not code for a protein, which are interspersed throughout DNA. There are only 15 % noncoding sequences in *E. coli*, 71 % in *C. elegans*, and 82 % in *Drosophila*. However, the noncoding sequences in human genome are 98 %, while only 2 % are coding regions. It is obvious the higher organisms have a relatively stable proteome and a relatively static number of protein-coding genes, which is not only much lower than expected but also varies by less than 30 % between the simple nematode worm *C. elegans* (which has only 10^3 cells) and humans (10^{14} cells), which have far greater developmental and physiological complexity. Moreover, only a minority of the genomes of multicellular organisms is occupied by protein-coding sequences, the proportion of which declines with increasing complexity, with a concomitant increase in the amount of noncoding intergenic and intronic sequences. Now it is widely believed that there are many units called noncoding genes in noncoding sequences which translate in noncoding RNA but not protein. Thus, there seems to be a progressive shift in transcriptional output between microorganisms and multicellular organisms from mainly protein-coding mRNAs to mainly noncoding RNAs, including intronic RNAs.

Noncoding RNA genes include highly abundant and functionally important RNA families such as transferRNA (tRNA) and ribosomal RNA (rRNA), as well as RNAs such as snoRNAs, microRNAs, siRNAs, piRNAs, and lastly long ncRNAs. Recent transcriptomic and bioinformatic studies suggest the existence of thousands of ncRNAs encoded within human genome. Numerous researches proved that noncoding RNAs have very important functions. For example, SINE elements serve as recombination hot spots allowing the exchange of genetic material between unrelated sequences and also can act as tissue-specific enhancers or silencers of the adjacent genes. The Xist gene lies within the X-inactivation center and is required to initiate X chromosome inactivation. Xist encodes a large, spliced, polyadenylated, and noncoding RNA that is expressed exclusively from the otherwise inactive X chromosome. NcRNAs, especially the miRNAs, have also been implicated in many diseases, including various cancers and neurological diseases. There are many estimated noncoding genes in sequenced genomes which may be much more than coding genes.

In 2004, the *C. elegans* genome was reported approximately 1,300 genes known to produce functional ncRNA transcripts including about 590 tRNAs, 275 rRNAs, 140 *trans*-spliced leader RNA genes, 120 miRNA genes, 70 spliceosomal RNA genes, and 30 snoRNA genes. Recently, with hundreds of noncoding RNA found, the number is increased drastically. Applying a novel cloning strategy, Wei Deng et al. have cloned 100 new and 61 known ncRNAs in *C. elegans*. Studying of genomic environment and transcriptional characteristics has shown that two-thirds of all ncRNAs, including many intronic snoRNAs, are independently transcribed under the control of ncRNA-specific upstream promoter elements. Furthermore, the percent of the transcription levels of the ncRNAs varying with developmental stages is at least 60 %. This work also found two new classes of ncRNAs, stem-bulge RNAs (sbRNAs) and snRNA-like RNAs (snlRNAs). They are all identified and featured distinct internal motifs, upstream elements, secondary structures, and high and developmentally variable expression. Most of the novel ncRNAs are conserved in *C. briggsae*, but only one homolog was found outside the nematodes. Preliminary estimates indicate the transcriptome may contain ~2,700 small noncoding RNAs, which are potentially acted as regulatory elements in nematode development. This estimation highly increased the number of noncoding RNA in *C. elegans*. Furthermore, combined microarray is designed to analyze the relationship between ncRNA and host gene expression. Results show that the expression of intronic ncRNA loci with conserved upstream motifs was not correlated to (and much higher than) expression levels of their host genes. Promoter-less intronic ncRNAs, which even show a clear correlation to host gene expression, also have a surprising amount of "expressional freedom" compared to host gene function. Taken together, the microarray analysis presents a more complete and detailed picture of a noncoding transcriptome than hitherto has been presented for any other multicellular organism.

By using a whole-genome tiling microarray, the *C. elegans* noncoding transcriptome is mapped. Three samples are designed and individually produced 108669, 97548, and 5738 transfrags which after removal of redundancies suggested the presence of at least 146,249 stably expressed regions with an average and median length of 156 and 103 nt, respectively. After combining overlapping transcripts, it is estimated the total transcription in *C. elegans* is at least 70 %. More tests show the experiments are very high precise and 90 % transcriptions are further confirmed by added experiments. These new findings are the same as other conservative summation of the mammalian sequences (human, mouse), which indicates that (at least) 60–70 % of the mammalian genome is transcribed on one or both strands.

In the past few years, considerable number of noncoding RNAs (ncRNAs) has been detected by using experiments and computations. Although the functions of the many recently identified ncRNAs remain mostly unknown, increasing evidence stands in support of the notion that ncRNAs represent a diverse and important functional output of most genomes. To fusion these information together, NONCODE presents an integrated knowledge database dedicated to ncRNAs and has many distinctive features. First, the ncRNAs in NONCODE include almost all the types of ncRNAs, except tRNAs and rRNAs. Second, all added ncRNA sequences and their related information have been confirmed manually by consulting relevant literature, and more than 80 % of the entries are based on experimental data. Third, based on the cellular process and function, which a given ncRNA is involved in, NONCODE introduced a novel classification system, labeled process function class, to integrate existing classification systems. In addition, ncRNAs have been grouped into nine other classes according to whether they are specific to gender or tissue or associated with tumors and diseases. NONCODE database is very powerful and gives much help in noncoding research.

There are rapidly research problems for ncRNAs, especially as the unknown function and mode of noncoding RNA, their complex structures, and so on. The functional genomics of ncRNAs will be a daunting task which may be equal or greater challenge than that we already face in working out the biochemical functions and biological roles of all of the known and predicted proteins. Most of the ncRNAs identified in genomic transcriptome studies have not been systematically studied and have yet to be ascribed as any functions. RNAs (including those derived from introns) appear to comprise a hidden layer of internal signals that control various levels of gene expression in physiology and development, including transcription, RNA splicing, editing, translation, and turnover. RNA regulatory networks may determine most of our complex characteristics and play a significant role in disease and constitute an unexplored world of genetic variation both within and between species. New methods and experiments will be designed to these new studies. In this area, bioinformatics will be a key, as it should be possible to use more information to identify transmitters and their receivers in RNA regulatory networks.

12.9   Identifying MicroRNAs and Their Targets

Nikolaus Rajewsky (✉)
Max-Delbrück-Center for Molecular Medicine, Berlin, Germany

MicroRNAs (miRNAs) are endogenous 22nt RNAs that can play important regulatory roles in animals and plants by targeting mRNAs for cleavage or translational repression. They are often conserved and with hairpin structures. The early idea about RNA started in the early 1960s when Rosalind Lee and Rhonda Feinbaum discovered that *lin*-4 does not code for a protein but instead produces a pair of small RNAs. The shorter *lin*-4 RNA is now recognized as the founding member of an abundant class of tiny regulatory RNAs called miRNAs. Hundreds of distinct miRNA genes are now known to exist and to be differentially expressed during development and across tissue types. Several detailed studies have shown that miRNA families can expand in plants and animals by the same processes of tandem, segmental, and whole-genome duplication as protein-coding genes. Many confidently identified miRNA genes are likely to regulate large number of protein-coding genes in these human and animals. These numbers will undoubtedly increase as high-throughput sequencing continues to be applied both to miRNA discovery and the validation of some of the many additional candidates proposed. The breadth and importance of miRNA-directed gene regulation are coming into focus as more miRNAs and their regulatory targets and functions are discovered.

Computational approaches have been developed to complement experimental approaches to miRNA gene identification. Homology searches are the early used methods and have revealed orthologs and paralogs of known miRNA genes. Gene-finding approaches that do not depend on homology or proximity to known genes have also been developed and applied to entire genomes. The two most sensitive computational scoring tools are MiRscan, which has been systematically applied to nematode and vertebrate candidates, and miRseeker which has been systematically applied to insect candidates. Both MiRscan and MiRseeker have identified dozens of genes that were subsequently (or concurrently) verified experimentally. They typically start by identifying conserved genomic segments that both fall outside of predicted protein-coding regions and potentially could form stem loops and then score these candidate miRNA stem loops for the patterns of conservation and pairing that characterize known miRNA genes.

Obviously, methods that rely on phylogenetic conservation of the structure and sequence of a miRNA cannot predict nonconserved genes. To overcome this problem, several groups have developed ab initio approaches to miRNA prediction that use only intrinsic structural features of miRNAs and not external information. With these ab initio prediction methods used, many nonconserved miRNAs have been discovered and experimentally verified in viruses and human. Very recently, by using deep sequencing technology, new method was developed to find miRNA effectively. Deep sequencing is a new biotechnology which may displace microarrays in the future and have already used for DNA sequencing, disease mapping, expressing profiling, and binding sites mapping. But analysis of output is highly nontrivial which needs new generation of computing power. MiRDeep is the first algorithm extract miRNA from deep sequencing and uses a probabilistic model of miRNA biogenesis to score compatibility of the position and frequency of sequenced RNA with the secondary structure of the miRNA precursor. MiRDeep can find overall miRNAs with a signal to noise ratio of at least 11:1 and sensitivity at least 90 %. MiRDeep finds 17 new miRNAs in *C. elegans* (signal over noise) at least of 3:1, and numerous novel miRNA genes are found in *C. elegans*, *planaria*, etc.

To understand biological miRNA function, it may be important to search for combinations of miRNA binding sites for sets of coexpressed miRNAs. The current predictions by TargetScan, PicTar, EMBL, and ElMMo all require stringent seed pairing and have a high degree of overlap. PicTar is a miRNA target-finding algorithm, which uses a probabilistic model to compute the likelihood that sequences are miRNA target sites when compared to the 3′ UTR background. This computational approach successfully identifies not only microRNA target genes for single microRNAs but also targets that are likely to be regulated by microRNAs that are coexpressed or act in a common pathway. Massive sequence comparisons using previously unavailable genome-wide alignments across eight vertebrate species strongly decreased the false positive rates of microRNA target predictions, allowed PicTar to predict (above noise), on average, ~200 targeted transcripts per microRNA. PicTar has been used to predict targets of vertebrate and *Drosophila* miRNAs. Vivo experimental validation suggests a high degree of accuracy (80–90 %) and sensitivity (60–70 %) for the PicTar algorithm in flies.

The most pressing question to arise from the discovery of the hundreds of different miRNAs is what all these tiny noncoding RNAs are doing. MiRNA is highly involved in posttranscriptional control. Standard Affymetrix analysis before and after knockdown with "antagomirs" results mRNA roughly regulated 300 gene go up and 300 down. For example, B cell numbers were reduced in the spleen of mice ectopically expressing *miR150*, just as the gene *cMyb* does. Is *cMyb* a direct target of *miR150*? It is really true. Experiments show ectopically expressed *miR150* downregulates *cMyb* protein. Further mutation experiments were done on *miR150* binding sites which were described by PicTar, and results show that some special sites are essential for downregulation of *cMyb* by *miR150*. There are many SNP databases available which give much useful information about evolution and disease and so on. SNP is at least 1 out of 100 bp in human genome. By population genetics methods to study function and evolution of short *cis*-regulatory sites using SNP, the binding sites of MiRNA can be further analyzed to find more evolution information. Many researches show ectopic, moderate experiment of a single miRNA can profoundly interfere with a developmental process by moderately repressing protein levels of a transcription factor. In evolution content, there may be broad rewiring of posttranscriptional control by miRNAs during metazoan evolution. Since many miRNAs and recognition motifs are deeply conserved but regulatory relationships only weekly conserved, it is interesting to further research whether lots of miRNA target rewiring during organism evolution. Many studies have suggested that TF is more conversed and their binding sites are low conserved, while miRNA's targets are more conserved. In this direction, the relationship and difference in evolution between TF and miRNA is also another charming topic. Although much progress is done in miRNA finding, target recognition, and special function description, there is much space to improve and lots of work to be continued.

12.10   Topics in Computational Epigenomics

Michael Q. Zhang (✉)
Department of Molecular and Cell Biology, The University of Texas at Dallas, 800 West Campbell Rd, RL11, Richardson, TX, USA 75080

Tsinghua National Laboratory for Information Science and Technology, Tsinghua University, Beijing 100084, China

This talk is divided into two parts: (1) introduction about epigenetics and epigenomics and (2) genome-wide study on CTCF binding sites.

Epigenetics refers to the inheritable changes in gene expression that cannot be attributed to changes in DNA sequences. There are two main mechanisms, RNAi and histone modification. Epigenomics is the genome-wide approach to studying epigenetics. The central goal is to define the DNA sequence features that direct epigenetic processes.

There are a number of epigenetic examples. In yeast, these include yeast silent mating-type loci and the silence of loci near telomeres and ribosomal DNA; in fly, the position-effect variegation (PEV) for eye color determination; and in plant, the first discovery of RNAi. In mammal, the well-known example is genomic imprinting, that is, epigenetic modification leading to differential expression of the two alleles of gene in somatic cells of the offspring.

The structure of chromatin plays important roles in epigenetic control. Two important histone modifications for transcriptional control are acetylation and methylation. The latter is mostly involved in the silencing of expression. Different combination of histone modification marker is referred to as the histone code.

The relationship between DNA methylation and histone modification is unclear. During development, methylations change. Different methylation patterns may determine cell types. Methylation pattern is largely fixed in post-development stage.

DNA methylation in CpG island and promoters also play roles in gene regulation. Now we show some of our works. In one study, we have experimentally identified whole-genome methylation markers and computationally identified sequence patterns associated with markers. Our studies show that unmethylated regions contain many CpG islands, more likely to be in promoter region and more conserved. Young Alu elements often localize close to unmethylated region. We also found that genes with methylated promoters are under-expressed. We also found sequence motif near unmethylated region. We have tried to build classifiers to predicted methylated and unmethylated regions and developed methods HMDFinder and MethCGI.

Recently, in collaboration with several labs, we tried to identify insulator (CTCF) binding sites in the human genome. CTCF is a DNA-binding domain zinc finger family, which accounts about one third of human TFs. CTCF is a ubiquitous, 11-zinc finger DNA-binding protein involved in transcriptional regulation, reading of imprinted sites, X chromosome inactivation, and enhancer blocking/insulator function. In cancer, CTCF appears to function as a tumor suppressor gene. Genome is organized into discrete functional and structural domains. Insulators are defined by function, block the action of distal enhancer when inserted between the enhancer and promoter, and serve as boundaries between heterochromatin and euchromatin. Mutations in insulators are implicated in human diseases. We performed whole-genome CTCF ChIP-chip experiment to identify CTCF binding sequences. We then validated by real-time quantitative PCR, showing a specificity >95 % and a sensitivity >80 %. We show CTCF sites are conserved, and CTCF binding correlates with gene number but not with chromosome length. On average CTCF binding sites are 8 kb away from the nearest 5' end, which is consistent with insulator function.

We also found two distinct patterns of CTCF distribution: (1) CTCF binding sites are depleted in superclusters of related genes, such as zinc finger protein clusters and olfactory receptor clusters. (2) CTCF binding sites are highly enriched in genes with a larger number of alternate transcripts, such as immunoglobulin locus. A vast majority of CTCF binding sites are characterized by a specific 20mer motif, present in >75 % of all CTCF binding sites. The motif is similar to preciously characterized motif and extends and refines at several key nucleotide positions.

In conclusion, we have defined a large number of CTCF binding sites in the human genome that exhibit unique distribution among sequence-specific transcriptional factors. CTCF binding sites can function as enhancer blockers and chromatin boundary elements. These CTCF binding sites are uniformly characterized by a unique 20mer motif, which are conserved and display interesting model of change throughout the vertebrate genomes, suggesting.

12.11   Understanding Biological Functions Through Molecular Networks

Jingdong Han (✉)
Chinese Academy of Sciences, Beijing, China

In this talk I'll show several studies based on molecular networks. Networks consist of nodes and edges connecting them. Some important concepts include degree, the number of edges of a node and hub, node with high degree, and characteristic path length, the average length of path between any two nodes in the network. There are mainly three types of molecular networks: protein network, regulatory network, and miRNA–mRNA network. We have been focused on protein networks.

Our first example is to map the structure of interactome networks of C. elegans. We use yeast two-hybrid systems. The resulted network shows some surprising features. For example, yeast orthologs are almost homogeneously distributed in the network, instead of as a core module. The network also shows a power-law degree distribution.

The next step is to annotate the mapped networks by data integration. Our first example is to extract molecular machines that function in C. elegans early embryogenesis, by integrating expression profiling, binary interactome mapping, and phenotypic profiling. We searched for fully connected subnetworks (cliques) and derived methods to prediction protein functions by completing incomplete cliques. Our second example is to generate breast cancer network using a similar approach. Based on this network, we predict new proteins involved in breast cancer. Proteins interacted with four seed proteins (known as breast cancer genes) are pulled out to generate a breast cancer gene network. A significant proportion of the network is not covered by existed literature. We also predicted a large number of high-quality interactions by Bayesian analysis. We integrated 27 omics datasets and predicted about 300,000 interactions among 11,000 human proteins. All data are deposited into the IntNetDB database.

With high-confidence network available, one can study other advanced topics. For example, we have shown that hubs can be further divided into two types: date hub and party hub, by the coexpression level hubs and their partners. In silico simulation of node removal shows that date hub is more important to the main CPL.

Recently network motifs are introduced to the study of molecular networks. One typical motif is feedback, including both positive feedback and negative feedback. Another interesting type is toggle switch. Please refer to the review of on *Nature*, 2002.

Another question we tried to answer is about aging, a complex process that involves many seemly unrelated biological processes. Could this reflect different aspects of concerted changes of the aging network? Is this network modularly designed? We proposed an algorithm to identify modules with an aging network and show existence of proliferation/differentiation switch at systems level. We also show that these modules are regulated like a toggle switch, that is, mutual inhibition.

The fifth question we have tried to answer is "can we infer the complete network from incomplete network?" The current interactome maps cover only a small fraction of the total interactome (3–15 %). Most networks show scale-free property. Will it be an artifact due to incomplete mapping? Our simulation studies show that limited sampling of networks of various topologies give rise to scale-free networks. Thus, at current coverage level, the scale-free topology of the maps cannot be extrapolated to the complete interactome network.

The next step of network study is to make prediction by network modeling at different levels, including differential equation model, Boolean model, Bayesian model, and statistical correlation models. Here we show a study using network statistics to predict disease-associated genes. The authors tried to find out the disease gene within a linkage loci containing more than 100 genes. The method first pulled out direct neighbors of the gene under study and then check whether these neighbors are involved in similar diseases. Genes with many partners involved in similar diseases are ranked higher.

12.12   Identification of Network Motifs in Random Networks

Rui Jiang (✉)
Tsinghua University, Beijing, China

The numerous persuasive evidences from many experimental investigations have proved that networks across many scientific disciplines share global statistical properties. The "small world" of short paths between nodes and highly clustered connections is one characteristic of natural networks. It has also been shown that many networks are "scale-free" networks, in which the node degrees follow a power-law distribution. However, recent studies have shown that the local different structures, termed "network motifs," are widespread in natural networks. Network motifs are these patterns that occur in networks at significantly higher frequency than the average in randomized networks and have been found in a wide variety of networks, ranging from the World Wide Web to the electronic circuits, from the transcriptional regulatory networks of *Escherichia coli* to the neural network of *Caenorhabditis elegans*. Therefore, network motifs are considered as the basic building blocks of most complex networks.

As in most previous studies, the networks and network motifs have been represented as deterministic graphs in which the connections between nodes include the presence and absence states. Nevertheless, most biological processes in organisms are in a dynamical equilibrium adapted to the rivalship between the internal conditional variations and the external environmental perturbations, rather than static. For example, in a living cell, DNA-binding proteins are believed to be in a balance between the bound and unbound states, thus introducing uncertainties in protein-DNA interactions. The evolution of regulatory pathways in cells is itself a stochastic process, and some protein-DNA interactions could change without affecting the functionality of the pathways. Functionally related network motifs are therefore not necessarily topologically identical. Therefore, networks and network motifs are intrinsic uncertain. Additionally, incomplete and/or incorrect observations due to experimental resolutions, systematic errors, and random noises also introduce considerable uncertainties into the observations. This situation prevails in biological interaction networks constructed by using data collected by high-throughput techniques such as the yeast two-hybrid assays and the chromatin immunoprecipitation method. With the intrinsic and experimental uncertainties, it is more suitable to describe the biological networks and network motifs as stochastic graphs in which connections between nodes are associated with probabilities and discuss the network motif identification problem in the circumstance of stochastic networks.

A stochastic network with N nodes could be represented by a probability matrix $\mathbf{P} = (\pi_{ij})_{N \times N}, 0 \leq \pi_{ij} \leq 1$. π_{ij} is the probability that the connection between node i and node j occurs in the stochastic network. By the probabilistic representation, a family of deterministic networks could be randomly drawn with respect to a stochastic network. The deterministic network is described by using the adjacency matrix $\mathbf{A} = (a_{ij})_{N \times N}$ which charactering the flow directions on the network. For directed graphs, $a_{ij} = 1$ if there is a directed edge pointing from node i to node j. For undirected graphs, $a_{ij} = 1$ if an undirected edge connects node i and node j. Network motifs are a small set of subgraphs embedded in stochastic networks with significant occurrences at number and could be also described by using probability matrices. A subgraph could be found in networks as various isomorphic structures with the same topology but dissimilar node orderings. These isomorphic structures could be mapped to each other by permuting their node labels. In the circumstance of stochastic networks, the network motif could be represented by a parametric probabilistic model, $\mathbf{\Theta}_f = (\theta_{ij})_{n \times n}, 0 \leq \theta_{ij} \leq 1$, which could be used to calculate the occurrence probability of a given n-node subgraph isomorphic structure P in the foreground. Another parametric probabilistic model is established to describe the suitable random ensemble drawn from the background and is characterized by a three-tuple $\mathbf{\Theta}_b = \{\mathbf{I}, \mathbf{O}, \mathbf{M}\}$, where $\mathbf{I}, \mathbf{O},$ and $\mathbf{M}$ are the distribution of the in, out, and mutual degrees for the stochastic network, respectively. Given the foreground and background mixture probabilistic model, a given n-node subgraph isomorphic structure P could be sampled from the population in the following procedure: randomly choose a generation distribution; for the foreground with the probability $\lambda_f = \lambda$, sample a subgraph according to the probability matrix Θ_f; for the background with the probability $\lambda_b = 1 - \lambda$, generate a subgraph under the parametric model Θ_b; and then select one isomorphic structure of the subgraph at random. Therefore, mathematically, the network motif identification problem could be formulated as searching for the mode of the likelihood function which describing the co-occurrence probability of the observations. Expectation maximization (EM) is an efficient statistical inference method to estimate the parameters which maximizing the likelihood of the mixture model through introducing some appropriate latent variables.

In experiments, this method has been applied to a wide range of available biological networks, including the transcriptional regulatory networks of *Escherichia coli* and *Saccharomyces cerevisiae*, as well as the protein-protein interaction networks of seven species [*E. coli, S. cerevisiae* (core), *C. elegans, Helicobacter pylori, Mus musculus, Drosophila melanogaster,* and *Homo sapiens*], and identifies several stochastic network motifs that are consistent with the current biological knowledge. Additionally, the ChIP-chip datasets for the *S. cerevisiae* regulatory network are also involved in the experiments. In the 3-node case, the feed-forward loop motifs are identified in the transcriptional regulatory networks for both species. Recent studies have shown that the feed-forward loop serves as a sensitive delay element in regulatory networks and could speed up the response time of the target gene's expression following stimulus steps in one direction (e.g., off to on) but not in the other direction (on to off). In 4-node case, the patterns with the solid biological evidences in regulatory networks are the stochastic bi-fan motifs in which the interaction between two regulators is of two states. Undirected stochastic motifs are also identified in protein-protein interaction networks via the available datasets.

The approach with the mixture probabilistic model and EM solver could be used to identify the stochastic network motifs embedded in the stochastic networks. The proposed unified parametric probabilistic model takes the intrinsic and experimental uncertainties into considerations, could capture the stochastic characteristics of network motifs, and could be applied to other types of networks conveniently.

12.13   Examples of Pattern Recognition Applications in Bioinformatics

Xuegong Zhang (✉)
Tsinghua University, Beijing, China

Concerning about the pattern recognition applications in bioinformatics, we give some examples on the basis of our major research interests. These interests are mainly surrounding the central dogma and focused on the molecular regulation systems in eukaryotes, especially in human beings. Most of the relevant interests are concentrated on the aspect of genome, including transcriptional regulation of genes, splicing regulation, regulation by and of noncoding RNAs, epigenomics, interaction of regulatory factors, and posttranslational modifications. Other interests include population genetics and traditional Chinese medicine.

First, on the issue of alternative splicing, two examples are mentioned. One of them is about how to identify the factors that make a splicing site alternative. From a pattern recognition point of view, this issue can be regarded as a classification problem to differentiate alternative splicing sites from constitutive splicing sites, and by training a machine to imitate the real splicing mechanism, the reasonableness of our supposed mechanism can be judged by the performances of the classification. The other problem is about the VSAS (very short alternative splicing events), and we feel interested to know what role the alternative splicing plays in this process. In this problem, the observed VSAS events can be classified into two groups depending on whether they insert new structure domains in the proteins, and they might be of different evolutionary status.

Next, we talk about RNA regulation, including computational recognition of noncoding RNAs especially microRNAs, study of their regulatory roles and target genes, mechanism of their transcription and regulation, and RNA editing. Most of the relevant problem can also be described or transformed into a pattern recognition problem; for example, the identification of microRNA can be solved as a classification problem of microRNA vs. pseudo microRNA.

Another point need to be noticed is the feature selection problem. Consider that we are doing a classification problem; for example, classifying the microarray data, there are two things that may need to be done, which are feature selection and classifier construction. There exist two kinds of methods to do feature gene selection, one kind of which is called two-step procedures (filtering methods), while the other kind of which is called recursive procedures (wrapper methods). For the filtering procedures, some criteria are first designed to select differentially expressed genes with certain stand-alone methods, and then classification is implemented using the selected genes. For the wrapper procedures however, the gene selection and classifier construction are done synchronously; that is, the entire genes are utilized to construct classifiers, and after that, useless genes are deserted, while useful genes are left. This gene selection process is recursive, and support vector machine (SVM) can be used in the recursive gene selection procedures. There is also an example about a story between SVM-RFE and R-SVM. The idea of the two methods is almost the same, both using the entire genes to train SVM (we just use linear SVM) and then selecting a subset of genes that gives the best performances. The advantage of the wrapper methods lies in that the following selection criteria are consistent with the classification criteria and possible collaborative effects of genes are considered. The main disadvantage lies in that the rankings are not comparable and there is no cutoff for selection. As a result, the problem existing in the wrapping methods is that it cannot evaluate the significance of rankings. In the application of the traditional statistical test method, the comparison between two samples is first transformed to the comparison of mean difference, and then the difference is ranked and measured by p-value, based on which we can draw conclusion accordingly. While using the machine learning method, the comparison between two samples is first transformed to the contribution in SVM, and then the contribution is ranked and the conclusion is drawn, but lacking the measures of the ranks, some p-value-like statistics as in the statistical test method.

On the issue of gene expression and transcriptional regulation, the main topics include discovery and recognition of transcription factor binding sites (TFBSs), prediction of transcription starting points (TSSs), tissue- and developmental stage-specific regulations, as well as analysis of regulatory pathways and networks. And some problem can also be described as a pattern recognition problem.

On the issue of epigenomics, we mainly do the following research: predicting the methylation status of CpG islands, genomic factors that decide on the methylation of CpG islands, and so on. As is well known, CpG islands are areas in the genome within which some Cs are methylated, while others are not. So the pattern recognition task in front of us is to classify the methylated from the unmethylated, and our classifier is constructed and trained using the known methylated and unmethylated CpG islands.

12.14 Considerations in Bioinformatics

Yanda Li (✉)
Tsinghua University, Beijing, China

On the issue of how to understand bioinformatics, different understandings may lead to different direction choice, focus, working methods, as well as final results of the study. Some people regard bioinformatics as serviceable and auxiliary means for biologists, and some regard it as applications of computer science on information processing or technique of pattern recognition and machine learning on molecular biology data analysis. From Prof. Yanda Li's point of view, however, bioinformatics plays an essential and central role in terms of the molecular biology or life sciences, the reason of which is explained by doing an analogy from understanding bioinformatics to understanding characteristics of a machine.

Since everything in this world is interconnected with each other, we may start from considering how we understand the characteristics of a machine or device. As is well known, the characteristics of a device are determined by the whole, but not the part. And though the whole is complicated on many aspects such as component structure, materials, driving force, and peripherals, its characteristics are mainly determined by the interaction of the components. The method of describing this interaction, which needs to omit minor factors, highlighting the dominating ones, is called *system modeling*. Through the mathematical (or computing) analysis toward the model, we can understand the response of the system to the outside world and also understand the overall characteristics of the system (static, dynamic) as well as various local features. However, no matter what type the system is (mathematical or of other type), the interaction of its components is actually characterized by relationship between informations. Therefore, information system modeling is the basic method to understand the overall characteristics of a system. It is even believed that understanding toward machines and that toward living creatures is substantially the same. Of course as an open problem as yet, some people oppose the above analogy between machines and living creatures, whereas Wiener and many posterior scholars hold similar views on behalf of the above analogy. This issue is then elevated to a philosophical height, say, recognizing living creatures can be understood in the same manner as machines means life is nothing but a physical interaction at an advanced stage, which has no insurmountable difference from low-level interactions in nature. There is no God and the soul. Hence, the fundamental task of bioinformatics is to find the internal information system model of the living creatures, that is, to understand the living creatures on the whole from an information system perspective.

Bioinformatics is calling for system theory and cybernetics. Since the biological system itself is a very complex system, its overall characteristics mainly come from interactions, and cybernetics is just a theory studying interactions. These interactions, to put it further, are actually characterized by information interactions. Therefore, studying bioinformatics from a perspective of information, rather than separately analyze the chemical composition, is a more promising way to simplify a complex system.

The development of bioinformatics can be traced from data management, sequence alignment, and biological molecule identification to various molecular interactions as well as biological molecular network modeling and analysis. Now the analysis of biological molecular network will be an important issue in front of bioinformatics. Analysis of the network includes structural analysis (simplification, clustering, and modularization) and the dynamic performance analysis (evolution and the response to the disturbance), by means of evolution and analysis of random networks.

Currently in the post-genome era, functional genomics and systems biology will be new studying directions we are facing. The gradually close integration of biological science, information science, controlling theory, and systems science enables the information analysis of complex systems become a core content in the post-genome era and bioinformatics. Besides, bioinformatics has a closer integration with diseases and medicine, and the internal coding system of living creatures may also meet breakthroughs, including the coding of regulation and cerebral neurobehavioral cognition. Specifically, the major challenges come from eight research area:

- Identification and characterization of pathogenic gene in complex diseases
- Noncoding DNA analysis and data mining
- Information analysis of oriented-differentiation problem in the stem cell research
- Analysis of protein-protein interaction networks
- Identification and prediction of foreign gene control (epigenetic gene control)
- Information analysis of brain-computer interface
- Analysis and formula optimization toward compound prescriptions of TCM by means of the complex system theory
- Character representation

ERRATUM

Chapter 1
Basics for Bioinformatics

Xuegong Zhang, Xueya Zhou, and Xiaowo Wang

R. Jiang et al. (eds.), Basics of Bioinformatics: Lecture Notes of the Graduate Summer
School on Bioinformatics of China, DOI 10.1007/978-3-642-38951-1,
©Tsinghua University Press, Beijing and Springer-Verlag Berlin Heidelberg 2013

DOI 10.1007/978-3-642-38951-1_13

Page 3
1.100 μm changed to
1–100 μm
1.10 nm changed to
1–10 nm

Page 4
Fig. 1.1 A prokaryotic cell changed to
Fig. 1.1 A prokaryotic cell (from Lodish et al: *Molecular Cell Biology, 4th edition*,
W.H. Freeman, 1999)
Fig. 1.2 An eukaryotic cell changed to
Fig. 1.2 A eukaryotic cell (from Lodish et al: *Molecular Cell Biology, 4th edition*,
W.H. Freeman, 1999)

Page 6
Fig. 1.4 The genetic codes changed to
Fig. 1.4 The genetic codes (from Purves et al: *Life: The Science of Biology, Sixth
Edition*, Sinauer Associates and W.H. Freeman, 2000)

Page 7
Fig. 1.5 The central dogma changed to
Fig. 1.5 The central dogma (from Purves et al: *Life: The Science of Biology, Sixth
Edition*, Sinauer Associates and W.H. Freeman, 2000)

The online version of the original chapter can be found at
http://dx.doi.org/10.1007/978-3-642-38951-1_1

Page 9, line 20 and line 24
Chr.1.22 changed to
Chr.1–22

Page 11
Fig. 1.7 Sequencing reaction changed to
Fig. 1.7 The first-generation sequencing technology (from Purves et al: *Life: The Science of Biology, Sixth Edition,* Sinauer Associates and W.H. Freeman, 2000)

Page 14
Fig. 1.9 changed to

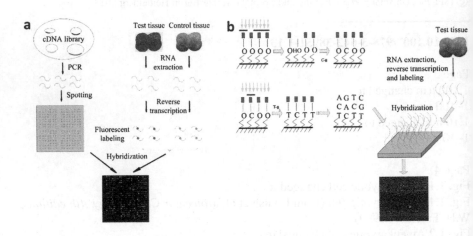

Page. 25 References changed to

1. Shendure J, Ji H (2008) Next-generation DNA sequencing. Nat Biotechnol 26:1135–1145
2. Fields S (2007) Site-seeing by sequencing. Science 316(5830):1441–1442
3. Lodish H, Berk A, Zipursky L, Matsudaira P, Baltimore D, Darnell J (1999) Molecular cell biology, 4th edn. W.H. Freeman, New York
4. Purves WK, Sadava D, Orians GH, Heller HC (2000) Life: the science of biology, 6th edn. Sinauer Associates/W.H. Freeman, Sunderland/New York
5. Hartwell L, Hood L, Goldberg M, Reynolds A, Silver L, Vere R (2006) Genetics: from genes to genomes, 3rd edn. McGraw-Hill, Boston
6. Hartl D, Clark A (2007) Principles of population genetics, 4th edn. Sinauer Associates, Sunderland
7. The International HapMap Consortium (2005) A haplotype map of the human genome. Nature 437:1299–1320

Printed in the United States
By Bookmasters